302

Current Topics in Microbiology and Immunology

R. N. Eisenman (Ed.)

The Myc/Max/Mad Transcription Factor Network

With 28 Figures and 3 Tables

Robert N. Eisenman, Prof. Dr., Ph.D.

Division of Basic Sciences
Fred Hutchinson Cancer Research Center
P.O. Box 19024
Seattle, Washington 98108-1024
USA

e-mail: eisenman@fhcrc.org

Cover Illustration: The cover figure depicts a Myc-Max protein heterodimer interacting with its CACGTG binding site in DNA, superimposed over an image of DAPI-stained cell nuclei. The large cell nuclei are located within a Drosophila larval salivary gland ectopically expressing Drosophila Myc (dMyc) while the small nuclei are from fat body cells lacking ectopic dMyc (from Pierce SB, Yost C, Britton JS, Loo LW, Flynn EM, Edgar BA, Eisenman RN (2004) dMyc is required for larval growth and endoreplication in Drosophila. Development 131: 2317-2327). See chapter by Nair and Burley, within, for review of structural studies on Myc-Max dimers.

Library of Congress Catalog Number 72-152360

ISSN 0070-217X
ISBN-10 3-540-23968-5 Springer Berlin Heidelberg New York
ISBN-13 978-3-540-23968-0 Springer Berlin Heidelberg New York

Springer is a part of Springer Science+Business Media
springeronline.com

Printed in Germany

Editor: Simon Rallison, Heidelberg
Desk editor: Anne Clauss, Heidelberg
Production editor: Nadja Kroke, Leipzig
Cover design: design & production GmbH, Heidelberg
Typesetting: LE-TEX Jelonek, Schmidt & Vöckler GbR, Leipzig
Printed on acid-free paper SPIN 11001973 27/3150/YL – 5 4 3 2 1 0

family proteins as well as with several bHLHZ proteins. BILLIN and AYER review the evidence for the Mlx network and describe its functions in energy metabolism.

The chapters included in this volume illustrate the complexities of the Myc/Max/Mad network and how its functions impinge on fundamental biological processes through regulation of transcription. I am grateful to all the authors for their efforts in putting together comprehensive and provocative chapters as well as for their patience during the long time it took for this volume to come to fruition. I thank Peter Vogt for suggesting the volume on the network and Ms. Anne Clauss for her help in assembling the volume. I hope that the exciting research described here will stimulate others to explore the functions of transcription factor networks.

Seattle, Washington, July 2005 *Robert N. Eisenman*

sequences. Binding of Myc-Max at E-boxes activates transcription from promoters in the vicinity of the binding sites. Underlying this transcriptional activity is the ability of Myc-Max to recruit several higher order chromatin modifying complexes to its binding sites, the major topic of the chapter by COLE and NIKIFOROV. Importantly Myc-Max have also been demonstrated to repress transcription of a number of genes, many of which are involved in cell cycle arrest and adhesion. The mechanism underlying Myc mediated repression is in part related to Myc's ability to interact with and inhibit the activity of other transcription factors, such as the BTB-POZ domain protein Miz-1. Repression by Myc and its biological consequences is described in the chapter by KLEINE-KOHLBRECHER, ADHIKARY, and EILERS.

The discovery of Max as an obligate dimerization partner for Myc led to the identification of other Max binding proteins. These include a novel group of bHLHZ proteins known as Mad proteins (now renamed Mxd), the closely related Mnt protein, and Mga. All these proteins have been associated with transcriptional repression at E-box binding sites. Indeed, Mad/Mxd and Mnt act as partial antagonists of Myc function. ROTTMANN and LÜSCHER review in detail the complex molecular and cellular biology of these proteins.

The Myc/Mad/Mad network then is defined by, and functions through, the interactions between individual Myc and Mad family proteins with Max as well as by interactions between Myc and Mad family proteins with higher order co-repressor and co-activator complexes. The structural biology of a number of these key interactions is reviewed in the chapter by NAIR and BURLEY who also discuss the basis for the high degree of specificity in complex formation. One concept that has emerged from these studies is the notion that a balance between Myc and Mad proteins may act to control key cellular events. A critical question that has haunted the Myc field for some time concerns the number and nature of the genes regulated by the network. This is the subject of the chapter by LEE and DANG who describe the approaches used the delineate target genes for Myc and how the thinking about Myc target genes has evolved.

Another major area of research interest relates to the biological consequences of both normal and abnormal Myc function. The chapter by WADE and WAHL describe evidence for the relationship between deregulated Myc protein expression and altered DNA repair and genomic instability. BLANCK, PIRITY and SCHREIBER-AGUS review the role of Myc/Max/Mad in embryonic development based on studies carried out in mice bearing targeted deletions of these genes. The chapter by GALLANT summarizes what research in invertebrate orthologs has taught us about the evolution of the Myc/Max/Mad network. Just as the Myc and Mad proteins do not function alone, the Myc/Max/Mad network is also unlikely to function in isolation. Mlx is a Max-like protein, which interacts with a subset of Max network

Preface

Scientists often look askance at their colleagues whose research appears too strongly focused on a single gene or gene product. We are supposed to be interested in the "big picture" and excessive zeal in pursuit of a single pixel might seem to border on an obsession that is likely to yield only details. However as this volume of *Current Topics in Microbiology and Immunology* demonstrates, this is certainly not the case for *myc*. Intense study of this enigmatic proto-oncogene over the last twenty years has only broadened our view of its functions and led to insights into mechanisms relating to transcriptional regulation as well as to cell growth, proliferation, differentiation, apoptosis and organismal development.

The *myc* gene originally came to light as a retroviral oncogene (v-*myc*) associated with a wide range of acute neoplasms. It was later shown to be a virally transduced cellular gene (c-*myc*) which is a member of family of oncogenes (c-*myc*, N-*myc*, L-*myc*). These family members are themselves subject to a bewildering assortment of genetic rearrangements associated with many different types of tumors derived from many different types of cells. These rearrangements (including chromosomal translocation, viral integration, and gene amplification) act to uncouple expression of the *myc* family genes from their normal physiological regulators. The chapter by LIU and LEVENS describes the key pathways leading to regulation of *myc* expression, showing that such regulation occurs at several different levels and through multiple mechanisms.

The early findings on *myc* regulation and its involvement in tumorigenesis suggested that *myc* plays a fundamental role in cell behavior and also served to attract a great deal of interest in understanding *myc*'s biological and molecular functions. One outcome of the strong research interest in *myc* was the realization that its encoded protein (Myc) does not function alone, but rather acts as part of a network, or module, of interacting proteins. Myc is a member of the ofbasic-helix-loop-helix-zipper (bHLHZ) class of proteins and forms a heterodimer with the bHLHZ protein Max. Myc-Max heterodimers recognize the sequence CACGTG and, with lower affinity, other related E-box

List of Contents

List of Contributors

(Addresses stated at the beginning of respective chapters)

CTMI (2006) 302:1–32

Making Myc

J. Liu · D. Levens (✉)

Gene Regulation Section, Laboratory of Pathology, NCI, DCS, Bldg. 10, Rm 2N106, Bethesda, MD 20892-1500, USA
levens@helix.nih.gov

Abstract Myc regulates to some degree every major process in the cell. Proliferation, growth, differentiation, apoptosis, and metabolism are all under Myc control. In turn, these processes feed back to adjust the level of *c-myc* expression. Although Myc is regulated at every level from RNA synthesis to protein degradation, *c-myc* transcription is particularly responsive to multiple diverse physiological and pathological signals. These signals are delivered to the *c-myc* promoter by a wide variety of transcription factors and chromatin remodeling complexes. How these diverse and sometimes disparate signals are processed to manage the output of the *c-myc* promoter involves chromatin, recruitment of the transcription machinery, post-initiation transcriptional regulation, and mechanisms to provide dynamic feedback. Understanding these mechanisms promises to add new dimensions to models of transcriptional control and to reveal new strategies to manipulate Myc levels.

1 *c-myc* Regulation

1.1 The Problem

As discussed elsewhere in this volume, Myc-Max heterodimers operating directly as a transcription factor recruit effector complexes to activate and repress transcription. Alternatively, by binding with E-boxes, Myc-Max competes with the other HLH-bZIP complexes to modify target gene action. Counted among Myc targets are genes essential for proliferation, growth, the cell cycle, apoptosis, metabolism, and both intra- and intercellular signaling. Thus the c-Myc network ensnares prey from virtually every important cellular activity (Grandori et al. 2000; Levens 2002, 2003). In turn, it would seem that *c-myc* expression should be coupled with direct or indirect feedback from many intra- and extracellular systems and subsystems. These systems regulate Myc at every level—from transcription, RNA processing, messenger (m)RNA half-life, and translation to protein turnover (Cole and Mango 1990; Wisdom and Lee 1991; Laird-Offringa 1992; Lavenu et al. 1995; Yeilding and Lee 1997; Brewer 1999; Creancier et al. 2001; Lemm and Ross 2002; Kim et al. 2003b). Although a number of factors bind to *c-myc* mRNA to influence its turnover and translation, it appears that most *c-myc* regulatory pathways are channeled through transcriptional control (though not necessarily exclusively so). Indeed, as will be discussed below, many important pathways reach the *c-myc* promoter through a variety of canonical, non-canonical, and atypical *cis*-elements. The central and elusive problem in *c-myc* regulation is discerning how multiple, and often disparate, signals are integrated to determine the final level of Myc.

1.2 Do Myc Levels Matter?

A number of observations indicate that cellular and organismal physiology and pathology are sensitive to slight alterations of Myc levels. The body sizes of mice bred to generate every diploid combination of normal, hypomorphic, and null *c-myc* alleles scaled with the amount of Myc; only the null-homozygotes were inviable, succumbing during development as reported (Davis et al. 1993; Trumpp et al. 2001). Somatic knockout of one or both *c-myc* alleles showed that the cell cycle length varies inversely with the dose of *c-myc* (Shichiri et al. 1993; Mateyak et al. 1997; Schorl and Sedivy 2003). These same studies indicate that there is no upregulation of the normal *c-myc* allele to compensate for the impaired or absent expression of its partner; there is no *c-myc* signal upregulating basal expression acting in *trans*. In contrast to underexpression, it is likely that Myc overexpression depresses expression from normal *c-myc* alleles (first appreciated by the silence of the unrearranged allele in Burkitt lymphoma, this phenomenon is not universal in all cases of the disease) (Siebenlist et al. 1984; Facchini et al. 1997). A variety of studies indicate that Myc-targeted gene expression varies quantitatively, if not qualitatively, as Myc levels are altered (Levens 2002, 2003). Chromosomal translocations, rearrangements, and viral insertions that deregulate *c-myc* expression without activating mutations within the Myc protein indicate that failing to confine Myc levels within physiological bounds is an important step in the carcinogenesis of many, if not most, tumors. Even in Burkitt lymphoma, the malignancy most closely associated with abnormal *c-myc* expression, the range of *c-myc* expression in some cases barely exceeds (1.47-fold) the levels found in normal tissues (Saez et al. 2003). It seems that Myc levels do matter.

The kinetic features of *c-myc* mRNA and protein indicate that the protein must be tightly regulated. Both the mRNA and the protein possess short half-lives (20–30 min), but may be stabilized in some pathological or physiological circumstances (Hann and Eisenman 1984; Dani et al. 1985; Rabbitts et al. 1985; Sears et al. 1999). In normal resting cells, *c-myc* mRNA levels are low: as low as one molecule per cell. It has been estimated that about 40% of the mRNAs in cells are single copy (Lockhart and Winzeler 2000). If these are Poisson distributed, as expected, at any given moment every cell has a unique expression profile. Rapidly dividing tissues support higher levels of *c-myc*, but even in embryonic cells, *myc* RNA levels have been estimated to be approximately five mRNAs per cell (Evingerhodges et al. 1988; Warrington et al. 2000). During mitogenic stimulation of normal cells or in tumors, *c-myc* transcripts may transiently rise to higher levels. During the G0–G1 transition, *c-myc* mRNA levels spike before dropping to steady-state levels (Dean et al. 1986). Apparently less Myc is required to sustain than to initiate proliferation; if so, fluctuating Myc expression might prove deleterious. However, because of rapid turnover and low abundance, it would seem that cells lack a sufficient reservoir to buffer the stochastic noise expected to buffet *c-myc* mRNA levels (Elowitz et al. 2002; Swain et al. 2002). Somehow all of the signals converging on *c-myc* must be integrated in a manner ensuring sufficient stability to prevent abnormal proliferation or apoptosis, yet responsive enough to allow Myc to fulfill its role as an immediately-early gene.

1.3 Promoters

Under most circumstances, *c-myc* transcription is initiated at two promoters, P1 and P2, with the latter supporting about 75% of c-*myc* transcripts (see Fig. 1). Minor amounts of *c-myc* RNA initiate at P0 and P3, and cryptic promoters may become activated following chromosomal translocation, under the influence of pathologically juxtaposed regulatory elements (Spencer and Groudine 1991; Marcu et al. 1992). The discovery of antisense transcription of the murine and human *c-myc* exon 1 presaged the more recent appreciation of opposite-strand transcription as a general phenomenon (Spencer and Groudine 1991, Rinn 2003; Marcu et al. 1992). Usually, only P2-initiated—and to a lesser extent P1-initiated—transcripts contribute significantly to the pool of *c-myc* mRNAs. Under some physiological circumstances (e.g., G0–G1 transition in lymphocytes—Broome et al. 1987) or pathological conditions (e.g., following translocations in Burkitt lymphoma), initiation at P1 nearly equals that at P2. If the minor promoter-initiated—or antisense—transcripts are physiologically relevant, they will most likely serve regulatory roles, contributing to RNA processing or stability [through alternative secondary structures or formation of microRNA (miRNA)] or altering transcription-driven chromatin remodeling and modification. P0-initiated transcription complexes must traverse through *cis*-elements upstream of P1 and P2, and so may potentially contribute to the structural reorganization and exchange of promoter-bound regulatory protein complexes. These possibilities remain largely unexplored. The P2 promoter itself is remarkably resistant to inactivating mutations and deletions. Eliminating both the TATA box and the transcription start-site preserves sufficient information for the transcription machinery to still locate the promoter and initiate transcription (Krumm et al. 1995). But at *c-myc*, most of the time, initiation is not the problem.

Unless the *c-myc* gene has been irreversibly silenced, a transcriptionally engaged RNA polymerase is paused in the promoter proximal region in most cells (Marcu et al. 1992). The presence of this polymerase dictates that *c-myc*

Fig. 1 Anatomy of c-myc promoter's DNase I hypersensitive sites and putative *trans*-factor binding sites: The DNA region from ~2.5 kb upstream of P2 to c-*myc* the second exon is shown. Locations of binding sites for over 40 factors directly binding to DNA are indicated relative to the human P2 promoter. Specific binding sites for each factor are listed in Table 1. Conventional double-stranded DNA-binding proteins are listed *below the line* and single-stranded nucleic DNA-binding proteins are listed *above the line*. Exon III and downstream hypersensitive sites are not shown

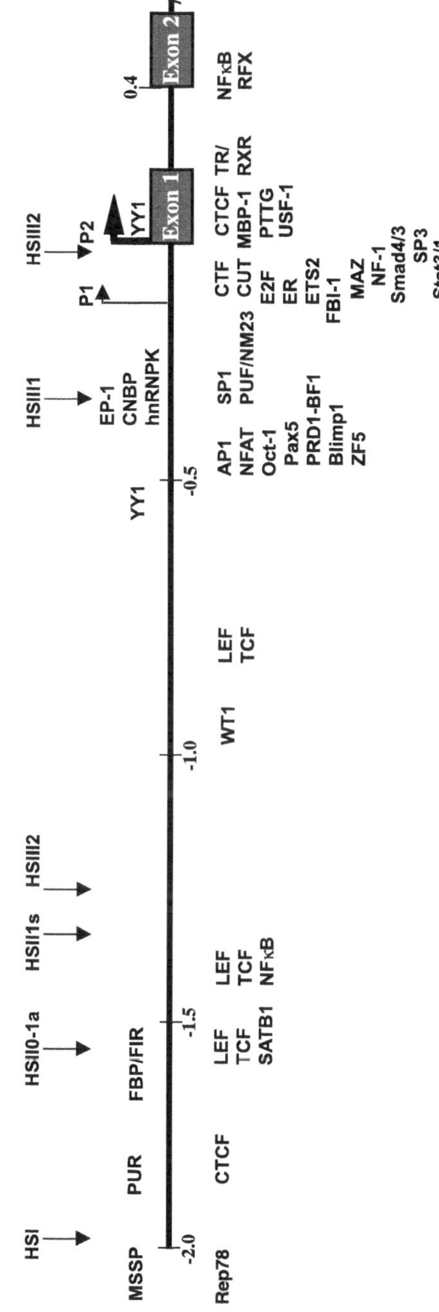
c-myc anatomy
HSI
HSII0-1a
HSII1s
HSII2
HSIII1
HSIII2
MSSP
PUR
FBP/FIR
YY1
EP-1
CNBP
hnRNPK
P1
P2
YY1
Exon 1
Exon 2
-2.0
-1.5
-1.0
-0.5
0.4
Rep78
CTCF
LEF
TCF
SATB1
LEF
TCF
NFκB
WT1
LEF
TCF
AP1
NFAT
Oct-1
Pax5
PRD1-BF1
Blimp1
ZF5
SP1
PUF/NM23
CTF
CUT
E2F
ER
ETS2
FBI-1
MAZ
NF-1
Smad4/3
SP3
Stat3/1
CTCF
MBP-1
PTTG
USF-1
TR/
RXR
NFκB
RFX

expression requires a cycle of promoter escape with the transition to elongation before reinitiation occurs. De novo activation from the silent state is not the usual situation for *c-myc* transcription. Paused RNA polymerases, such as first identified at the *hsp70* promoter, customarily have been considered poised for a rapid response (Rougvie and Lis 1988). Although pre-positioning the RNA polymerase might accelerate Myc induction, recent analysis of transcription factor and polymerase dynamics in vivo indicates that DNA–protein interactions are rapid and not usually rate limiting (McNally et al. 2000). In vitro, even in the absence of activators, neither preinitiation complex formation nor elongation is often the rate-limiting step in the transcription cycle. On some promoters the transition from initiation to elongation—promoter escape—is rate limiting (Kugel and Goodrich 1998). The pausing of RNA polymerase at promoters has additional implications for gene expression. Unless activated or removed, a paused polymerase trumps the action of all factors operating to recruit the basal machinery. So factors acting to delay promoter escape provide a check against spurious activation of a vacant promoter.

The site(s) of polymerase pausing on the *c-myc* promoter has not been rigorously defined and the nascent transcripts sprouting from the paused polymerase have never been isolated. Nuclear run-on experiments employing different combinations of nucleotide triphosphates to advance the polymerase incrementally revealed variable pause sites ranging over approximately the first 50 nucleotides (Wolf et al. 1995). Furthermore, the *cis*-elements imposing the pause have also not been rigorously identified. Whether local promoter sequences define intrinsic pause sites (as occurs with RNA polymerase alone), or distant sequences recruit pause-controlling *trans*-factors, is not known (Pal et al. 2001). The control of pausing may be linked with conformational changes demanded by the transition form initiation to elongation. The requirement for DNA melting at all promoters is self-evident. How this melting occurs has not been fully revealed. The structures of all RNA polymerase-template complexes that have been solved (whether phage, bacterial, or polymerase II) reveal a sharp bend with the active site (Zhang et al. 1999; Cramer et al. 2000; Gnatt et al. 2001; Tahirov et al. 2002; Yin and Steitz 2002). Bent DNA melts more easily, and so this feature of transcription complexes may help to open the duplex to permit pairing with the incoming ribonucleotide triphosphates in preparation for phosphodiester bond formation (Kahn et al. 1994). Whether the melted region at the *c-myc* start site is composed only of a transcription bubble sequestered entirely within the active site of RNA polymerase II is not known. When the last general transcription factor, TFIIH, joins the preinitiation complex it carries along the XPB/p89/ERCC3 3′-5′ and XPD/p80/ERCC2 5′-3 helicases; the former is essential for initiation and promoter escape, whereas the latter, though inessential for transcription, facilitates promoter escape (Zawel et al. 1995; Hoeijmakers et al. 1996; Ohkuma 1997; Coin and Egly 1998; Frit et al. 1999; Tirode et al. 1999; Akoulitchev et al. 2000). The helicases of the TFIIH core function during nucleotide excision repair to expose damaged bases for removal. TFIIH has been proposed to act as a molecular wrench modifying DNA conformation at start sites from a downstream location, though alternative models for TFIIH action at transcription start sites exist (Robert et al. 1998; Douziech et al. 2000; Kim et al. 2000). The roles of the helicases in modifying promoter structure have been relatively less studied than the role of the TFIIH CAK (cyclin-activating kinase/cdk7) subcomplex. CDK7, the kinase within CAK, plays a major role phosphorylating the hepta-residue repeat comprising the carboxyl terminal domain (CTD) of the large subunit of RNA polymerase II. Depending on the state of CTD phosphorylation, additional factors involved in transcription and RNA processing (including capping) are recruited to early transcription complexes. Several factors regulating *c-myc* transcription [e.g., FUSE-binding protein (FBP), FBP-interacting repressor (FIR), estrogen receptor E2] interact with TFIIH (Pearson and Greenblatt 1997; Chen et al. 2000; Liu et al. 2000, 2001; Keriel et al. 2002). Shortly after initiation, factors such as DRB sensitivity-inducing factor (DSIF) and negative elongation factor (NELF) are recruited and contribute to pausing, at least to some promoters such as the *Drosophila HSP70* promoter (Wu et al. 2003). It should be stressed that the mechanistic role of TFIIH, DSIF, and NELF, as well as other factors involved in promoter escape, have been explored only on a very small number of promoters; whether the details of the disposition of these factors during initiation and promoter escape can be generalized to all promoters from these few cases is not known.

A role of the nascent *c-myc* RNA in regulating transcript growth, either by directly manipulating the transcription apparatus or through the recruitment of sequence or structure-specific RNA-binding proteins [as for human immunodeficiency virus (HIV) via the TAR sequence in the nascent transcript, the TAT protein recruits PTEF, cyclin T-CDK9, which phosphorylates the CTD and stimulates elongation], has not been reported, but should be considered (Garber et al. 1998; Zhou et al. 1998).

1.4 Chromatin Changes

c-myc was one of the first genes analyzed by indirect labeling for DNase I hypersensitive sites (Siebenlist et al. 1984). Constitutive and regulated hypersensitive sites upstream, downstream, and within the *c-myc* gene have been mapped (Fig. 1). Constitutive DNase hypersensitive site (HS) I is associated with the binding of CCCTC-binding factor (CTCF). This remarkable multi-zinc-finger protein associates with several regions of the *c-myc* gene, including the promoter, under repressed conditions. Since CTCF has enhancer-blocking activity, it is likely to play a role in eliminating or selectively gating the influence of more distant regulatory proteins on the *c-myc* promoter. In cases of Burkitt lymphoma with far upstream translocations, the influence of the immunoglobulin enhancer must penetrate HS I by an unknown mechanism to negate CTCF's barrier function. CTCF also makes protein–protein interactions with other *c-myc* regulators, such as YB1 (Filippova et al. 1996; Chernukhin et al. 2000; Ohlsson et al. 2001; Qi et al. 2003). Although HS II_1 and II_2 map to upstream regions binding various factors, the agents responsible for conferring hypersensitivity have not been unambiguously assigned. In the case of site II_2, cleavage occurs within the CT-element, an entangled mess of overlapping and inter-nested binding sites for conventional and single-strand selective factors; these sites and factors are not easily functionally or biochemically deconvoluted. Sites III_1 and III_2 overlap the P1 and P2 promoters; again, the agent of hypersensitivity has not been ascribed to any single protein or complex, although the ME1a1 site is implicated (Albert et al. 2001). Additional HS sites map 3′ of *c-myc* (Mautner et al. 1995). Whereas some hypersensitive sites are enhanced when *c-myc* is expressed, other sites, such as HS I, persist even in cells with irreversibly silenced *c-myc*.

Actively transcribed *c-myc* genes carry 15 nucleosomes in a 3.6-kb array stretching from upstream of the promoter into intron 1. When inactive, *c-myc* genes harbor an additional four nucleosomes masking segments near HS I, the CT-element, P0, and P1. The major P2 start-site is nucleosome-free irrespective of gene activity (Michelotti et al. 1996b; Pullner et al. 1996; Albert et al. 1997; Schuhmacher et al. 1999). The hypersensitive sites and nucleosome arrangement of *c-myc* promoters embedded in episomal vectors recapitulate those of endogenous *c-myc* genes (Michelotti et al. 1996b; Albert et al. 1997; Madisen et al. 1998; Albert et al. 2001).

The role of chromatin-modifying and remodeling complexes in regulating *c-myc* expression is complicated and confusing. Whereas histone deacetylase (HDAC) inhibitors augment expression from transfected or transgenic *c-myc* promoters consistent with the notion that increased acetylation supports increased transcription, HDAC inhibition paradoxically depresses endogenous *c-myc* in most situations (for example, see Van Lint et al. 1996; Chambers et al. 2003; and many others). The dynamics of HAT, HDAC, and remodeling complex recruitment and dismissal during *c-myc* gene induction and shut-off are incompletely described and likely to prove complicated.

1.5
Where Are *c-myc* Regulatory Elements?

A Hind III site at −2329 relative to P1 has often served as the operational upstream boundary of the *c-myc* promoter; a Pvu II site at the end of exon 1 usually delimits the downstream segment, although sometimes the 5′ portion of intron 1 is also included. Most characterized *cis*-elements and candidate *cis*-elements have been mapped to this interval. Although convenient, these arbitrary choices are poorly justified. *c-myc* promoter-driven reporter genes—whether transiently or stably transfected, integrated, or episomal, as well as transgenes passaged through the germline in mice—have failed to recapitulate proper *c-myc* expression, although certain features of *c-myc* transcription have been coarsely mimicked. Embedding *c-myc* reporters and transgenes in 30 kb or even 50 kb of natural flanking sequence has proved insufficient to confer physiological regulation (Lavenu et al. 1994; Mautner et al. 1996). Why is proper *c-myc* regulation so difficult to achieve? Perhaps important *cis*-elements reside at vast distances from the coding sequence, or perhaps the *c-myc* promoter is particularly sensitive to perturbation of its natural chromosomal context and so requires proper boundary elements to define chromatin and topological domains. Supporting the argument that context is key for *c-myc* governance is the extreme vulnerability of the locus to chromosomal damage, even from vast distances. In the case of Burkitt lymphoma, translocations hundreds of kilobases upstream, or downstream, as well as within the gene, deregulate transcription. In these cases, the cytogenetically juxtaposed, but molecularly remote, immunoglobulin enhancer overrides or usurps all of the locally acting elements with their associated factors to enforce *c-myc* expression. So either context is paramount or vital and remote elements operate on the promoter from either side. Besides translocations, viral insertions, gene amplification, and mutations all deregulate *c-myc* expression.

2
cis-Elements and Transacting-Factors Regulating *c-myc* Expression

2.1
cis-Elements

There is no evidence for a compact enhancer that confers or explains the physiological patterns of *c-myc* expression. There is no evidence for the assembly of a precisely arranged enhanceosome composed of multiple transcription factors and architectural DNA-binding proteins, as well as chromatin remodeling and modifying complexes (Thanos and Maniatis 1995). Virtually every major signal transduction pathway impacts directly or indirectly the *c-myc* promoter (Table 1). Some of the *cis*-elements receiving the signals have been well characterized, whereas others have been revealed only in silico. The *c-myc* promoter generally lacks canonical binding elements while relying on atypical binding sites to recruit many of the *c-myc* regulatory proteins. Generally, non-canonical sites are suboptimal for binding transcription factors. Most of the activators and repressors that bind *c-myc cis*-elements recruit coactivators or corepressors, at least in vitro, and in some cases, chromatin immunoprecipitation studies have demonstrated these effector complexes at *c-myc* in vivo. Reliance upon non-canonical *cis*-elements to recruit these effectors has several implications. First, the weak binding may contribute to the observation that *c-myc* levels are adjusted several fold by many, perhaps even most, agents, but very few single signals impel changes in Myc levels of sufficient degree to constitute an on–off switch. Second, higher concentrations of each factor may be required to achieve *cis*-element occupancy, and so a strong or sustained stimulus might be required to activate expression. Third, the stabilization of weak binding factors to their *cis*-elements through cooperatively interacting partners in principle serves to cross-couple signals and promote synergy. Fourth, fractional *cis*-element occupancy also confounds in vivo footprinting and other protection studies that work best at saturation.

A number of *cis*-elements may be densely inter-nested with the promoter. Sequences responsible for negative autoregulation (probably occurring through both direct and indirect mechanisms), as well as sites for binding CTCF, MBP-1 (a protein related to enolase) (Ray and Miller 1991; Subramanian and Miller 2000; Lee et al. 2002), and other factors occur so close to start sites that cohabitation is difficult to imagine; sequential or alternative action at the promoter seems more likely.

2.2
Traditional *trans*-Acting Factors

The literature describing the pathways delivering signals to the *c-myc* promoter via conventional transcription factors constitutes a veritable compendium of gene regulatory phenomena. Although in any one setting or cell line a particular pathway may dominate, in other situations the influence of that same pathway may be minimal or irrelevant. Signaling pathways striking *c-myc* include: MAP kinase, JAK/STAT, Ras, IFN-γ PI3-K, Fas, Wnt, TGF-β, interleukins, cytokines, lymphokines, steroid and peptide hormones, pharmacologic agents, NF-κB-activating pathways, E2F-activating pathways,

Table 1 List of putative transcription factors regulating *c-myc* transcription. Transcription factors directly binding to the *c-myc* promoter are listed alphabetically. Their binding sites are relative to the human P2 transcription start site. Also listed are signal transduction pathways influencing the binding or dissociation of factors to the *c-myc* promoter. A considerable number of factors bind to non-canonical sites

Factor	Binding site (relative to human P2)	Signaling	Canonical	Reference(s)
AP1	(−467/−461)	MAPK, JAK/Stat, TGF-β, FasL/CD95	NC	Takimoto et al. 1989; Hay et al. 1987; Shaulian and Karin 2002; Cippitelli et al. 2003
CEBPα	Through E2F			Johansen et al. 2001
CNBP	CT (−316/−278)		N/A	Michelotti et al. 1995
CTCF	TRE (+2/+48)	TGF-β, MAPK	N/A	Lutz et al. 2002; Ohlsson et al. 2001
CTF	(−61/−47)	TGF-β, JAK/Stat, MAPK	NC	Gronostajski 2000
CUT	ME1a1 (−61/−47)		C	Dufort and Nepveu 1994
E2F	(−77/−69)	Rb, TGF-β, MAPK, focal adhesin signaling	C	Thalmeier et al. 1989; Ogawa et al. 2002; Iakova et al. 2003; Kowalik 2002
ER	(−138/−22)	Ras/Raf	NC	Dubik and Shiu 1992; Shang et al. 2000
ETS2	(−77/−69)	Ras/Raf/MEK/ERK, JAK/Stat	C	Roussel et al. 1994; de Nigris et al. 2001
FBI-1	(−197/−183)		N/A	Hernandez and Pessler 2003
FBP/FIR	FUSE (−1741/−1652)	Wnt	N/A	Michelotti et al. 1996b; Liu et al. 2000, 2001; Braddock et al. 2002
hnRNPK	CT (−316/−278)		N/A	Michelotti et al. 1996a,b

Table 1 (continued)

Factor	Binding site (relative to human P2)	Signaling	Canonical	Reference(s)
LEF	TBE (−1528/−1415, −1330/−1321, −763/−754)	Wnt, PI3 K	NC(?)	Reya et al. 2000; Peifer 2002
MAZ	ME1a1 (−61/−47), ME1a2 (−95/−81)		C	Bossone et al. 1992; Song et al. 2001
MBP1	(−20/−15)		N/A	Ray and Miller 1991; Ray et al. 1995
MSSP1/2	(−2355/−2348)		N/A	Fujimoto et al. 2001
MYB	(strong site at −1154, multiple weak sites)		C and NC	Zobel et al. 1991; Schmidt et al. 2000
NF-1	(−61/−47)	TGF-β, JAK/Stat, MAPK	NC	Gronostajski 2000
NF-κB	(−1392/−1299, +285/+295)	NF-κB signaling, MAPK, INF-γ, BCR Signaling JAK/Stat, SAPK/JNK		Grumont et al. 2002; Shaffer et al. 2001; Arcinas et al. 2001; Jeay et al. 2001
NFAT	(−467/−461)	Ras, JNK/Stat		Rao et al. 1997
Oct-1	(−467/−461)	MAPK, JAK/Stat, Ras, DNA damage	C	Takimoto et al. 1989; Zhao et al. 2000
Pax5	(−453/−428)		N/A	Pasqualucci et al. 2001
PRD1-BF1/Blimp1	(−453/−428)	Jak/Stat Pathway, Fas/CD95	C	Lin et al. 1997; Gupta et al. 2001
PTTG	(−4/+20)	MEK1	N/A	Pei 2001
PuF/NM23	CT (−316/−278)		N/A	Michelotti et al. 1997
PUR	(−1808/−1789)		N/A	Bergemann and Johnson 1992
Rep78	<−2 kb		N/A	Hermonat 1994
RFX	MIE1 (+350/+363)	Jak/Stat,	C	Zhang 1993

Table 1 (continued)

Factor	Binding site (relative to human P2)	Signaling	Canonical	Reference(s)
SATB1	(−1531/−1507)		N/A	Cai et al. 2003
Smad4/3	TIE (−84/−75)	TGF-β	NC	Kowalik 2002; Chen et al. 2002
Sp1	CT (−316/−278), ME1a1 (−61/−47)	MAPK, INF-γ	NC	Harris et al. 2000
Sp3	ME1a1 (−61/−47)		C	Majello et al. 1997
Stat3/1	(−81/−73)	Jak/Stat Pathway, p38 MAPK, FasL		Bowman et al. 2001; Jenab and Morris 1997
TR/RXR	(Human: −2.2 kb, mouse: +100/+133)	MAPK, PKC	NC	Lutz et al. 2003; Perez-Juste et al. 2000
TCF	TBE (−1528/−1415, −1330/−1321, −763/−754)	Wnt, PI3 K	NC(?)	Peifer 2002; Reya et al. 2000; He et al. 1998
USF-1	(−20/+28)	MAPK	NC	Kiermaier 1999
VDR				Simpson et al. 1987; Katayama et al. 2003
WT1	(−918/−911)	Ras, ERK		Hewitt 1995
YY1	(−553/−423, −20/+28)	TGF-β, MAPK	NC	Riggs et al. 1993; Kurisaki et al. 2003
ZF5	(−453/−403)		N/A	Numoto et al. 1993; Kaplan and Calame 1997

C, canonical;
N/A, not available;
NC, non-canonical

etc. Many of these signals branch and influence more than one *trans*-factor. Each of the *c-myc trans*-factors merits a separate review; some of them are listed in Table 1.

2.3
Funny DNA: The Role of Topology and Conformation

Work from multiple laboratories has contributed to the notion that nonstandard transcription factors binding at atypical binding sites participate in the transcriptional regulation of *c-myc*. Several regions of the *c-myc* gene are associated with non-B-DNA conformation. To understand how these *cis*-elements and their *trans*-factors operate it is important to consider what drives the formation of non-B-DNA.

Conceptually, several processes may directly or indirectly drive conformational changes occurring at *c-myc cis*-elements. First, unwinding torsional stress (negative supercoiling) destabilizes duplex DNA, and certain regions of DNA, especially segments with high A–T content, preferentially melt when the unwinding torque is high enough. Within a topological domain, each hotspot for melting competes with every other hotspot, so the response to torque is inextricably coupled with the creation and destruction of topological boundaries (Benham 1992; Fye and Benham 1999). Protein–protein interactions between DNA-bound factors or attachments to immobile structures restricting rotation of DNA along its helical axis impose topological borders. Topological domains may be nested. Wrapping and fixation of DNA around a nucleosome restrains approximately one supercoil, and unless the grip of the nucleosome is breeched, this DNA constitutes a separate topological domain (Sinden 1994). As long as this DNA is firmly held, the entire protein-DNA assembly may be rotated en bloc, transmitting stress to the unrestrained linker regions. Loosening of histone tails secondary to chromatin modifications such as acetylation would be predicted to expand the amount of linker DNA available to accommodate torsion (Norton et al. 1990; Morales and Richard-Foy 2000). Thus the particular chromatin arrangement within DNA loops may help to focus torsional strain onto DNA segments predisposed to melting or forming alternative structures such as Z-DNA (Rich and Zhang 2003). Second, helicases expend energy from ATP to open bound segments of duplex. As noted previously, the helicase of TFIIH contributes to transcription regulation; whether TFIIH might also contribute to DNA melting at elements other than start sites (directly or as torque generator acting from the promoter) has not been explored except during DNA repair. Several dozen helicase-like open-reading frames reside in the human genome. While many of these are generally presumed to be RNA, rather than DNA, helicases, it is premature to ascribe a molecule with certainty to one class or the other in the absence of experimental information (Caruthers and McKay 2002). So the possibility that helicases may be recruited to *cis*-elements to facilitate the transition to non-B-DNA remains plausible. Third, the machinery of transcription, replication, recombination, and repair demands transient single-stranded regions at the sites of catalytic activity, apart from supercoil-induced melting driven by complexes translocating along DNA. Auxiliary single-stranded (ss) DNA-binding proteins, chromatin remodeling machines, and topoisomerases accompany each of these genetic transactions, and so all have the potential to reconfigure those *cis*-elements prone to altered states. Fourth, although it is assumed that homeostatic mechanisms maintain a monotonous chemical and physical intranuclear environment, changes in parameters that alter the stability of B-DNA such as ionic strength, pH, divalent cations, polyamines, and temperature may all conspire to alter DNA structure. Utilization of elements responsive to these parameters in principle would directly couple *c-myc* transcription with the maintenance of intranuclear homeostasis (note that the sensitivity of PCR to slight variations of these parameters illustrates that it may not be too far fetched to conceive of physiological or pathological changes in DNA structure due to changes of the intranuclear milieu). Cells embedded in tissues are also subject to considerable mechanical force. If these forces were transmitted to DNA via anchored chromatin, DNA structure could be affected at susceptible sequences. So in principle DNA elements that adopt non-B conformations may serve as *cis*-acting stress sensors acting concertedly with conventional *trans*-acting stress-sensing pathways.

2.4 Strange Factors

Besides the panoply of well-recognized, well-characterized transcription factors binding duplex DNA and operating through conventional mechanisms, *c-myc* promoter recruits a menagerie of strange gene regulators binding to elements assuming unusual DNA structures and conformations. Two regions of *c-myc* sequence are particularly associated with altered DNA structures. First, the CT-element, found 100 to 145 bp upstream of the P1 promoter, has been reported to adopt H-DNA, tetraplex, and single-stranded conformations in addition to the standard B-form duplex (Kinniburgh 1989; Postel 1992; Michelotti et al. 1996b; Simonsson et al. 1998). Each of these states is associated, at least in vitro, with a set of conformation-sensitive binding proteins. The extent of regulatory input in vivo conferred by a particular conformational state and its associated factors is not known. Altered structures or conformation in the region of the CT-element is compatible with the absence of phased nucleosomes in this region; nucleosomes constrain B-DNA but have not been shown to engage non-B helices or melted DNA.

Candidate *trans*-factors operating through the CT-element are:

Sp1 The predominant duplex-binding protein interacting with the CT-element and a site further downstream. The essential role of Sp1 in regulating housekeeping genes is well-established (DesJardins and Hay 1993; Michelotti et al. 1996b). Acting locally on the *c-myc* promoters, Sp1 seems to be the conduit through which the immunoglobulin enhancer mediates the activation of P1 relative to P2 often occurring in Burkitt lymphoma (Geltinger et al. 1996).

hnRNP K A prototype for the KH-motif, bearing three repeats of this nucleic-acid binding module. Belying its name, this protein binds more tightly and sequence specifically with ssDNA than with RNA (Tomonaga and Levens 1995; Braddock et al. 2002a). Although there is some indication that hnRNP K recognizes duplex CT-elements, the structure of the ssCT-element complexed with hnRNP K provides no insight as to how this interaction might occur (Braddock et al. 2002a). hnRNP K interacts with TFIID as well as numerous other signaling and gene regulatory proteins. Its ability to stabilize single-stranded loops introduces torsional and flexural hinges into promoters, facilitating interactions between flanking sites (Takimoto et al. 1993; Tomonaga and Levens 1995; Geltinger et al. 1996; Michelotti et al. 1996a; Tomonaga et al. 1998). Between sculpting DNA and recruiting diverse partners, hnRNP K seems to be an adapter gating the interactions of other molecules with greater intrinsic transcription effector activity (Bomsztyk et al. 1997). Although hnRNP K is associated with increased *c-myc* expression, on other genes it may play a negative role. Whether hnRNP K plays a positive or negative role is likely to be context dependent in that it would be determined by the intrinsic activities of the more potent effectors it serves.

CNBP Cellular nucleic acid-binding protein. This multi-zinc-finger protein also binds avidly and in a sequence-specific manner with ssDNA and RNA (Rajavashisth et al. 1989; Michelotti et al. 1996b; Pellizzoni et al. 1997; Crosio et al. 2000). Evidence indicates that this protein may function in the translational regulation of some mRNAs. CNBP binds the purine-rich strand of the CT-element. Knockout of CNBP in mice diminishes *c-myc* expression in those zones of the embryo (forebrain most prominently) where CNBP is abundant; *c-myc* levels are unaffected in regions where CNBP is absent indicating either that other factors substitute for CNBP or that alternate mechanisms bypass the CT-element (Chen et al. 2003). Expressing CNBP in the $CNBP^{-/-}$ cells augments the expression of a transfected *c-myc*-reporter. Importantly, these results dramatize the differential utilization of transcription factors to regulate *c-myc* expression in different cells. The mechanism of transcription activation by CNBP has not been elucidated.

nm23/NDPK Has been associated with a variety of enzymatic and regulatory activities. Initially identified as a transcript downregulated in metastatic cells, nm23 was shown to have tumor-suppressor activity. It was subsequently discovered in an expression screen designed to identify CT-element (also termed NHE—nuclease hypersensitive element) binding factors. This same protein has been associated with nucleoside diphosphate kinase activity, histidine-kinase activity, and both sequence-specific and generalized DNase activities (Hartsough and Steeg 2000; Postel et al. 2000; Roymans et al. 2002). Recently nm23 has surfaced as a subunit of an S-phase octamer-co-activating complex (OCA-S) (Zheng et al. 2003). The protein lacks intrinsic transcription activating function and may not possess sufficient DNA binding specificity to find its physiological targets in vivo unless complexed with partner proteins (Michelotti et al. 1997).

MAZ A multi-zinc-finger protein first identified binding with the *c-myc* promoter. It interacts at several sites within the vicinity of the promoter including the CT-element and a site further downstream inter-nested with Sp1 and E2F binding sites (Bossone et al. 1992, Sakatsume 1996).

YB-1/NSEP Identified as a component of a *c-myc* promoter binding ribonucleoprotein complex (the RNA component has not been characterized further). YB-1 has also been identified as a ssDNA-binding protein interacting with other promoters in vitro, with supporting evidence for a regulatory role for several genes in vivo (Davis et al. 1989; Kinniburgh 1989). YB-1-related proteins bear cold shock domains and have been reported to bind mRNAs and regulate translation in addition to recognizing DNA (Kloks et al. 2002). Recent studies indicate that signaling pathways potentially leading to specific proteolysis and cytoplasmic-nuclear shuttling might determine whether YB-1 binds to RNA or DNA.

The second region associated with altered structures and DNA conformations is a segment far upstream of P1 and P2 which possesses a peculiar sensitivity to torsional strain: the FUSE that binds FBP resides in an AT-rich segment that is easily melted by application of supercoiling forces (Michelotti et al. 1996b; He et al. 2000). This same region of DNA is hypersensitive to single-strand selective oxidation by potassium permanganate in vivo in cells expressing *c-myc*, but not in cells with silent *c-myc* genes. FUSE in nuclei of *c-myc* expressing cells is also sensitive to S1 nuclease. When *c-myc* is silent, a regular nucleosome array runs through the FUSE region (Michelotti et al. 1996b; Pullner et al. 1996; Albert et al. 1997; Albert et al. 2001). When *c-myc* is expressed, this array is disturbed in the vicinity of FUSE. A BRG1-containing complex has been proposed to participate in the remodeling of FUSE chromatin. Just as Brahma-related gene 1 (BRG1) action has also been implicated in the generation or stabilization of a Z-DNA segment in the human colony-stimulating factor (CSF)-1 promoter, it is plausible that it may act similarly on *c-myc* (Chi 2003). Immediately upstream of the FUSE region is one of three Z-DNA-forming segments in the *c-myc* gene (Wittig et al. 1992; Wolfl et al. 1997). Antibodies recognizing Z-DNA can be cross-linked to this Z-DNA-forming region in nuclei. Just as negative supercoiling favors melting of FUSE, so conversion of right-handed B-DNA into left-handed Z-DNA is also driven by torsional stress. Melting at FUSE versus Z-DNA formation would compete to absorb torsional stress. Because nucleosomes do not accommodate non-B-DNA structures, both melting of the duplex and Z-DNA formation may contribute to the disturbance of the regular nucleosomal ladder in this region when *c-myc* is expressed. An origin of replication has been mapped to the FUSE/Z-DNA region. As occurs with other origins, nascent strand synthesis maps to a broader zone beyond the FUSE region (Tao et al. 2000; Liu et al. 2003). Functional coordination or direct mechanisms linking DNA synthesis and *c-myc* expression have not been explored.

2.5 FUSE-Binding Protein

FBP engages FUSE through four KH-motifs with each KH domain engaging 4–6 nucleotides (Braddock et al. 2002b). The cognate sequence segments are separated by spacer DNA (due to the intrinsic flexibility of ssDNA, there is no obligatory helical phasing between the segments engaged by each motif). FBP binds tightly with ssDNA and supercoiled DNA, but forms no stable complex with relaxed duplexes. The carboxyl terminus of FBP bears a tyrosine-rich motif that engages TFIIH to activate transcription (Tomonaga and Levens 1995; Duncan et al. 1996; Liu et al. 2001; Braddock et al. 2002a). FBP's activation domain stimulates the 3′-5′ helicase activity of the XPB/ERCC3/p89 subunit of TFIIH and facilitates initiation and advancement to promoter escape (see Sect. 2.8). The amino-terminus of FBP confers repressor activity when transferred to heterologous DNA binding domains (Duncan et al. 1996). FBP has two closely related sibs, FBP2 and FBP3. FBP2 and 3 bind to FUSE through four KH-motifs highly homologous to those in FBP and possess even more potent carboxyl terminal activation domains (Davis-Smyth et al. 1996). An adenovirus vector over-expressing FBP augments *c-myc* mRNA levels, whereas the same vector expressing a dominant-interfering FBP (central DNA binding domain only, devoid of amino and carboxyl terminal effector domains) depresses *c-myc* RNA (He et al. 2000). FBP itself is downregulated by the direct Myc target p38/JTV-1 (Kim et al. 2003a). Though a core protein in a multi-transfer (t)RNA aminoacyl-synthetase complex, knockout of p38 surprisingly does not impair protein synthesis (Kim et al. 2002). Rather, mice lacking p38 die in the immediate neonatal period with hyperplastic internal organs and increased *c-myc* levels. p38 targets FBP for ubiquitination and degradation. So, normally FBP augments Myc levels, Myc augments p38/JTV-1, and p38 downregulates FBP, closing a homeostatic feedback loop.

2.6 FBP-Interacting Repressor

The central DNA binding domain and the amino terminus of FBP bind FIR. FBP, FIR, and FUSE may form a ternary complex possessing both activation and repression moieties. The amino-terminus of FIR engages TFIIH and depresses, but does not abolish, the same XPB/ERCC3/p89 helicase activity augmented by FBP. FIR does not block initiation, but retards the advance of the transcription complex to promoter escape (see Sect. 2.8). *Drosophila* FIR (*puf60, hfp, dFIR*) was first reported to participate in the developmental regulation of alternative splicing (Van Buskirk and Schupbach 2002). More recently, *dFIR* was found to repress *Drosophila c-myc (dmyc)* at the RNA level, and dFIR was implicated in the regulation of cell cycle progression, influencing both G1/S and G2/M progression (Quinn et al. 2004).

2.7 Special AT-Rich Binding Protein 1

Immediately downstream of FUSE is an A–T rich segment. This segment is especially prone to melt at low levels of supercoiling that may nucleate the destabilization of FUSE. This same segment has the properties of a base-unwinding-region (BUR) and binds with special AT-rich binding protein 1 (SATB1), an atypical homeobox protein that nucleates higher order chromatin organization, especially chromatin loops (Cai et al. 2003, Dickinson et al. 1997; Yasui et al. 2002). SATB1 may contribute to cell-type specific folding of *c-myc* chromatin or the partitioning of *c-myc* upstream sequences into subdomains. SATB1 recruits chromatin remodeling and modifying complexes. SATB1 does not bind to ssDNA, and so FBP and SATB1 actions are likely to prove mutually exclusive, if not antagonistic (Dickinson et al. 1992; Yasui et al. 2002).

Proteins binding "generic" Z-DNA exist, but Z-DNA-binding proteins that are also sequence-specific have not been described, so whether the upstream Z-DNA segment of *c-myc* plays physical roles such as excluding nucleosomes or as a capacitor storing torsional energy, rather than serving as a platform to recruit special *trans*-factors, is not yet known (Rich and Zhang 2003).

2.8
A Scheme to Regulate *c-myc* Transcription

Where, when, and how much *c-myc* to transcribe must be explained in order to fully understand Myc biology. The cell uses several levels of molecular organization to answer these questions. First, it seems that special mechanisms park *c-myc* loci at particular intranuclear sites. Recent data reveal that *c-myc* genes are non-randomly distributed within nuclei; moreover, frequent-translocation partners with *c-myc*—such as the immunoglobulin heavy chain in Burkitt lymphoma—dwell closer on average to *c-myc* than do cytogenetically indifferent loci (Roix et al. 2003). Whether *c-myc* loci are deployed to stations conducive for proper expression, or whether this localization reflects a more passive partitioning of silent and expressed genes is not known. In terminally differentiated cells *c-myc* expression is irreversibly silenced, but cells retaining proliferative potential preserve the capacity to express *c-myc*. Most of these latter cells have a paused polymerase. The initial events in *c-myc* induction have not been defined. Binding of transcription factors to *c-myc* regulatory sequences and chromatin remodeling are likely to occur concomitantly and are probably interdependent. Depending on the variety and magnitude of signals, transcription factors of all sorts flicker on and off the *c-myc* gene. While some of these factors recruit chromatin remodeling and modifying complexes, the immediate issue for the induction of *c-myc* transcription is to restart the paused polymerase. Operating through TFIIH and perhaps other basal transcription components, the paused polymerase is spurred by activators through a series of otherwise slow transitions. The density of bound factors and the frequency of interactions between their activation domains and the promoter-bound apparatus control progression to the point of promoter escape. Only following escape would the promoter be available for reinitiation. In this scheme, a single intense signal acting repetitively through a responsive transcription factor, or multiple weak signals acting through diverse *cis*-elements, would ratchet the pre-promoter escape transcription complex through its various stages. Signal integration would occur through multiple sequential (but not necessarily ordered) activating events. Utilization of multiple, kinetically equivalent, pre-promoter escape intermediates would reduce the temporal variance and damp stochastic fluctuations when compared with a process regulated at a single rate-limiting step that yields only very small numbers of product. Following the delivery of an activating signal (note that experimentally this has often involved employing a single agent—few studies have dealt with synergy or antagonism between *c-myc*-regulating signals delivered in combination) *c-myc* transcription stereotypically peaks between 1 and 2 h after stimulation, and declines rapidly thereafter. Upon achieving log-phase, *c-myc* mRNA levels stabilize above resting levels but well below peak levels. Thus, once cells have experienced a pulse of *c-myc* transcription, a lower level sustains proliferation. Overlying the switches that upregulate *c-myc*, a molecular cruise control system may operate to constrain and prevent chaotic fluctuations of Myc levels. A scheme can be jury-rigged from the features of the FBP–FIR–FUSE system to superimpose dynamic, real-time feedback onto the *c-myc* promoter. Upon activation, transcription pumps torsional stress into the DNA upstream of the promoter. Loops between *trans*-factors and the translocating transcription apparatus at least transiently accumulate torsional energy (until either the loop breaks or a topoisomerase landing within the loop relieves all the tension). If the stress is focused mainly into the linker regions, transcribing even a short distance has the capacity, in principle, to drive structural transitions at sensitive sites within the loop. Hence the ability to recruit, hold, and functionally engage topology and/or conformation-sensitive factors would be linked to ongoing gene activity. The effector domains of some of these proteins (such as FBP and FIR) could reach back to the pre-promoter escape transcription complex and influence further transcription, or they might influence the rate of reinitiation on the next round of transcript synthesis. Such a mode of regulation would occur in real-time in response to ongoing RNA synthesis irrespective of the particular pathway activated to drive transcription. In contrast, conventional feedback requiring synthesis of a *c-myc* primary transcript, splicing, processing, mRNA transport to the cytoplasm, translation, dimerization, transport back to the nucleus, protein modification, incorporation into larger chromatin modifying and remodeling complexes, binding at target sites, and finally—if Myc autoregulation is direct via the P2 promoter—transcriptional repression of *c-myc*. For indirect autoregulation, a second cycle of expression would delay feedback repression even further. These delays would limit the ability of end-product feedback to impose tight homeostasis on a rapidly fluctuating system.

Chromatin immunoprecipitation studies have revealed temporal evolution of the spectrum of factors bound to the *c-myc* promoter during induction and shutoff (Shang et al. 2000). These shifting patterns either represent the superimposition of independent pathways activated and repressed with distinct kinetics (as occurs in yeast), or, alternatively, they represent the dynamic progression of a molecular machine through different stages of operation (Bryant and Ptashne 2003). Distinguishing between these alternatives is an experimental challenge with fundamental consequences for understanding gene expression and with practical implications for devising strategies to reregulate a rogue *c-myc* oncogene.

References

Akoulitchev S, Chuikov S, Reinberg D (2000) TFIIH is negatively regulated by cdk8-containing mediator complexes. Nature 407:102–106

Albert T, Mautner J, Funk JO, Hortnagel K, Pullner A, Eick D (1997) Nucleosomal structures of c-myc promoters with transcriptionally engaged RNA polymerase II. Mol Cell Biol 17:4363–4371

Albert T, Wells J, Funk JO, Pullner A, Raschke EE, Stelzer G, Meisterernst I, Farnham PJ, Eick D (2001) The chromatin structure of the dual c-myc promoter P1/P2 is regulated by separate elements. J Biol Chem 276:20482–20490

Arcinas M, Heckman CA, Mehew JW, Boxer LM (2001) Molecular mechanisms of transcriptional control of bcl-2 and c-myc in follicular and transformed lymphoma. Cancer Res 61:5202–5206

Benham CJ (1992) Energetics of the strand separation transition in superhelical DNA. J Mol Biol 225:835–847

Bergemann A, Johnson E (1992) The HeLa Pur factor binds single-stranded DNA at a specific element conserved in gene flanking regions and origins of DNA replication. Mol Cell Biol 12:1257–1265

Bomsztyk K, Van Seuningen I, Suzuki H, Denisenko O, Ostrowski J (1997) Diverse molecular interactions of the hnRNP K protein. FEBS Lett 403:113–115

Bossone SA, Asselin C, Patel AJ, Marcu KB (1992) MAZ, a zinc finger protein, binds to c-MYC and C2 gene sequences regulating transcriptional initiation and termination. Proc Natl Acad Sci U S A 89:7452–7456

Bowman T, Broome MA, Sinibaldi D, Wharton W, Pledger WJ, Sedivy JM, Irby R, Yeatman T, Courtneidge SA, Jove R (2001) Stat3-mediated Myc expression is required for Src transformation and PDGF-induced mitogenesis. Proc Natl Acad Sci U S A 98:7319–7324

Braddock DT, Baber JL, Levens D, Clore GM (2002a) Molecular basis of sequence-specific single-stranded DNA recognition by KH domains: solution structure of a complex between hnRNP K KH3 and single-stranded DNA. EMBO J 21:3476–3485

Braddock DT, Louis JM, Baber JL, Levens D, Clore GM (2002b) Structure and dynamics of KH domains from FBP bound to single-stranded DNA. Nature 415:1051–1056

Brewer G (1999) Evidence for a 3′-5′ decay pathway for c-myc mRNA in mammalian cells. J Biol Chem 274:16174–16179

Broome HE, Reed JC, Godillot EP, Hoover RG (1987) Differential promoter utilization by the c-Myc gene in mitogen-2-stimulated and interleukin-2-stimulated human-lymphocytes. Mol Cell Biol 7:2988–2993

Bryant GO, Ptashne M (2003) Independent recruitment in vivo by Gal4 of two complexes required for transcription. Mol Cell 11:1301–1309

Cai S, Han HJ, Kohwi-Shigematsu T (2003) Tissue-specific nuclear architecture and gene expression regulated by SATB1. Nat Genet 34:42–51

Caruthers JM, McKay DB (2002) Helicase structure and mechanism. Curr Opin Struct Biol 12:123–133

Chambers AE, Banerjee S, Chaplin T, Dunne J, Debernardi S, Joel SP, Young BD (2003) Histone acetylation-mediated regulation of genes in leukaemic cells. Eur J Cancer 39:1165–1175

Chen CR, Kang YB, Siegel PM, Massagué J (2002) E2F4/5 and p107 as Smad cofactors linking the TGFβ receptor to c-myc repression. Cell 110:19–32

Chen DS, Riedl T, Washbrook E, Pace PE, Coombes RC, Egly JM, Ali S (2000) Activation of estrogen receptor alpha by S118 phosphorylation involves a ligand-dependent interaction with TFIIH and participation of CDK7. Mol Cell 6:127–137

Chen W, Liang Y, Deng W, Shimizu K, Ashique AM, Li E, Li YP (2003) The zinc-finger protein CNBP is required for forebrain formation in the mouse. Development 130:1367–1379

Chernukhin IV, Shamsuddin S, Robinson AF, Carne AF, Paul A, El-Kady AI, Lobanenkov VV, Klenova EM (2000) Physical and functional interaction between two pluripotent proteins, the Y-box DNA/RNA-binding factor, YB-1, and the multivalent zinc finger factor, CTCF. J Biol Chem 275:29915–29921

Chi TH, Wan M, Lee P, Akashi K, Metzger D, Chambon P, Wilson CB, Crabtree GR (2003) Sequential roles of Brg, the ATPase subunit of BAF chromatin remodeling complexes, in thymocyte development. Immunity 19:169–182

Cippitelli M, Fionda C, Di Bona D, Lupo A, Piccoli M, Frati L, Santoni A (2003) The cyclopentenone-type prostaglandin 15-deoxy-delta-12,14-prostaglandin J2 Inhibits CD95 ligand gene expression in T lymphocytes: interference with promoter activation via peroxisome proliferator-activated receptor-gamma-independent mechanisms. J Immunol 170:4578–4592

Coin F, Egly JM (1998) Ten years of TFIIH. Cold Spring Harb Symp Quant Biol 63:105–110

Cole MD, Mango SE (1990) cis-Acting determinants of c-myc mRNA stability. Enzyme 44:167–180

Cramer P, Bushnell DA, Fu JH, Gnatt AL, Maier-Davis B, Thompson NE, Burgess RR, Edwards AM, David PR, Kornberg RD (2000) Architecture of RNA polymerase II and implications for the transcription mechanism. Science 288:640–649

Creancier L, Mercier P, Prats AC, Morello D (2001) c-myc Internal ribosome entry site activity is developmentally controlled and subjected to a strong translational repression in adult transgenic mice. Mol Cell Biol 21:1833–1840

Crosio C, Boyl PP, Loreni F, Pierandrei-Amaldi P, Amaldi F (2000) La protein has a positive effect on the translation of TOP mRNAs in vivo. Nucleic Acids Res 28:2927–2934

Dani C, Mechti N, Piechaczyk M, Lebleu B, Jeanteur P, Blanchard JM (1985) Increased rate of degradation of c-Myc messenger-RNA in interferon-treated daudi cells. Proc Natl Acad Sci U S A 82:4896–4899

Davis AC, Wims M, Spotts GD, Hann SR, Bradley A (1993) A null c-Myc mutation causes lethality before 10.5 days of gestation in homozygotes and reduced fertility in heterozygous female mice. Genes Dev 7:671–682

Davis TL, Firulli AB, Kinniburgh AJ (1989) Ribonucleoprotein and protein factors bind to an H-DNA-forming c-myc DNA element: possible regulators of the c-myc gene. Proc Natl Acad Sci U S A 86:9682–9686

Davis-Smyth T, Duncan RC, Zheng T, Michelotti G, Levens D (1996) The far upstream element-binding proteins comprise an ancient family of single-strand DNA-binding transactivators. J Biol Chem 271:31679–31687

de Nigris F, Mega T, Berger N, Barone MV, Santoro M, Viglietto G, Verde P, Fusco A (2001) Induction of ETS-1 and ETS-2 Transcription Factors Is Required for Thyroid Cell Transformation. Cancer Res 61:2267–2275

Dean M, Levine RA, Ran W, Kindy MS, Sonenshein GE, Campisi J (1986) Regulation of c-myc transcription and mRNA abundance by serum growth factors and cell contact. J Biol Chem 261:9161–9166

DesJardins E, Hay N (1993) Repeated CT elements bound by zinc finger proteins control the absolute and relative activities of the two principal human c-myc promoters. Mol Cell Biol 13:5710–5724

Dickinson LA, Joh T, Kohwi Y, Kohwishigematsu T (1992) A tissue-specific Mar/Sar DNA-binding protein with unusual binding-site recognition. Cell 70:631–645

Dickinson LA, Dickinson CD, Kohwi-Shigematsu T (1997) An atypical homeodomain in SATB1 promotes specific recognition of the key structural element in a matrix attachment region. J Biol Chem 272:11463–11470

Douziech M, Coin F, Chipoulet JM, Arai Y, Ohkuma Y, Egly JM, Coulombe B (2000) Mechanism of promoter melting by the xeroderma pigmentosum complementation group B helicase of transcription factor IIH revealed by protein-DNA photo-cross-linking. Mol Cell Biol 20:8168–8177

Dubik D, Shiu RPC (1992) Mechanism of estrogen activation of c-myc oncogene expression. Oncogene 7:1587–1594

Dufort D, Nepveu A (1994) The human cut homeodomain protein represses transcription from the c-myc promoter. Mol Cell Biol 14:4251–4257

Duncan R, Collins I, Tomonaga T, Zhang T, Levens D (1996) A unique transactivation sequence motif is found in the carboxyl-terminal domain of the single-strand-binding protein FBP. Mol Cell Biol 16:2274–2282

Elowitz MB, Levine AJ, Siggia ED, Swain PS (2002) Stochastic gene expression in a single cell. Science 297:1183–1186

Evingerhodges MJ, Bresser J, Brouwer R, Cox I, Spitzer G, Dicke K (1988) Myc and Sis expression in acute myelogenous leukemia. Leukemia 2:45–49

Facchini LM, Chen S, Marhin WW, Lear JN, Penn LZ (1997) The Myc negative autoregulation mechanism requires Myc-Max association and involves the c-myc P2 minimal promoter. Mol Cell Biol 17:100–114

Filippova GN, Fagerlie S, Klenova EM, Myers C, Dehner Y, Goodwin G, Neiman PE, Collins SJ, Lobanenkov VV (1996) An exceptionally conserved transcriptional repressor, CTCF, employs different combinations of zinc fingers to bind diverged promoter sequences of avian and mammalian c-myc oncogenes. Mol Cell Biol 16:2802–2813

Frit P, Bergmann E, Egly JM (1999) Transcription factor IIH: a key player in the cellular response to DNA damage. Biochimie 81:27–38

Fujimoto M, Matsumoto K-i, Iguchi-Ariga SMM, Ariga H (2001) Disruption of MSSP, c-myc single-strand binding protein, leads to embryonic lethality in some homozygous mice. Genes Cells 6:1067–1075

Fye RM, Benham CJ (1999) Exact method for numerically analyzing a model of local denaturation in superhelically stressed DNA. Phys Rev E 59:3408–3426

Garber ME, Wei P, KewalRamani VN, Mayall TP, Herrmann CH, Rice AP, Littman DR, Jones KA (1998) The interaction between HIV-1 Tat and human cyclin T1 requires zinc and a critical cysteine residue that is not conserved in the murine CycT1 protein. Genes Dev 12:3512–3527

Geltinger C, Hortnagel K, Polack A (1996) TATA box and Sp1 sites mediate the activation of c-myc promoter P1 by immunoglobulin kappa enhancers. Gene Expr 6:113–127

Gnatt AL, Cramer P, Fu JH, Bushnell DA, Kornberg RD (2001) Structural basis of transcription: an RNA polymerase II elongation complex at 3.3 angstrom resolution. Science 292:1876–1882

Grandori C, Cowley SM, James LP, Eisenman RN (2000) The Myc/Max/Mad network and the transcriptional control of cell behavior. Annu Rev Cell Dev Biol 16:653–699

Gronostajski RM (2000) Roles of the NFI/CTF gene family in transcription and development. Gene 249:31–45

Grumont RJ, Strasser A, S G (2002) B cell growth is controlled by phosphatidylinosotol 3-kinase-dependent induction of Rel/NF-kappaB regulated c-myc transcription. Mol Cell 10:1283–1294

Gupta S, Anthony A, Pernis AB (2001) Stage-specific modulation of IFN-regulatory factor 4 function by kruppel-type zinc finger proteins. J Immunol 166:6104–6111

Hann SR, Eisenman RN (1984) Proteins encoded by the human c-Myc oncogene—differential expression in neoplastic-cells. Mol Cell Biol 4:2486–2497

Harris VK, Coticchia CM, List H-J, Wellstein A, Riegel AT (2000) Mitogen-induced expression of the fibroblast growth factor-binding protein is transcriptionally repressed through a non-canonical E-box element. J Biol Chem 275:28539–28548

Hartsough MT, Steeg PS (2000) Nm23/nucleoside diphosphate kinase in human cancers. J Bioenerg Biomembr 32:301–308

Hay N, Bishop JM, Levens D (1987) Regulatory elements that modulate expression of human c-Myc. Genes Dev 1:659–671

He LS, Liu JH, Collins I, Sanford S, O'Connell B, Benham CJ, Levens D (2000) Loss of FBP function arrests cellular proliferation and extinguishes c-myc expression. EMBO J 19:1034–1044

He T-C, Sparks AB, Rago C, Hermeking H, Zawel L, da Costa LT, Morin PJ, Vogelstein B, Kinzler KW (1998) Identification of c-MYC as a target of the APC pathway. Science 281:1509–1512

Hermonat PL (1994) Down-regulation of the human c-fos and c-myc proto-oncogene promoters by adeno-associated virus Rep78. Cancer Lett 81:129–136

Hernandez N, Pessler F (2003) Flexible DNA binding of the BTB/POZ-domain protein FBI-1. J Biol Chem 278:29327–29335

Hewitt SM HS, McDonnell TJ, Rauscher FJ 3rd, Saunders GF (1995) Regulation of the proto-oncogenes bcl-2 and c-myc by the Wilms' tumor suppressor gene WT1. Cancer Res 55:5386–5389

Hoeijmakers JH, Egly JM, Vermeulen W (1996) TFIIH: a key component in multiple DNA transactions. Curr Opin Genet Dev 6:26–33

Iakova P, Awad SS, Timchenko NA (2003) Aging reduces proliferative capacities of liver by switching pathways of C/EBPalpha growth arrest. Cell 113:495–506

Jeay S, Sonenshein GE, Kelly PA, Postel-Vinay MC, Baixeras E (2001) Growth hormone exerts antiapoptotic and proliferative effects through two different pathways involving nuclear factor-kappaB and phosphatidylinositol 3-kinase. Endocrinology 142:147–156

Jenab S, Morris PL (1997) Transcriptional regulation of Sertoli cell immediate early genes by interleukin-6 and interferon-gamma is mediated through phosphorylation of STAT-3 and STAT-1 proteins. Endocrinology 138:2740–2746

Johansen LM, Iwama A, Lodie TA, Sasaki K, Felsher DW, Golub TR, Tenen DG (2001) c-Myc is a critical target for C/EBPalpha in granulopoiesis. Mol Cell Biol 21:3789–3806

Kahn JD, Yun E, Crothers DM (1994) Detection of localized DNA flexibility. Nature 368:163–166

Kaplan J, Calame K (1997) The ZiN/POZ domain of ZF5 is required for both transcriptional activation and repression. Nucleic Acids Res 25:1108–1116

Katayama ML, Pasini FS, Folgueira MA, Snitcovsky IM, Brentani MM (2003) Molecular targets of 1,25(OH)2D3 in HC11 normal mouse mammary cell line. J Steroid Biochem Mol Biol 84:57–69

Keriel A, Stary A, Sarasin A, Rochette-Egly C, Egly JM (2002) XPD mutations prevent TFIIH-dependent transactivation by nuclear receptors and phosphorylation of RAR alpha. Cell 109:125–135

Kiermaier A GJ, Desbarats L, Saffrich R, Ansorge W, Farrell PJ, Eilers M, Packham G (1999) DNA binding of USF is required for specific E-box dependent gene activation in vivo. Oncogene 18:7200–7211

Kim JY, Kang YS, Lee JW, Kim HJ, Ahn YH, Park H, Ko YG, Kim S (2002) p38 is essential for the assembly and stability of macromolecular tRNA synthetase complex: implications for its physiological significance. Proc Natl Acad Sci U S A 99:7912–7916

Kim MJ, Park BJ, Kang YS, Kim HJ, Park JH, Kang JW, Lee SW, Han JM, Lee HW, Kim S (2003a) Downregulation of FUSE-binding protein and c-myc by tRNA synthetase cofactor p38 is required for lung cell differentiation. Nat Genet 34:330–336

Kim SY, Herbst A, Tworkowski KA, Salghetti SE, Tansey WP (2003b) Skp2 regulates Myc protein stability and activity. Mol Cell 11:1177–1188

Kim TK, Ebright RH, Reinberg D (2000) Mechanism of ATP-dependent promoter melting by transcription factor IIH. Science 288:1418–1421

Kinniburgh AJ (1989) A cis-acting transcription element of the c-myc gene can assume an H-DNA conformation. Nucleic Acids Res 17:7771–7778

Kloks CP, Spronk CA, Lasonder E, Hoffmann A, Vuister GW, Grzesiek S, Hilbers CW (2002) The solution structure and DNA-binding properties of the cold-shock domain of the human Y-box protein YB-1. J Mol Biol 316:317–326

Kowalik TF (2002) Smad about E2F. TGFbeta repressionof c-Myc via a Smad3/E2F/p107 complex. Mol Cell 10:7–8

Krumm A, Hickey LB, Groudine M (1995) Promoter-proximal pausing of RNapolymerase-Ii defines a general rate-limiting step after transcription initiation. Genes Dev 9:559–572

Kugel JF, Goodrich JA (1998) Promoter escape limits the rate of RNA polymerase II transcription and is enhanced by TFIIE, TFIIH, and ATP on negatively supercoiled DNA. Proc Natl Acad Sci U S A 95:9232–9237

Kurisaki K, Kurisaki A, Valcourt U, Terentiev AA, Pardali K, ten Dijke P, Heldin C-H, Ericsson J, Moustakas A (2003) Nuclear factor YY1 inhibits transforming growth factor beta- and bone morphogenetic protein-induced cell differentiation. Mol Cell Biol 23:4494–4510

Laird-Offringa IA (1992) What determines the instability of c-myc proto-oncogene mRNA? Bioessays 14:119–124

Lavenu A, Pournin S, Babinet C, Morello D (1994) The cis-acting elements known to regulate c-Myc expression ex-vivo are not sufficient for correct transcription in-vivo. Oncogene 9:527–536

Lavenu A, Pistoi S, Pournin S, Babinet C, Morello D (1995) Both coding exons of the c-myc gene contribute to its posttranscriptional regulation in the quiescent liver and regenerating liver and after protein synthesis inhibition. Mol Cell Biol 15:4410–4419

Lee H, Guo Y, Ohta M, Xiong LM, Stevenson B, Zhu JK (2002) LOS2, a genetic locus required for cold-responsive gene transcription encodes a bi-functional enolase. EMBO J 21:2692–2702

Lemm I, Ross J (2002) Regulation of c-myc mRNA decay by translational pausing in a coding region instability determinant. Mol Cell Biol 22:3959–3969

Levens D (2002) Disentangling the MYC web. Proc Natl Acad Sci U S A 99:5757–5759

Levens DL (2003) Reconstructing MYC. Genes Dev 17:1071–1077

Lin Y, Wong K-k, Calame K (1997) Repression of c-myc transcription by Blimp-1, an inducer of terminal B cell differentiation. Science 276:596–599

Liu G, Malott M, Leffak M (2003) Multiple functional elements comprise a mammalian chromosomal replicator. Mol Cell Biol 23:1832–1842

Liu JH, He LS, Collins I, Ge H, Libutti D, Li JF, Egly JM, Levens D (2000) The FBP interacting repressor targets TFIIH to inhibit activated transcription. Mol Cell 5:331–341

Liu JH, Akoulitchev S, Weber A, Ge H, Chuikov S, Libutti D, Wang XW, Conaway JW, Harris CC, Conaway RC, Reinberg D, Levens D (2001) Defective interplay of activators and repressors with TFIIH in xeroderma pigmentosum. Cell 104:353–363

Lockhart DJ, Winzeler EA (2000) Genomics, gene expression and DNA arrays. Nature 405:827–836

Lutz M, Burke LJ, LeFevre P, Myers FA, Thorne AW, Crane-Robinson C, Bonifer C, Filippova GN, Lobanenkov V, Renkawitz R (2003) Thyroid hormone-regulated enhancer blocking: cooperation of CTCF and thyroid hormone receptor. EMBO J 22:1579–1587

Lutz W, Leon J, Eilers M (2002) Contributions of Myc to tumorigenesis. Biochim Biophys Acta 1602:61–71

Madisen L, Krumm A, Hebbes TR, Groudine M (1998) The immunoglobulin heavy chain locus control region increases histone acetylation along linked c-myc genes. Mol Cell Biol 18:6281–6292

Majello B, De Luca P, Lania L (1997) Sp3 is a bifunctional transcription regulator with modular independent activation and repression domains. J Biol Chem 272:4021–4026

Marcu KB, Bossone SA, Patel AJ (1992) Myc function and regulation. Annu Rev Biochem 61:809–860

Mateyak MK, Obaya AJ, Adachi S, Sedivy JM (1997) Phenotypes of c-myc-deficient rat fibroblasts isolated by targeted homologous recombination. Cell Growth Differ 8:1039–1048

Mautner J, Joos S, Werner T, Eick D, Bornkamm GW, Polack A (1995) Identification of two enhancer elements downstream of the human c-myc gene. Nucleic Acids Res 23:72–80

Mautner J, Behrends U, Hortnagel K, Brielmeier M, Hammerschmidt W, Strobl L, Bornkamm GW, Polack A (1996) c-myc expression is activated by the immunoglobulin kappa-enhancers from a distance of at least 30 kb but not by elements located within 50 kb of the unaltered c-myc locus in vivo. Oncogene 12:1299–1307

McNally JG, Muller WG, Walker D, Wolford R, Hager GL (2000) The glucocorticoid receptor: rapid exchange with regulatory sites in living cells. Science 287:1262–1265

Michelotti EF, Tomonaga T, Krutzsch H, Levens D (1995) Cellular nucleic-acid binding-protein regulates the Ct element of the human c-Myc protooncogene. J Biol Chem 270:9494–9499

Michelotti EF, Michelotti GA, Aronsohn AI, Levens D (1996a) Heterogeneous nuclear ribonucleoprotein K is a transcription factor. Mol Cell Biol 16:2350–2360

Michelotti EF, Sanford S, Freije JM, MacDonald NJ, Steeg PS, Levens D (1997) Nm23/PuF does not directly stimulate transcription through the CT element in vivo. J Biol Chem 272:22526–22530

Michelotti GA, Michelotti EF, Pullner A, Duncan RC, Eick D, Levens D (1996b) Multiple single-stranded cis elements are associated with activated chromatin of the human c-myc gene in vivo. Mol Cell Biol 16:2656–2669

Morales V, Richard-Foy H (2000) Role of histone N-terminal tails and their acetylation in nucleosome dynamics. Mol Cell Biol 20:7230–7237

Norton VG, Marvin KW, Yau P, Bradbury EM (1990) Nucleosome linking number change controlled by acetylation of histones H3 and H4. J Biol Chem 265:19848–19852

Numoto M, Niwa O, Kaplan J, Wong KK, Merrell K, Kamiya K, Yanagihara K, Calame K (1993) Transcriptional repressor ZF5 identifies a new conserved domain in zinc finger proteins. Nucleic Acids Res 21:3767–3775

Ogawa H, Ishiguro K, Gaubatz S, Livingston DM, Nakatani Y (2002) A complex with chromatin modifiers that occupies E2F- and Myc-responsive genes in G0 cells. Science 296:1132–1136

Ohkuma Y (1997) Multiple functions of general transcription factors TFIIE and TFIIH in transcription: possible points of regulation by trans-acting factors. J Biochem (Tokyo) 122:481–489

Ohlsson R, Renkawitz R, Lobanenkov V (2001) CTCF is a uniquely versatile transcription regulator linked to epigenetics and disease. Trends Genet 17:520–527

Pal M, McKean D, Luse DS (2001) Promoter clearance by RNA polymerase II is an extended, multistep process strongly affected by sequence. Mol Cell Biol 21:5815–5825

Pasqualucci L, Neumeister P, Goossens T, Nanjangud G, Chaganti RS, Kuppers R, Dalla-Favera R (2001) Hypermutation of multiple proto-oncogenes in B-cell diffuse large-cell lymphomas. Nature 412:341–346

Pearson A, Greenblatt J (1997) Modular organization of the E2F1 activation domain and its interaction with general transcription factors TBP and TFIIH. Oncogene 15:2643–2658

Pei L (2001) Identification of c-myc as a Down-stream target for pituitary tumor-transforming gene. J Biol Chem 276:8484–8491

Peifer M (2002) Developmental biology: colon construction. Nature 420:274–275

Pellizzoni L, Lotti F, Maras B, Pierandrei-Amaldi P (1997) Cellular nucleic acid binding protein binds a conserved region of the 5′ UTR of Xenopus laevis ribosomal protein mRNAs. J Mol Biol 267:264–275

Perez-Juste G, Garcia-Silva S, Aranda A (2000) An element in the region responsible for premature termination of transcription mediates repression of c-myc gene expression by thyroid hormone in neuroblastoma cells. J Biol Chem 275:1307–1314

Postel EH (1992) Modulation of c-myc transcription by triple helix formation. Ann N Y Acad Sci 660:57–63

Postel EH, Berberich SJ, Rooney JW, Kaetzel DM (2000) Human NM23/nucleoside diphosphate kinase regulates gene expression through DNA binding to nuclease-hypersensitive transcriptional elements. J Bioenerg Biomembr 32:277–284

Pullner A, Mautner J, Albert T, Eick D (1996) Nucleosomal structure of active and inactive c-myc genes. J Biol Chem 271:31452–31457

Qi CF, Martensson A, Mattioli M, Dalla-Favera R, Lobanenkov VV, Morse HC (2003) CTCF functions as a critical regulator of cell-cycle arrest and death after ligation of the B cell receptor on immature B cells. Proc Natl Acad Sci U S A 100:633–638

Quinn LM, Dickins RA, Coombe M, Hime GR, Bowtell DD, Richardson H (2004) Drosophila Hfp negatively regulates dmyc and stg to inhibit cell proliferation. Development 131:1411–1423

Rabbitts PH, Watson JV, Lamond A, Forster A, Stinson MA, Evan G, Fischer W, Atherton E, Sheppard R, Rabbitts TH (1985) Metabolism of C-Myc gene-products—C-Myc messenger-RNA and protein expression in the cell-cycle. EMBO J 4:2009–2015

Rajavashisth TB, Taylor AK, Andalibi A, Svenson KL, Lusis AJ (1989) Identification of a zinc finger protein that binds to the sterol regulatory element. Science 245:640–643

Rao A, Luo C, Hogan PG (1997) Transcription factors of the NFAT family: regulation and function. Annu Rev Immunol 15:707–747

Ray R, Miller DM (1991) Cloning and characterization of a human c-myc promoter-binding protein. Mol Cell Biol 11:2154–2161

Ray R, Steele R, Seftor E, Hendrix M (1995) Human breast carcinoma cells transfected with the gene encoding a c-myc promoter-binding protein (MBP-1) inhibits tumors in nude mice. Cancer Res 55:3747–3751

Reya T, O'Riordan M, Okamura R, Devaney E, Willert K, Nusse R, Grosschedl R (2000) Wnt signaling regulates B lymphocyte proliferation through a LEF-1 dependent mechanism. Immunity 13:15–24

Rich A, Zhang S (2003) Timeline: Z-DNA: the long road to biological function. Nat Rev Genet 4:566–572

Riggs K, Saleque S, Wong K, Merrell K, Lee J, Shi Y, Calame K (1993) Yin-yang 1 activates the c-myc promoter. Mol Cell Biol 13:7487–7495

Robert F, Douziech M, Forget D, Egly JM, Greenblatt J, Burton ZF, Coulombe B (1998) Wrapping of promoter DNA around the RNA polymerase II initiation complex induced by TFIIF. Mol Cell 2:341–351

Roix JJ, McQueen PG, Munson PJ, Parada LA, Misteli T (2003) Spatial proximity of translocation-prone gene loci in human lymphomas. Nat Genet 34:287–291

Rougvie AE, Lis JT (1988) The RNA polymerase-Ii molecule at the 5′ end of the uninduced Hsp70 gene of drosophila-melanogaster is transcriptionally engaged. Cell 54:795–804

Roussel MF, Davis JN, Cleveland JL, Ghysdael J, Hiebert SW (1994) Dual control of myc expression through a single DNA binding site targeted by ets family proteins and E2F-1. Oncogene 9:405–415

Roymans D, Willems R, Van Blockstaele DR, Slegers H (2002) Nucleoside diphosphate kinase (NDPK/NM23) and the waltz with multiple partners: possible consequences in tumor metastasis. Clin Exp Metastasis 19:465–476

Saez AI, Artiga MJ, Romero C, Rodriguez S, Cigudosa JC, Perez-Rosado A, Fernandez I, Sanchez-Beato M, Sanchez E, Mollejo M, Piris MA (2003) Development of a real-time reverse transcription polymerase chain reaction assay for c-myc expression that allows the identification of a subset of c-myc plus diffuse large B-cell lymphoma. Lab Invest 83:143–152

Sakatsume O, Tsutsui H, Wang Y, Gao H, Tang X, Yamauchi T, Murata T, Itakura K, Yokoyama KK (1996) Binding of THZif-1, a MAZ-like zinc finger protein to the nuclease-hypersensitive element in the promoter region of the c-MYC protooncogene. J Biol Chem 271:31322–31333

Schmidt M, Nazarov V, Stevens L, Watson R, Wolff L (2000) Regulation of the resident chromosomal copy of c-myc by c-Myb is involved in myeloid leukemogenesis. Mol Cell Biol 20:1970–1981

Schorl C, Sedivy JM (2003) Loss of protooncogene c-Myc function impedes G(1) phase progression both before and after the restriction point. Mol Biol Cell 14:823–835

Schuhmacher M, Staege MS, Pajic A, Polack A, Weidle UH, Bornkamm GW, Eick D, Kohlhuber F (1999) Control of cell growth by c-Myc in the absence of cell division. Curr Biol 9:1255–1258

Sears R, Leone G, DeGregori J, Nevins JR (1999) Ras enhances Myc protein stability. Mol Cell 3:169–179

Shaffer AL, Rosenwald A, Hurt EM, Giltnane JM, Lam LT, Pickeral OK, Staudt LM (2001) Signatures of the immune response. Immunity 15:375–385

Shang Y, Hu X, DiRenzo J, Lazar MA, Brown M (2000) Cofactor dynamics and sufficiency in estrogen receptor-regulated transcription. Cell 103:843–852

Shaulian E, Karin M (2002) AP-1 as a regulator of cell life and death. Nat Cell Biol 4:E131–E136

Shichiri M, Hanson KD, Sedivy JM (1993) Effects of C-Myc expression on proliferation, quiescence, and the G0 to G1 transition in nontransformed cells. Cell Growth Differ 4:93–104

Siebenlist U, Hennighausen L, Battey J, Leder P (1984) Chromatin structure and protein-binding in the putative regulatory region of the c-Myc gene in Burkitt-lymphoma. Cell 37:381–391

Simonsson T, Pecinka P, Kubista M (1998) DNA tetraplex formation in the control region of c-myc. Nucleic Acids Res 26:1167–1172

Simpson RU, Hsu T, Begley DA, Mitchell BA, Alizadeh BN (1987) Transcriptional regulation of the c-myc protooncogene by 1,25-dihydroxyvitamin D3 in HL-60 promyelocytic leukemia cells. J Biol Chem 262:4101–4108

Sinden RR (1994) DNA supercoiling. In: Sinden RR (ed) DNA structure and function. Academic Press, San Diego, pp 95–133

Song J, Ugai H, Kanazawa I, Sun K, Yokoyama KK (2001) Independent repression of a GC-rich housekeeping gene by Sp1 and MAZ involves the same cis-elements. J Biol Chem 276:19897–19904

Spencer CA, Groudine M (1991) Control of C-Myc regulation in normal and neoplastic-cells. Adv Cancer Res 56:1–48

Subramanian A, Miller DM (2000) Structural analysis of alpha-enolase. Mapping the functional domains involved in down-regulation of the c-myc protooncogene. J Biol Chem 275:5958–5965

Swain PS, Elowitz MB, Siggia ED (2002) Intrinsic and extrinsic contributions to stochasticity in gene expression. Proc Natl Acad Sci U S A 99:12795–12800

Tahirov TH, Temiakov D, Anikin M, Patlan V, McAllister WT, Vassylyev DG, Yokoyama S (2002) Structure of a T7 RNA polymerase elongation complex at 2.9 angstrom resolution. Nature 420:43–50

Takimoto M, Quinn J, Farina A, Staudt L, Levens D (1989) fos/jun and octamer-binding protein interact with a common site in a negative element of the human c-myc gene. J Biol Chem 264:8992–8999

Takimoto M, Tomonaga T, Matunis M, Avigan M, Krutzsch H, Dreyfuss G, Levens D (1993) Specific binding of heterogeneous ribonucleoprotein particle protein-K to the human c-Myc promoter, in-vitro. J Biol Chem 268:18249–18258

Tao L, Dong Z, Leffak M, Zannis-Hadjopoulos M, Price G (2000) Major DNA replication initiation sites in the c-myc locus in human cells. J Cell Biochem 78:442–457

Thalmeier K, Synovzik H, Mertz R, Winnacker EL, Lipp M (1989) Nuclear factor E2F mediates basic transcription and trans-activation by E1a of the human MYC promoter. Genes Dev 3:527–536

Thanos D, Maniatis T (1995) Virus induction of human IFN beta gene expression requires the assembly of an enhanceosome. Cell 83:1091–1100

Tirode F, Busso D, Coin F, Egly JM (1999) Reconstitution of the transcription factor TFIIH: assignment of functions for the three enzymatic subunits, XPB, XPD, and cdk7. Mol Cell 3:87–95

Tomonaga T, Levens D (1995) Heterogeneous nuclear ribonucleoprotein-K Is a DNA-binding transactivator. J Biol Chem 270:4875–4881

Tomonaga T, Michelotti GA, Libutti D, Uy A, Sauer B, Levens D (1998) Unrestraining genetic processes with a protein-DNA hinge. Mol Cell 1:759–764

Trumpp A, Refaeli Y, Oskarsson T, Gasser S, Murphy M, Martin GR, Bishop JM (2001) c-Myc regulates mammalian body size by controlling cell number but not cell size. Nature 414:768–773

Van Buskirk C, Schupbach T (2002) Half pint regulates alternative splice site selection in Drosophila. Dev Cell 2:343–353

Van Lint C, Emiliani S, Verdin E (1996) The expression of a small fraction of cellular genes is changed in response to histone hyperacetylation. Gene Expr 5:245–253

Warrington JA, Nair A, Mahadevappa M, Tsyganskaya M (2000) Comparison of human adult and fetal expression and identification of 535 housekeeping/maintenance genes. Physiol Genomics 2:143–147

Wisdom R, Lee W (1991) The protein-coding region of c-myc mRNA contains a sequence that specifies rapid mRNA turnover and induction by protein synthesis inhibitors. Genes Dev 5:232–243

Wittig B, Wolfl S, Dorbic T, Vahrson W, Rich A (1992) Transcription of human c-myc in permeabilized nuclei is associated with formation of Z-DNA in three discrete regions of the gene. EMBO J 11:4653–4663

Wolf DA, Strobl LJ, Pullner A, Eick D (1995) Variable pause positions of RNA-polymerase-Ii lie proximal to the c-Myc promoter irrespective of transcriptional activity. Nucleic Acids Res 23:3373–3379

Wolfl S, Wittig B, Dorbic T, Rich A (1997) Identification of processes that influence negative supercoiling in the human c-myc gene. Biochim Biophys Acta 1352:213–221

Wu CH, Yamaguchi Y, Benjamin LR, Horvat-Gordon M, Washinsky J, Enerly E, Larsson J, Lambertsson A, Handa H, Gilmour D (2003) NELF and DSIF cause promoter proximal pausing on the hsp70 promoter in Drosophila. Genes Dev 17:1402–1414

Yasui D, Miyano M, Cai ST, Varga-Weisz P, Kohwi-Shigematsu T (2002) SATB1 targets chromatin remodelling to regulate genes over long distances. Nature 419:641–645

Yeilding NM, Lee WM (1997) Coding elements in exons 2 and 3 target c-myc mRNA downregulation during myogenic differentiation. Mol Cell Biol 17:2698–2707

Yin YW, Steitz AA (2002) Structural basis for the transition from initiation to elongation transcription in T7 RNA polymerase. Science 298:1387–1395

Zawel L, Kumar KP, Reinberg D (1995) Recycling of the general transcription factors during RNA polymerase II transcription. Genes Dev 9:1479–1490

Zhang GY, Campbell EA, Minakhin L, Richter C, Severinov K, Darst SA (1999) Crystal structure of Thermus aquaticus core RNA polymerase at 3.3 angstrom resolution. Cell 98:811–824

Zhang XY, Jabrane-Ferrat N, Asiedu CK, Samac S, Peterlin BM, Ehrlich M (1993) The major histocompatibility complex class II promoter-binding protein RFX (NF-X) is a methylated DNA-binding protein. Mol Cell Biol 13:6810–6818

Zhao H, Jin S, Fan F, Fan W, Tong T, Zhan Q (2000) Activation of the transcription factor Oct-1 in response to DNA damage. Cancer Res 60:6276–6280

Zheng L, Roeder RG, Luo Y (2003) S phase activation of the histone H2B promoter by OCA-S, a coactivator complex that contains GAPDH as a key component. Cell 114:255–266

Zhou Q, Chen D, Pierstorff E, Luo KX (1998) Transcription elongation factor P-TEFb mediates Tat activation of HIV-1 transcription at multiple stages. EMBO J 17:3681–3691

Zobel A, Kalkbrenner F, Guehmann S, Nawrath M, Vorbrueggen G, Moelling K (1991) Interaction of the v- and c-Myb proteins with regulatory sequences of the human c-myc gene. Oncogene 6:1397–1407

CTMI (2006) 302:33–50

Transcriptional Activation by the Myc Oncoprotein

M. D. Cole (✉) · M. A. Nikiforov

Departments of Pharmacology and Genetics, Dartmouth Medical School,
Lebanon, NH 03756, USA
michael.cole@dartmouth.edu

Abstract The Myc transcription factor functions as a downstream effector of most mitogenic signals. Myc is synthesized rapidly in response to extracellular mitogenic signals, and blocking Myc induction abolishes or at least severely attenuates any mitogenic response. Furthermore, ectopic Myc expression can often bypass the requirement for extracellular signals for entry into S phase. Thus, the Myc transcription factor is both necessary and in many ways sufficient to promote the growth of diverse cell types. Given this potent biological activity, it is not surprising that mutations in the *myc* gene are among the most frequent in human and animal cancers. Understanding the molecular basis of Myc function has been a central issue in the fields of cancer biology and signal transduction for 20 years.

1
Early Myc Transcription Factor Connections

The earliest studies showing that v-Myc protein was localized in the nucleus hinted that Myc could be a transcription factor (Hann et al. 1983), and studies began to explore the possibility that it could regulate specific genes (Dean et al. 1987; Prendergast and Cole 1989). The first concrete link between Myc and transcription factors came from the finding that Myc contained both leucine zipper and helix-loop-helix motifs (Landschulz et al. 1988; Murre et al. 1989). Both of these motifs were first documented in sequence-specific DNA binding proteins, providing a clear indication that Myc should also be a DNA binding protein. However, it was not until the identification of the Max protein that Myc gained real credibility as a sequence-specific transcription factor (Blackwood and Eisenman 1991). Max is a small, ubiquitously expressed protein that can itself homodimerize and bind to DNA. However, Max homodimers are inhibited from binding DNA in vivo by phosphorylation (Berberich and Cole 1992). Myc must heterodimerize with Max to bind to DNA (Blackwood et al. 1992), yet DNA binding by the Myc/Max heterodimer is not inhibited by phosphorylation at sites that are comparable to those that inhibit Max homodimers (Berberich and Cole 1992). Max can also dimerize with other proteins, but these proteins and the structure of the DNA binding domain will be discussed elsewhere in this volume.

Sequence-specific transcription factors are usually modular, with a well-defined, evolutionarily conserved DNA binding domain and a more loosely defined effector domain that either activates or represses transcription when tethered near a basal promoter. Before Myc was definitively shown to be a DNA binding protein, it was found that fusion of the Myc N-terminus to the Gal4 DNA binding domain created a potent transactivator (>100-fold; Kato et al. 1990). However, this work raised two puzzles that remain unresolved. First, there was not a good correlation between domains within the N-terminus that were biologically important with those that promoted transcriptional activation (Kato et al. 1990). Second, once the Max protein and consensus DNA binding sites were available, it was found that Myc/Max heterodimers were much less potent at transcriptional activation (typically 3- to 4-fold) than Gal4-Myc fusion proteins, even using concatamerized binding sites (Kretzner et al. 1992). The latter, relatively weak activity has now been confirmed in vivo through the analysis of chromosomally localized target genes which have Myc/Max binding sites (Bush et al. 1998). In fact, several broad studies of Myc target genes using microarrays find an average transcriptional activation of chromosomal targets of approximately twofold, for example (Coller et al. 2000; O'Connell et al. 2003). Only a small number of the Myc responsive genes identified in these studies are verified as in vivo targets, but the average response of verified targets is quite comparable (2- to 3-fold). Hence, while virtually all studies agree that Myc is a direct activator of transcription, its activity is inevitably quite modest and pales in comparison to potent transactivators like NF-κB (100- to 1,000-fold in comparable assays).

2
Biologically Functional Myc Domains

A fruitful approach toward understanding Myc function has been to map the functional domains in various assays (Fig. 1). Distinct evolutionarily conserved regions called Myc homology boxes I and II (MBI and MBII) within the N-terminus were required for the ability of Myc to cooperate with an H-*ras* oncogene in the transformation of primary rodent fibroblasts and for transformation of a sensitized rodent fibroblast cell line (Stone et al. 1987). Similar domains were found to be required for inducing apoptosis and blocking differentiation (Evan et al. 1992; Freytag et al. 1990). All of these Myc-induced phenotypes are dependent on expression of Myc at levels higher than usually observed under physiologically normal settings.

Until recently, almost no information was available concerning the domains of Myc that function to promote normal cell proliferation since virtually

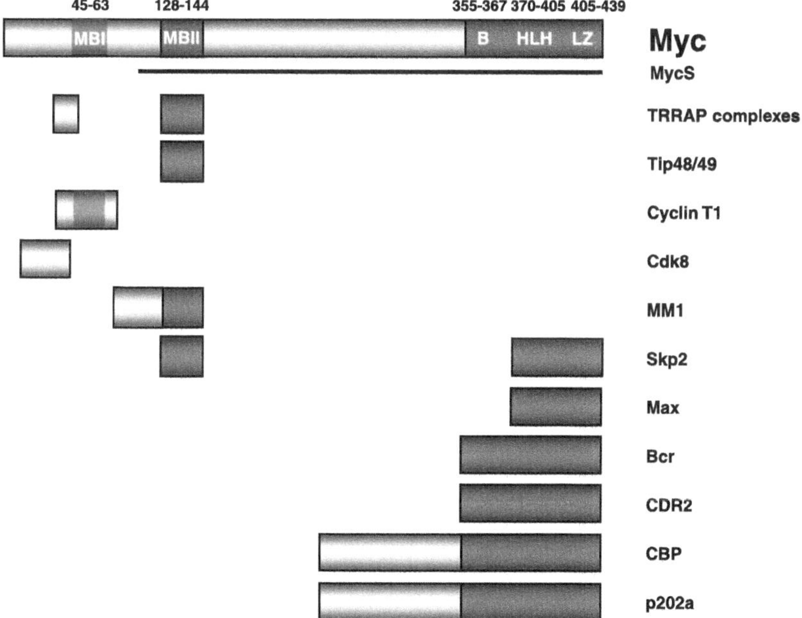

Fig. 1 Functional domains of the Myc protein. The conserved Myc homology box I (*MBI*), Myc homology box II (*MBII*) and DNA binding domains are highlighted. The DNA binding domain consists of basic (*B*), helix-loop-helix (*HLH*), and leucine zipper (*LZ*) domains. Amino acid numbers corresponds to c-Myc. The regions of c-Myc that interact with different nuclear cofactors are shown in the *lower section* of the figure

all cells express an endogenous level of c-myc. The availability of a *myc*-null cell line made it possible to assess the role of Myc-dependent transactivation in fibroblast proliferation (Mateyak et al. 1997). The first surprise from these assays is that the conserved MBI region of c-Myc is not required for fibroblast proliferation, based on the activity of the MycS protein. MycS is a naturally occurring protein variant that originates from translation initiation at either of two methionines at amino acids 103 or 111 in relation to the initial methionine of the most abundant Myc2 protein (Spotts et al. 1997; Fig. 1). MycS thus lacks the N-terminal 100 amino acids of c-Myc, including MBI, but retains the C-terminal 260 amino acids spanning MBII and the DNA binding domain. MycS is devoid of all transactivation activity assayed with reporter constructs, but ectopic expression of MycS can induce anchorage-independent growth and apoptosis as well as rescue the cell cycle delay of Myc-deficient fibroblasts (Xiao et al. 1998). Thus, transactivation by the Myc N terminus (as defined by transient reporter assays) is neither necessary nor sufficient in many biological assays. However, MycS is completely defective for cooperation with H-*ras* in the transformation of primary rodent cells (McMahon et al. 1998). A comparison of target genes that respond to MycS should prove valuable in deciphering the Myc proliferative versus transforming activity. Even though MycS lacks detectable transactivation in transient reporter assays, activation of specific cellular promoters may be required for MycS function, and the apparent dichotomy between MycS and c-Myc is only a consequence of the inability of transient reporter assays to recapitulate the regulation of chromosomal targets. Alternately, the biological functions of Myc and MycS may be linked to gene repression rather than activation (Xiao et al. 1998). The mechanism of Myc repression will not be discussed in this review.

A second surprise from fibroblast proliferation assays is that MBII is not absolutely essential for proliferation. The first assay for MBII mutants in c-Myc reported a cell doubling time intermediate between wild-type (wt) c-Myc and vector controls, a rescue of approximately 50% of the growth defect (Bush et al. 1998). More recent studies of both c-Myc△MBII and N-Myc△MBII mutants demonstrate a more substantial rescue of the slow growth phenotype (Nikiforov et al. 2002; Oster et al. 2003). In support of an uncoupling of proliferation from oncogenic transformation, L-*myc* can also fully rescue *myc*-null cell growth even though it is defective for transactivation in transient assays and for transformation in cooperation with H-*ras* (Barrett et al. 1994; Birrer et al. 1988; Landay et al. 2000; Nikiforov et al. 2002).

One possible interpretation for the difference in Myc domain requirements between the fibroblast proliferation and H-*ras* cooperation assays is the absolute level of "Myc function" required. It has been estimated that as few as 450 molecules of Myc per cell are sufficient for normal proliferation (Mehmet et al. 1997), compared to 5,000 molecules per cell after serum stimulation and over 100,000 molecules per cell in some tumor lines (Moore et al. 1987). Loss of MBI in the MycS protein or "weakened" Myc proteins such as MycΔMBII or L-Myc may still have enough activity to promote cell proliferation in a *myc*-null background. On the other hand, the H-*ras* cooperation and other assays require Myc levels above those found in normal fibroblasts, and hence these assays may be more sensitive to any loss of wt Myc function. However, it is also possible that different forms of Myc have qualitatively distinct biological activities mediated by their ability to activate or repress select target genes.

An alternate model for the role of MBII in proliferation is based on the difference between immortalized versus primary cells. The most stringent MBII-dependent assay for Myc function is the transformation of primary rat embryo fibroblasts in cooperation with H-*ras* (Land et al. 1983). An MBII-dependence is also observed for the proliferation of primary cerebellar granule neurons (Kenney et al. 2003). Immortalization of cells in culture often requires the disruption of the p53 pathway through loss of p53 itself or of p19ARF, a modulator of the p53/mdm2 regulatory loop (Lundberg et al. 2000). Because of this and other undefined genetic or epigenetic changes that accompany immortalization, immortalized cells may have a qualitatively different dependence on MBII for target gene regulation. For example, as discussed in more detail below, the telomerase reverse transcriptase (*TERT*) gene is profoundly dependent on MBII for activation in primary human fibroblasts, but only minimally dependent on MBII in an immortalized line (Nikiforov et al. 2002). Thus, the response of target genes to Myc overexpression may vary between primary and immortalized cells, accounting for the differences in biological activity observed in different systems.

3
Nuclear Cofactors of Myc

The function of most sequence-specific transcription factors is to recruit nuclear cofactors to specific promoters. These cofactors can be general transcription factors (GTFs) that facilitate the entry and movement of RNA polymerase or they can modify local chromatin structure to either enhance or repress transcription or the binding of other factors. Numerous proteins have been identified as potential Myc cofactors, which have been reviewed recently (Sakamuro and Prendergast 1999). For the purpose of this review, we will focus on recent work characterizing cofactors that have been demonstrated to play a functional role in Myc activities and which have also been shown to coprecipitate with Myc in vivo (Fig. 2).

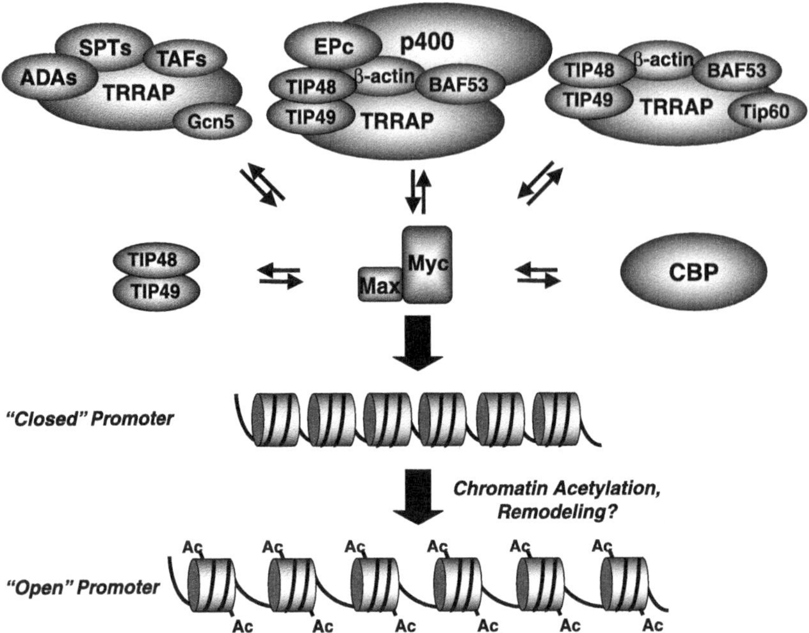

Fig. 2 Model for Myc recruitment of nuclear cofactors that promote chromatin modification. The Myc protein recruits several complexes that can promote localized modification and remodeling of chromatin. These complexes may alter the acetylation around Myc target genes or perturb chromatin in some other undefined way

3.1
Myc and Chromatin Modification

The MBII domain has recently been shown to facilitate Myc binding to a novel large nuclear cofactor called TRRAP (*tr*ansactivation/*t*ransformation-domain *a*ssociated *p*rotein), which was purified by affinity chromatography using the c-Myc N-terminal transactivation domain (McMahon et al. 1998). TRRAP is a 3,830-amino-acid protein with limited homology to the phosphoinositide (PI)-3 kinase/ATM family, although TRRAP lacks the kinase catalytic residues present in other members of the family (McMahon et al. 1998). TRRAP binding to the N-terminus is directly correlated with Myc oncogenic activity, since deletions or mutations in Myc that disrupt TRRAP binding are transformation-defective, and the weakly transforming L-Myc protein exhibits poor TRRAP binding (McMahon et al. 1998; Nikiforov et al. 2002). Furthermore, the disruption of endogenous TRRAP pools using antisense and ectopic expression of TRRAP fragments with dominant inhibitory activity severely impair Myc-mediated oncogenic transformation (McMahon et al. 1998). These data imply that the recruitment of TRRAP to cellular promoters is essential for Myc-mediated oncogenic transformation.

The identification of TRRAP as an essential cofactor provided an important mechanistic insight into the function of the Myc N-terminal domain when TRRAP was found to be part of the SAGA complex (Grant et al. 1998a; Saleh et al. 1998; Vassilev et al. 1998). SAGA (*S*PT/*A*DA/*G*CN5/*a*cetyltransferase) is a 1.8-MDa complex containing approximately 20 proteins which have been implicated in transcriptional regulation, primarily through genetic screens in yeast (Grant et al. 1997). Myc binds directly to a small internal domain of TRRAP that is similar in location to the binding site on Tra1p for transcription factors in yeast (Brown et al. 2001; Park et al. 2001). Several recent studies have demonstrated that, in addition to TRRAP, many other components of the SAGA complex are also highly conserved from yeast to humans (Martinez et al. 2001; Ogryzko et al. 1998; Smith et al. 1998). Among the many proteins contained in SAGA, the only one with a clearly defined biochemical function is the histone acetyltransferase GCN5 (Georgakopoulos and Thireos 1992; Marcus et al. 1994; Wang et al. 1997). Histone acetylation by transcription cofactors has frequently been associated with gene activation (Grant et al. 1998b), making this an attractive mechanism for Myc-mediated transactivation. This model is appealing since the alternate Max heterodimeric partners, Mad and Mxi, can antagonize Myc function through recruitment of Sin3A/Sin3B and the histone deacetylases HDAC1/2 (Ayer 1999; Knoepfler and Eisenman 1999).

A series of studies has found several other complexes that contain TRRAP in addition to the human ortholog of the yeast SAGA complex. The SAGA complex is related to the STAGA and TFTC coactivator complex defined in transcription assays (Brand et al. 1999; Martinez et al. 2001). TRRAP is also found in a complex with the TIP60 H2A/H4 HAT (Ikura et al. 2000), and this complex shares many subunits with a complex containing the Swi2/Snf2-related p400 protein (Fuchs et al. 2001). However, the latter complex lacks HAT activity. The p400 complex binds to c-Myc in U2OS extracts, presumably through TRRAP, but the functional consequences of this recruitment remain unclear. Alternately, it is possible that TRRAP-containing complexes may directly repress target gene expression, even though histone acetylation is usually linked to gene activation. In support of this concept, the ARG1 gene in yeast was shown to be repressed through recruitment of GCN5 HAT activity and the SAGA complex (Ricci et al. 2002). Mutants of GCN5 that lack HAT activity fail to repress ARG1, and the SAGA complex is recruited directly to ARG1 through the ArgR/Mcm1 repressor complex. It is not clear if it is the acetylation of histones or some other substrate that mediates repression, but the direct involvement of TRRAP/TRA1-linked HAT activity to repression could explain the MBII-dependence of Myc repression (Claassen and Hann 1999).

A general model of Myc-mediated histone acetylation as the basis of oncogenic transformation raises a number of important questions that must be resolved in the future. On one hand, Myc recruits HAT activity (McMahon et al. 2000), and TRRAP is recruited to the promoters of several Myc-responsive genes following serum stimulation in association with induction of H4 but not always H3 acetylation (Bouchard et al. 2001; Frank et al. 2001). It was recently shown that the silent TERT gene acquires both H3 and H4 acetylation in the course of being activated by c-Myc in primary human fibroblasts (Nikiforov et al. 2002). On the other hand, another study reported that activation of the Myc target genes *cad* and *TERT* in cell lines occurs without concomitant increases in histone H3 or H4 acetylation (Eberhardy et al. 2000). Furthermore, MBII mutants that fail to recruit HAT activity can still induce several Myc target genes in their native chromosomal context (Nikiforov et al. 2002). The MycS protein does not bind to TRRAP, but it can rescue the growth defect in *myc*-null fibroblasts (McMahon et al. 1998; Xiao et al. 1998). This suggests that MycS may interact with other cofactors to mediate cell cycle progression, anchorage-independent growth, and apoptosis. Finally, although the TIP60 H4 HAT is recruited to c-Myc target genes and promotes localized histone acetylation, abolishing TIP60 HAT activity had no impact on Myc target gene expression (Frank et al. 2003). Thus, it remains possible that localized HAT recruitment may correlate well with Myc target gene activation because this activity is recruited as part of TRRAP complexes. However, HAT activity itself may be dispensable and the critical function recruited by Myc is provided by TRRAP itself or other associated proteins.

Another set of cofactors recruited by Myc are evolutionarily conserved proteins called TIP49 and TIP48, which contain ATPase motifs (Wood et al. 2000). These proteins are found as part of the TRRAP:TIP60 HAT complex in mammalian cells (Ikura et al. 2000), but some mutations in Myc retain TIP49/48 binding while losing TRRAP binding, suggesting that these proteins may interact with Myc independently (Wood et al. 2000). An ATPase-defective mutant of TIP49 was a potent inhibitor of Myc oncogenic transformation but had little effect in proliferation (Wood et al. 2000). This same mutation enhanced Myc-mediated apoptosis (Dugan et al. 2002), suggesting that apoptosis and transformation may result from different pathways. The biochemical function of the TIP49/48 proteins remains unclear. They are not components of the analogous H4 histone acetyltransferase complex in yeast (Allard et al. 1999), although they are found in other yeast chromatin remodeling complexes (Shen et al. 2000).

3.2
TRRAP-Independent Target Gene Activation

While the recruitment of TRRAP complexes and the localized modification of chromatin are likely to be part of Myc-dependent gene activation, considerable evidence suggests that Myc has other functions in activating target genes. Little or no change in histone acetylation was found during the activation of the *cad* and *TERT* promoters in some cell types (Eberhardy et al. 2000; Eberhardy and Farnham 2001), and Myc mutants that are defective in TRRAP and HAT recruitment can still activate many target genes nearly as well as wt Myc in log phase cells (Nikiforov et al. 2002). Further exploration into the mechanism of this TRRAP-independent gene activation has led to the finding that RNA PolII remains engaged at the *cad* promoter even in the absence of Myc protein binding (Eberhardy and Farnham 2001). No change in PolII binding to the promoter was found in either serum-stimulated fibroblasts or in differentiating U937 cells despite large changes in Myc binding. On the other hand, PolII binding to the 3′ end of the *cad* gene was regulated in concert with Myc binding to the promoter and transcription of the gene, suggesting that Myc might regulate promoter clearance or elongation of a stalled PolII preinitiation complex (Eberhardy and Farnham 2001). Significant changes in PolII binding were also found in an intronic region of another Myc-regulated gene, nucleolin. Stimulation of PolII elongation is thought to be controlled by phosphorylation of the carboxy-terminal domain (CTD) by protein kinases. Among the kinases that phosphorylate the CTD is P-TEFb (positive transcription elongation factor b) which is a complex of cdk9 and cyclin T1 (Price 2000). In in vitro binding assays, the Myc transactivation domain can bind to P-TEFb, whereas the USF transactivation domain cannot (Eberhardy and Farnham 2002). Furthermore, an artificial *cad* promoter containing Gal4 binding sites was stimulated by a Gal4-Myc fusion and by Gal4-cyclinT1 cotransfected with a cdk9 expression vector. These data are consistent with a model in which Myc regulates promoter clearance, independent of its role in chromatin modification (Fig. 3). However, it remains to be shown that Myc/Max heterodimers recruit P-TEFb or any other CTD kinase in vivo.

3.3
Myc, Ubiquitylation and Transcriptional Activation

The c-Myc protein turns over rapidly in the cell with a half-life of 15–20 min (Hann and Eisenman 1984). c-Myc turnover is linked to ubiquitylation (Salghetti et al. 1999), and the half-life is modulated by the Ras pathway and by mutations that are commonly found in Burkitt's lymphomas (Gregory and Hann 2000; Salghetti et al. 1999; Sears et al. 1999). The turnover pathway has

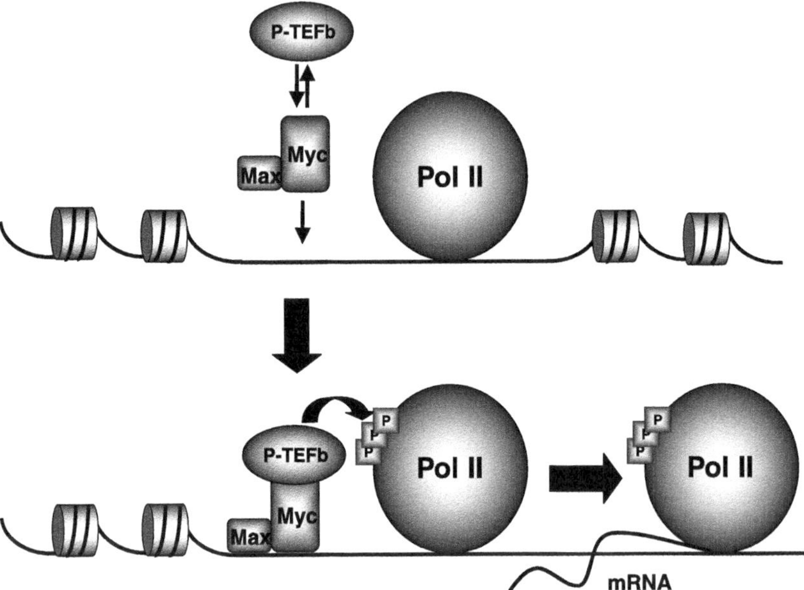

Fig. 3 Model for Myc recruitment of basal transcription factors. The Myc protein can promote a stalled RNA polymerase to transcribe mRNAs by recruiting the P-TEFb complex which phosphorylates the carboxy terminal domain of the polymerase

taken an unexpected twist lately with the discovery that Myc ubiquitylation may be directly linked to transcriptional activation. The Myc "degron" maps within the transactivation domain and it has been proposed that the ubiquitylation of many transcription factors is required for their activity (Conaway et al. 2002; Salghetti et al. 2001; Salghetti et al. 1999). It has recently been shown that the ubiquitin ligase component Skp2 binds to Myc and regulates its turnover (Kim et al. 2003; von der Lehr et al. 2003). Moreover, Skp2 actually stimulates Myc-dependent transcriptional activation for both transiently expressed and endogenous target genes. The mechanism by which Skp2 stimulates Myc transactivation remains unclear, but this observation may offer a partial explanation for the oncogenic activity of Skp2 (Gstaiger et al. 2001; Latres et al. 2001).

3.4 Myc and CBP

Another nuclear cofactor found to activate transcription in conjunction with c-Myc is CREB-binding protein (CBP) (Vervoorts et al. 2003), which is a close relative of the E1A-associated p300 protein. Both CBP and p300 have inherent HAT activity, and both can be recruited to many transcription factors. CBP binds to c-Myc in vivo, and cotransfection of CBP with c-Myc provides a substantial stimulation of Myc-dependent transactivation (Vervoorts et al. 2003). Curiously, CBP binds to c-Myc through a C-terminal region, yet transfection of N-terminal deletion mutants of Myc that contain the CBP interaction domain do not activate transcription or have other biological activity. This implies that Myc recruitment of endogenous CBP pools is not sufficient to activate transcription, even though CBP is recruited to Myc-regulated promoters. One possibility is that CBP acetylates c-Myc itself, perhaps leading to a change in Myc protein ubiquitination (Vervoorts et al. 2003).

3.5
Inhibitors of Myc Activity

The cofactors discussed above are thought to mediate the activation of Myc target genes. A number of other proteins that interact with Myc domains have the opposite effect: They inhibit Myc function and/or block Myc-mediated transactivation. The majority of these Myc inhibitors have been identified through yeast two-hybrid screens and we will discuss recently characterized interactions that have been demonstrated to occur in vivo with endogenous proteins. One inhibitor of Myc to emerge is the cdr2 protein that is normally expressed in cerebellar Purkinje neurons (Okano et al. 1999). Cdr2 binds to c-Myc, but not Max, through the leucine zipper, and this binding sequesters c-Myc into the cytoplasm. Cdr2 is an antigen associated with perineoplastic cerebellar degeneration (PCD), a disorder in which the onconeural antigen cdr2 is expressed in breast and ovarian cancers and the anti-cdr2 antibodies promote neural degeneration (Okano et al. 1999). PCD antisera block the cdr2-Myc interaction in vitro and could theoretically free Myc to promote unscheduled cell-cycle entry in Purkinje neurons, subsequently leading to cell death. However, only 20% of Purkinje neurons express both c-Myc and cdr2, whereas all Purkinje neurons express cdr2 itself (Okano et al. 1999), making it unclear if targeting of only a subset of neurons could account for the disease.

Two other inhibitors of c-Myc function have also been described. The differentiation and interferon inducible p202a protein can inhibit c-Myc transcriptional activity and dimerization with Max (Wang et al. 2000). p202 is induced in differentiated cells and overexpression inhibits cell proliferation. p202a can interact with (and inhibit) a number of other transcription factors besides Myc, such as c-Fos, c-Jun, AP2, E2F, myoD, and NF-κB (Wang et al. 2000). Overexpression of p202a can both reduce dimerization of c-Myc with

Max and inhibit Myc-dependent transactivation. Furthermore, the overexpression of p202 can reduce the expression of Myc target genes, although this may also be an indirect consequence of growth inhibition and/or its effects on other transcription factors.

Another inhibitor of c-Myc function is the breakpoint cluster region (BCR) protein. BCR is the fusion partner for the c-Abl tyrosine kinase in chronic myelogenous leukemia. The BCR protein scored an interaction with c-Myc in a yeast two-hybrid screen that was validated by in vivo co-immunoprecipitation (Mahon et al. 2003). c-Myc does not interact with the oncogenic fusion protein BCR-ABL since the binding domain within BCR is C-terminal to the junction. BCR appears to suppress c-Myc activity by competing with Max for binding to the C-terminal B/HLH/LZ (basic region/helix-loop-helix/leucine zipper) domain. BCR can suppress the ability of c-Myc to activate the expression of an artificial reporter construct as well as the endogenous cyclin D2 gene. It can also suppress the ability of c-Myc to cooperate with H-Ras(G12V) in the transformation of NIH3T3 cells.

The preceding three c-Myc inhibitors raise interesting questions that remain unresolved. What fraction of the endogenous c-Myc protein is inhibited by any of these proteins at native levels of expression? Since inhibitor binding is mutually exclusive with Myc/Max dimerization, one might expect a variable pool of c-Myc protein that was complexed with inhibitors rather than Max. However, in our hands, immunoprecipitation of Max can remove virtually all detectable c-Myc from lysates of cells that overexpress Myc proteins (M.D. Cole, unpublished observations), suggesting the all of the c-Myc protein is in Max complexes. On the other hand, if the inhibitor-Myc complexes promoted an enhanced c-Myc turnover, this complex might not accumulate to a significant extent (Mahon et al. 2003). Further work will be required to resolve the dynamics of c-Myc inhibition in these systems.

3.6
Other Myc Interacting Proteins

Another repressor of c-Myc is the MM-1%STOP protein (Mori et al. 1998; Satou et al. 2001). MM-1 can interact with TIF1β/KAP1 and HDACs, although no binding of the latter cofactors has been described with endogenous c-Myc protein. Several other nuclear proteins have been described as Myc-interacting proteins, but further discussion of their role in Myc-dependent activities will require more thorough documentation that they interact with endogenous Myc protein in vivo.

4 The Myc Conundrum

The biggest puzzle for understanding Myc function is how a single transcription factor can have such profound biological activity while its ability to modulate any specific target gene expression is so muted. Does Myc promote cellular growth and oncogenic transformation by the twofold induction of thousands of target genes? Or are there specific target genes, perhaps like TERT, whose activation is proportionately much larger, since, for example, TERT expression is virtually undetectable in primary cells? Even minimal levels of TERT expression may enhance the long-term growth of tumor cells. Therefore, low but significant levels of other Myc target genes may have similar effects on cellular growth properties. Genes such as these might be below the threshold of current microarray experiments. An even larger mystery is the role of Myc in tumor cell growth versus the growth of normal cells. One might predict that the high levels of Myc found after chromosomal translocation, gene amplification, or even hyperstimulation of the signaling pathway leading to elevated endogenous *myc* expression would transcriptionally activate novel targets that were not normally Myc regulated. Yet, despite years of searching, no Myc targets have been discovered that are uniquely activated in tumor cells, with the possible exception of TERT. Even for TERT, there is little evidence that its activation in human tumor cells is actually Myc-dependent. This ultimately presents a quandary. Lots of genes have Myc binding sites, lots of genes respond weakly to changing Myc levels, but it has proved exceedingly difficult to link specific genes to Myc function. The weak transactivation activity of Myc and the target gene conundrum raise the possibility that Myc is not primarily a conventional transcription factor at all, but serves some other function in chromosome structure or dynamics. Further studies of the nuclear factors that interact with Myc should provide more insight into this enigmatic oncoprotein.

References

Allard S, Utley RT, Savard J, Clarke A, Grant P, Brandl CJ, Pillus L, Workman JL, Cote J (1999) NuA4, an essential transcription adaptor/histone H4 acetyltransferase complex containing Esa1p and the ATM-related cofactor Tra1p. EMBO J 18:5108–5119

Ayer DE (1999) Histone deacetylases: transcriptional repression with SINers and NuRDs. Trends Cell Biol 9:193–198

Barrett J, Birrer MJ, Kato GJ, Dosaka-Akita H, Dang CV (1994) Activation domains of L-Myc and c-Myc determine their transforming potencies in rat embryo cells. Mol Cell Biol 12:3130–3137

Berberich SJ, Cole MD (1992) Casein kinase II inhibits the DNA-binding activity of Max homodimers but not Myc/Max heterodimers. Genes Dev 6:166–176

Birrer MJ, Segal S, Degreve JS, Kaye F, Sausville EA, Minna JD (1988) L-myc cooperates with ras to transform primary rat embryo fibroblasts. Mol Cell Biol 8:2668–2673

Blackwood EM, Eisenman RN (1991) Max: a helix-loop-helix zipper protein that forms a sequence-specific DNA-binding complex with Myc. Science 251:1211–1217

Blackwood EM, Kretzner L, Eisenman RN (1992) Myc and Max function as a nucleoprotein complex. Curr Opin Genet Dev 2:227–235

Bouchard C, Dittrich O, Kiermaier A, Dohmann K, Menkel A, Eilers M, Luscher B (2001) Regulation of cyclin D2 gene expression by the Myc/Max/Mad network: Myc-dependent TRRAP recruitment and histone acetylation at the cyclin D2 promoter. Genes Dev 15:2042–2047

Brand M, Yamamoto K, Staub A, Tora L (1999) Identification of TATA-binding protein-free TAFII-containing complex subunits suggests a role in nucleosome acetylation and signal transduction. J Biol Chem 274:18285–18289

Brown CE, Howe L, Sousa K, Alley SC, Carrozza MJ, Tan S, Workman JL (2001) Recruitment of HAT complexes by direct activator interactions with the ATM-related Tra1 subunit. Science 292:2333–2337

Bush A, Mateyak M, Dugan K, Obaya A, Adachi S, Sedivy J, Cole MD (1998) c-myc null cells misregulate cad and gadd45 but not other proposed c-Myc targets. Genes Dev 12:3797–3802

Claassen G, Hann S (1999) Myc-mediated transformation: the repression connection. Oncogene 18:2925–2933

Coller HA, Grandori C, Tamayo P, Colbert T, Lander ES, Eisenman RN, Golub TR (2000) Expression analysis with oligonucleotide microarrays reveals that MYC regulates genes involved in growth, cell cycle, signaling, and adhesion. Proc Natl Acad Sci USA 97:3260–3265

Conaway RC, Brower CS, Conaway JW (2002) Emerging roles of ubiquitin in transcription regulation. Science 296:1254–1258

Dean M, Cleveland JL, Rapp UR, Ihle JN (1987) Role of myc in the abrogation of IL3 dependence of myeloid FDC-P1 cells. Oncogene Res 1:279–296

Dugan KA, Wood MA, Cole MD (2002) TIP49, but not TRRAP, modulates c-Myc and E2F1 dependent apoptosis. Oncogene 21:5835–5843

Eberhardy S, D'Cunha C, Farnham P (2000) Direct examination of histone acetylation on Myc target genes using chromatin immunoprecipitation. J Biol Chem 275:33798–33805

Eberhardy SR, Farnham PJ (2001) c-Myc mediates activation of the cad promoter via a post-RNA polymerase II recruitment mechanism. J Biol Chem 276:48562–48571

Eberhardy SR, Farnham PJ (2002) Myc recruits P-TEFb to mediate the final step in the transcriptional activation of the cad promoter. J Biol Chem 277:40156–40162

Evan GI, Wyllie AH, Gilbert CS, Littlewood TD, Land H, Brooks M, Waters CM, Penn LZ, Hancock DC (1992) Induction of apoptosis in fibroblasts by c-myc protein. Cell 69:119–128

Frank SR, Schroeder M, Fernandez P, Taubert S, Amati B (2001) Binding of c-Myc to chromatin mediates mitogen-induced acetylation of histone H4 and gene activation. Genes Dev 15:2069–2082

Frank SR, Parisi T, Taubert S, Fernandez P, Fuchs M, Chan HM, Livingston DM, Amati B (2003) MYC recruits the TIP60 histone acetyltransferase complex to chromatin. EMBO Rep 4:575–580

Freytag SO, Dang CV, Lee WMF (1990) Definition of the activities and properties of c-myc required to inhibit cell differentiation. Cell Growth Differ 1:339–343

Fuchs M, Gerber J, Drapkin R, Sif S, Ikura T, Ogryzko V, Lane WS, Nakatani Y, Livingston DM (2001) The p400 complex is an essential E1A transformation target. Cell 106:297–307

Georgakopoulos T, Thireos G (1992) Two distinct yeast transcriptional activators require the function of the GCN5 protein to promote normal levels of transcription. EMBO J 11:4145–4152

Grant PA, Duggan L, Cote J, Roberts SM, Brownell JE, Candau R, Ohba R, Owen-Hughes T, Allis CD, Winston F, Berger SL, Workman JL (1997) Yeast Gcn5 functions in two multisubunit complexes to acetylate nucleosomal histones: characterization of an Ada complex and the SAGA (Spt/Ada) complex. Genes Dev 11:1640–1650

Grant PA, Schieltz D, Pray-Grant MG, Yates JRR, Workman JL (1998a) The ATM-related cofactor Tra1 is a component of the purified SAGA complex. Mol Cell 2:863–867

Grant PA, Sterner DE, Duggan LJ, Workman JL, Berger SL (1998b) The SAGA unfolds: convergence of transcription regulators in chromatin-modifying complexes. Trends Cell Biol 8:193–197

Gregory MA, Hann SR (2000) c-Myc proteolysis by the ubiquitin-proteasome pathway: stabilization of c-Myc in Burkitt's lymphoma cells. Mol Cell Biol 20:2423–2435

Gstaiger M, Jordan R, Lim M, Catzavelos C, Mestan J, Slingerland J, Krek W (2001) Skp2 is oncogenic and overexpressed in human cancers. Proc Natl Acad Sci U S A 98:5043–5048

Hann SR, Eisenman RN (1984) Proteins encoded by the human c-myc oncogene: differential expression in neoplastic cells. Mol Cell Biol 4:2486–2497

Hann SR, Abrams HD, Rohrschneider LR, Eisenman RN (1983) Proteins encoded by v-myc and c-myc oncogenes: identification and localization in acute leukemia virus transformants and bursal lymphoma cell lines. Cell 34:789–798

Ikura T, Ogryzko VV, Grigoriev M, Groisman R, Wang J, Horikoshi M, Scully R, Qin J, Nakatani Y (2000) Involvement of the TIP60 histone acetylase complex in DNA repair and apoptosis. Cell 102:463–473

Kato GJ, Barrett J, Villa-Garcia M, Dang CV (1990) An amino-terminal c-myc domain required for neoplastic transformation activates transcription. Mol Cell Biol 10:5914–5920

Kenney AM, Cole MD, Rowitch DH (2003) Nmyc upregulation by sonic hedgehog signaling promotes proliferation in developing cerebellar granule neuron precursors. Development 130:15–28

Kim SY, Herbst A, Tworkowski KA, Salghetti SE, Tansey WP (2003) Skp2 regulates Myc protein stability and activity. Mol Cell 11:1177–1188

Knoepfler PS, Eisenman RN (1999) Sin meets NuRD and other tails of repression. Cell 99:447–450

Kretzner L, Blackwood EM, Eisenman RN (1992) Myc and Max proteins possess distinct transcriptional activities. Nature 359:426–429

Land H, Parada LF, Weinberg RA (1983) Tumorigenic conversion of primary embryo fibroblasts requires at least two cooperating oncogenes. Nature 304:596–602

Landay M, Oster SK, Khosravi F, Grove LE, Yin X, Sedivy J, Penn LZ, Prochownik EV (2000) Promotion of growth and apoptosis in c-myc nullizygous fibroblasts by other members of the myc oncoprotein family. Cell Death Differ 7:697–705

Landschulz WH, Johnson PF, McKnight SL (1988) The leucine zipper: a hypothetical structure common to a new class of DNA binding proteins. Science 240:1759–1764

Latres E, Chiarle R, Schulman BA, Pavletich NP, Pellicer A, Inghirami G, Pagano M (2001) Role of the F-box protein Skp2 in lymphomagenesis. Proc Natl Acad Sci U S A 98:2515–2520

Lundberg AS, Hahn WC, Gupta P, Weinberg RA (2000) Genes involved in senescence and immortalization. Curr Opin Cell Biol 12:705–709

Mahon GM, Wang Y, Korus M, Kostenko E, Cheng L, Sun T, Arlinghaus RB, Whitehead IP (2003) The c-Myc Oncoprotein Interacts with Bcr. Curr Biol 13:437–441

Marcus GA, Silverman N, Berger SL, Horiuchi J, Guarente L (1994) Functional similarity and physical association between GCN5 and ADA2: putative transcriptional adaptors. EMBO J 13:4807–4815

Martinez E, Palhan VB, Tjernberg A, Lymar ES, Gamper AM, Kundu TK, Chait BT, Roeder RG (2001) Human STAGA complex is a chromatin-acetylating transcription coactivator that interacts with pre-mRNA splicing and DNA damage-binding factors in vivo. Mol Cell Biol 21:6782–6795

Mateyak MK, Obaya AJ, Adachi S, Sedivy JM (1997) Phenotypes of c-Myc-deficient rat fibroblasts isolated by targeted homologous recombination. Cell Growth Differ 8:1039–1048

McMahon SB, Van Buskirk HA, Dugan KA, Copeland TD, Cole MD (1998) The novel ATM-related protein TRRAP is an essential cofactor for the c-Myc and E2F oncoproteins. Cell 94:363–374

McMahon SB, Wood MA, Cole MD (2000) The essential cofactor TRRAP recruits the histone acetyltransferase hGCN5 to c-Myc. Mol Cell Biol 20:556–562

Mehmet H, Littlewood TD, Sinnett-Smith J, Moore JP, Evan GI, Rozengurt E (1997) Large induction of c-Myc is not essential for the mitogenic response of Swiss 3T3 fibroblasts. Cell Growth Differ 8:187–193

Moore JP, Hancock DC, Littlewood TD, Evan GI (1987) A sensitive and quantitative enzyme-linked immunosorbence assay for the c-myc and N-myc oncoproteins. Oncogene Res 2:65–80

Mori K, Maeda Y, Kitaura H, Taira T, Iguchi-Ariga SM, Ariga H (1998) MM-1, a novel c-Myc-associating protein that represses transcriptional activity of c-Myc. J Biol Chem 273:29794–29800

Murre C, McCaw PS, Baltimore D (1989) A new DNA binding and dimerization motif in immunoglobulin enhancer binding, daughterless, MyoD, and myc proteins. Cell 56:777–783

Nikiforov MA, Chandriani S, Park J, Kotenko I, Matheos D, Johnsson A, McMahon SB, Cole MD (2002) TRRAP-dependent and TRRAP-independent transcriptional activation by Myc family oncoproteins. Mol Cell Biol 22:5054–5063

O'Connell BC, Cheung AF, Simkevich CP, Tam W, Ren X, Mateyak MK, Sedivy JM (2003) A large scale genetic analysis of c-Myc-regulated gene expression patterns. J Biol Chem 278:12563–12573

Ogryzko VV, Kotani T, Zhang X, Schiltz RL, Howard T, Quin J, Nakatani Y (1998) Histone-like TAFs within the PCAF histone acetylase complex. Cell 94:35–44

Okano HJ, Park WY, Corradi JP, Darnell RB (1999) The cytoplasmic Purkinje onconeural antigen cdr2 down-regulates c-Myc function: implications for neuronal and tumor cell survival. Genes Dev 13:2087–2097

Oster SK, Mao DY, Kennedy J, Penn LZ (2003) Functional analysis of the N-terminal domain of the Myc oncoprotein. Oncogene 22:1998–2010

Park J, Kunjibettu S, McMahon SB, Cole MD (2001) The ATM-related domain of TRRAP is required for histone acetyltransferase recruitment and Myc-dependent oncogenesis. Genes Dev 15:1619–1624

Prendergast GC, Cole MD (1989) Posttranscriptional regulation of cellular gene expression by the c-myc oncogene. Mol Cell Biol 9:124–134

Price DH (2000) P-TEFb, a cyclin-dependent kinase controlling elongation by RNA polymerase II. Mol Cell Biol 20:2629–2634

Ricci AR, Genereaux J, Brandl CJ (2002) Components of the SAGA histone acetyltransferase complex are required for repressed transcription of ARG1 in rich medium. Mol Cell Biol 22:4033–4042

Sakamuro D, Prendergast G (1999) New Myc-interacting proteins: a second Myc network emerges. Oncogene 18:2942–2954

Saleh A, Schieltz D, Ting N, McMahon SB, Litchfield DW, Yates III JR, Lees-Miller SP, Cole MD, Brandl CJ (1998) Tra1p is a component of the yeast Ada.Spt transcriptional regulatory complexes. J Biol Chem 273:26559–26570

Salghetti SE, Kim S, Tansey WP (1999) Destruction of Myc by ubiquitin-mediated proteolysis: cancer-associated and transforming mutations stabilize Myc. EMBO J 18:717–726

Salghetti SE, Caudy AA, Chenoweth JG, Tansey WP (2001) Foreign body aspiration in children: value of radiography and complications of bronchoscopy. Science 293:1651–1653

Satou A, Taira T, Iguchi-Ariga SM, Ariga H (2001) A novel transrepression pathway of c-Myc. Recruitment of a transcriptional corepressor complex to c-Myc by MM-1, a c-Myc-binding protein. J Biol Chem 276:46562–46567

Sears R, Leone G, DeGregori J, Nevins J (1999) Ras enhances Myc protein stability. Mol Cell 3:169–179

Shen X, Mizuguchi G, Hamiche A, Wu C (2000) A chromatin remodelling complex involved in transcription and DNA processing. Nature 406:541–545

Smith ER, Eisen A, Gu W, Sattah M, Pannuti A, Zhou J, Cook RG, Lucchesi JC, Allis CD (1998) ESA1 is a histone acetyltransferase that is essential for growth in yeast. Proc Natl Acad Sci U S A 95:3561–3565

Spotts GD, Patel SV, Xiao Q, Hann SR (1997) Identification of downstream-initiated c-Myc proteins which are dominant-negative inhibitors of transactivation by full-length c-Myc proteins. Mol Cell Biol 17:1459–1468

Stone J, De Lange T, Ramsay G, Jakobovits E, Bishop JM, Varmus H, Lee W (1987) Definition of regions in human c-myc that are involved in transformation and nuclear localization. Mol Cell Biol 7:1697–1709

Vassilev A, Yamauchi J, Kotani T, Prives C, Avantaggiati ML, Qin J, Nakatani Y (1998) The 400 kDa subunit of the PCAF histone acetylase complex belongs to the ATM superfamily. Mol Cell 2:869–875

Vervoorts J, Luscher-Firzlaff JM, Rottmann S, Lilischkis R, Walsemann G, Dohmann K, Austen M, Luscher B (2003) Stimulation of c-MYC transcriptional activity and acetylation by recruitment of the cofactor CBP. EMBO Rep 4:484–490

von der Lehr N, Johansson S, Wu S, Bahram F, Castell A, Cetinkaya C, Hydbring P, Weidung I, Nakayama K, Nakayama KI, Soderberg O, Kerppola TK, Larsson LG (2003) The F-box protein Skp2 participates in c-Myc proteosomal degradation and acts as a cofactor for c-Myc-regulated transcription. Mol Cell 11:1189–1200

Wang H, Liu C, Lu Y, Chatterjee G, Ma XY, Eisenman RN, Lengyel P (2000) The interferon- and differentiation-inducible p202a protein inhibits the transcriptional activity of c-Myc by blocking its association with Max. J Biol Chem 275:27377–27385

Wang L, Mizzen C, Ying R, Candau R, Barlev N, Brownell J, Allis CD, Berger S (1997) Histone acetyltransferase activity is conserved between yeast and human GCN5 and is required for complementation of growth and transcriptional activation. Mol Cell Biol 17:519–527

Wood MA, McMahon SB, Cole MD (2000) An ATPase/helicase complex is an essential cofactor for oncogenic transformation by c-Myc. Mol Cell 5:321–330

Xiao Q, Claassen G, Shi J, Adachi S, Sedivy J, Hann S R (1998) Transactivation-defective c-MycS retains the ability to regulate proliferation and apoptosis. Genes Dev 12:3803–3808

CTMI (2006) 302:51–62

Mechanisms of Transcriptional Repression by Myc

D. Kleine-Kohlbrecher · S. Adhikary · M. Eilers (✉)

Institute for Molecular Biology and Tumor Research, University of Marburg, 35033 Marburg, Germany
eilers@imt.uni-marburg.de

Abstract Myc proteins are nuclear proteins that exert their biological functions at least in part through the transcriptional regulation of large sets of target genes. Recent microarray analyses show that several percent of all genes may be directly regulated by Myc. A large body of data shows that Myc proteins both positively and negatively affect transcription. The basic mechanism underlying Myc's activation of transcription is well understood, but the mechanisms through which Myc negatively regulates or represses transcription are far less understood. In this chapter, we will review our current knowledge about this less-well-understood topic.

1 Introduction

Myc proteins are nuclear proteins that exert their biological functions at least in part through the transcriptional regulation of large sets of target genes. Recent microarray analyses show that several percent of all genes may be directly regulated by Myc. A large body of data shows that Myc proteins both positively and negatively affect transcription. The basic mechanism underlying Myc's activation of transcription is well understood and is reviewed in detail in other parts of this volume (see M.D. Cole and M.A. Nikiforov, this volume). In contrast, the mechanisms through which Myc negatively regulates or represses transcription are far less understood. In this chapter, we will review our current knowledge about this topic.

2
Targets of Myc-Mediated Repression

Multiple targets of gene repression have been identified, from both microarray analyses and directed searches. These analyses have used a number of different approaches, including inducible alleles of Myc (Coller et al. 2000), Myc knockout cells (O'Connell et al. 2003), or comparing primary tumors that do or do not express an amplified Myc gene (Berwanger et al. 2002). An updated list of target genes together with data on their validation as Myc targets can be found at www.myc-cancer-gene.org. While repressed genes, like induced genes, fall into multiple functional classes, most mechanistic work has focused on a relatively small number of target genes.

The first class of genes encodes proteins that are selectively expressed in quiescent cells or that directly or indirectly inhibit cell proliferation. This group encompasses the cell cycle inhibitors p21Cip1 (Claassen and Hann, 2000; Gartel et al. 2001; Herold et al. 2002; Seoane et al. 2002; van de Wetering et al. 2002; Wu et al. 2003), p27kip1 (Yang et al. 2001), p15ink4b (Seoane et al. 2001; Staller et al. 2001; Warner et al. 1999), p18ink4c (Knoepfler et al. 2002), and p57kip2 (Dauphinot et al. 2001), as well as the differentiation-inducing proteins C/EBP-α (Freytag and Geddes, 1992; Yang et al. 1993), the growth-arrest proteins gas1 (Lee et al. 1997) and gas2 (see below), and the Myc-antagonist Mad4 (Kime and Wright 2003). This long list points to a role for Myc-mediated gene repression in the control of cellular differentiation and in the response to growth arrest signals. In some cases, repression has been shown to occur in response to several stimuli. For example, Myc represses induction of p21Cip1 in response to DNA damage or addition of transforming growth factor (TGF)-β, and in response to induction of differentiation, arguing that repression does not merely reflect the loss of a particular signal transduction pathway.

A role for Myc-mediated repression in the response to cellular stress is further supported by the suppression of a group of GADD genes (growth arrest and DNA damage), most notably *gadd45* (Amundson et al. 1998). Gadd45 has been implicated in multiple responses to stress and in G2/M checkpoint control (Wang et al. 1999). Potentially due to loss of this checkpoint, Gadd45$^{-/-}$ cells are genomically unstable. Whether the G2/M checkpoint is generally compromised in Myc-transformed cells is not completely clear (Li and Dang

1999). Similar to overexpressed Myc, loss of *gadd45* facilitates transformation of primary mouse embryo fibroblasts by Ras and abolishes Ras-induced senescence.

The general importance of transcriptional repression of "arrest genes" may be hard to determine (in contrast to individual protein/protein interactions, see below). However, it appears clear from the literature that the repression of individual genes significantly contributes to the phenotype of Myc-transformed cells. For example, repression of C/EBP-α is required for inhibition of adipogenesis by Myc (Freytag and Geddes 1992; Yang et al. 1993). Repression of p15Ink4b is a key element of the resistance of Myc-transformed cells to growth inhibition by TGF-β, and repression of p21Cip1 is important for the inability of Myc-transformed cells to arrest in the G1 phase of the cell cycle upon exposure to DNA damage (Herold et al. 2002; Seoane et al. 2002). Although the underlying mechanism is not completely clear, it should be pointed out that repression of ferritin expression by Myc has been shown to be required for transformation (Wu et al. 1999).

A second class of genes that is often repressed by Myc encodes proteins involved in cell adhesion, including a number of integrins (Inghirami et al. 1990). Altered cell adhesion is a hallmark of many Myc-transformed cells and has been observed in different cell types (Coller et al. 2000). In stem cells of epithelia and of the hematopoietic system, Myc has been suggested to regulate the balance between self-renewal and exit from the stem cell compartment by regulation of adhesive interactions between stem cells and the local microenvironments. This appears to occur through Myc-mediated downregulation of integrins and alterations in cell adhesion (Frye et al. 2003; Waikel et al. 2001; Wilson et al. 2004).

Finally, Myc promotes angiogenesis through suppression of thrombospondin. In transgenic mice, deregulated expression of Myc strongly promotes angiogenesis when apoptosis is suppressed (Pelengaris et al. 2002). Suppression of thrombospondin plays a causative role in the induction of angiogenesis by Myc (Tikhonenko et al. 1996). While the mechanism of repression in unclear, phosphorylation of the Myc-amino-terminus by Ras-dependent events plays a central role in regulating thrombospondin expression (Watnick et al. 2003).

3 Target Sites of Repression

Unlike transcriptional activation by Myc, which is mediated by binding of Myc to the E-box sequence CACGTG and related sequences, no simple consensus sequence for transcriptional repression by Myc has emerged. This opens the possibility that transcriptional repression is simply an indirect consequence of the altered physiological (e.g., transformed) state of a cell that is induced by Myc. For example, many published examples of "p53-repressed" genes appear to be regulated indirectly as a consequence of p21Cip1 activation by p53 and subsequent Cdk2 inhibition (Lohr et al. 2003). Likewise, it appears possible that repression of cell-cycle inhibitors might result from the fact that Myc enhances cell proliferation. Indeed, there is evidence in the literature for such indirect mechanisms of gene repression by Myc: For example, inhibition of nuclear factor (NF)-κB-dependent transcription by Myc is a consequence of high levels of E2F1-protein in Myc-transformed cells, since E2F1 binds and inhibits the p65 subunit of NF-κB (Tanaka et al. 2002). Similarly, inhibition of MyoD-dependent transcription by Myc seems indirect and may be a consequence of deregulated cyclin E/Cdk2 kinase activity (Crescenzi et al. 1994).

One argument against the notion that all repression is similarly indirect was the identification of mutants of Myc that distinguish transcriptional activation from repression and the detailed analysis of the resulting phenotypes. For example, gene repression, but not activation, is enhanced in a lymphoma-derived allele of Myc, suggesting that repression and activation are regulated independently; transformation by this mutant is also enhanced, pointing to a role of repression in transformation (Lee et al. 1996). More recently, we have described a single point mutant (*MycV394D*) that is unable to bind to Miz1 and that uncouples Miz1-dependent repression from gene activation and other forms of repression. This mutant is fully capable of inducing growth and cell cycle progression of established fibroblast cell lines, arguing that specific pathways of repression indeed exist (Herold et al. 2002).

Early analyses mainly of the adenovirus major late and the C/EBP-α promoters suggested that repression by Myc was mediated by the "initiator" element of both promoters (Li et al. 1994). This view was further supported by the analysis of the ferritin promoter, another target of repression by Myc (Wu et al. 1999). Initiators are defined as sequence elements on DNA, which are capable of positioning the start site of transcription independent of a tumor-associated transplantation antigen (TATA) element. From this, the concept of an "initiator"-dependent pathway of transcriptional repression has emerged. Since repression of core promoter activity by Myc was observed in multiple studies, the notion that Myc generally represses through "initiator" elements has persisted to the present time.

There are several reasons to suspect that this notion is wrong. First, initiator elements are now thought to be recognized by a subunit of the TFIID complex, p150, and to our knowledge no interaction of Myc with this or any other component of TFIID (with the exception of TBP) has been documented (Kaufmann et al. 1998; Verrijzer et al. 1995). Second, several core promoters that are repressed by Myc contain TATA elements and have no documented initiators. One example is the cell cycle inhibitor, p21Cip1. Third, array analyses have found multiple repressed genes, but by no means are all TATA-less genes targets for repression by Myc (*cyclin A* being just one of many examples). Also, it is likely that many early studies reporting repression of "core promoters" looked at non-physiological phenomena like squelching of transactivation domains. Squelching of a promoter will become more pronounced when a reporter plasmid is truncated so that the promoter under study is stripped of enhancer elements.

In our view, "initiator-dependent" repression simply reflects the fact that at least two of the transcription factors Myc interacts with, Sp1 and Miz1, have a preference for binding at core promoters. Not surprisingly, other sites of repression have been reported; for example, repression of the GADD45 gene occurs through a WT1/EGR1 binding site (Amundson et al. 1998), repression of the platelet-derived growth factor (PDGF) receptor β chain occurs through NF-Y binding sites (Izumi et al. 2001) and repression of Smad-function contributes to repression of the p15ink4b gene (Feng et al. 2002).

4 Role of DNA-Binding in Myc-Induced Repression

Repression by Myc through the Smad- and the NF-Y binding sites has been ascribed to direct protein/protein interactions between Myc and Smad2 and NF-Y, respectively (Feng et al. 2002; Izumi et al. 2001); Myc is thought to be recruited to DNA through the respective interaction.

In contrast, several interactions have been proposed to account for the repression through core promoter elements. Repression of p27Kip1 has been ascribed to direct binding of Myc/Max complexes to the start site of the promoter (Yang et al. 2001). However, other core promoters do not bind Myc/Max complexes in the absence of recruiting proteins: Examples are the core of the p21Cip1 and the p15Ink4b promoter (Seoane et al. 2001; Wu et al. 2003).

Second, repression of core promoter activity has been suggested to result from interactions between Myc and YY1 (Shrivastava et al. 1993), TFII-I (Roy et al. 1993), Sp1 (Gartel et al. 2001), and Miz1 (Peukert et al. 1997). Little or no follow-up work has been published on the initial reports of the Myc/YY1 and Myc/TFII-I interactions, making it hard to judge the relevance of these interactions. However, by current standards, they cannot be seen as fully validated interactions since key experiments such as chromatin immunoprecipitations demonstrating the presence of these factors on Myc-regulated genes in vivo have not been performed.

In contrast, a number of different experimental approaches including chromatin-immunoprecipitation support a role for the interaction between Myc and Miz1 in the regulation of the p15Ink4b, p21Cip1, and Mad4 promoters (op.cit). Miz1 is a zinc-finger protein that contains 13 zinc fingers (see Fig. 1). At its amino-terminus, Miz1 carries a BTB/POZ-domain, which is a protein/protein interaction domain found in multiple zinc-finger proteins; POZ-domains are involved in multiple protein/protein interactions, including homo- and heterodimerization and recruitment of transcriptional co-repressors such as N-CoR.

The identification of a specific point mutant of Myc, MycV394D, which is unable to bind to Miz1 but fully capable of transcriptional activation, has made it possible to identify those genes and processes that are regulated by Myc

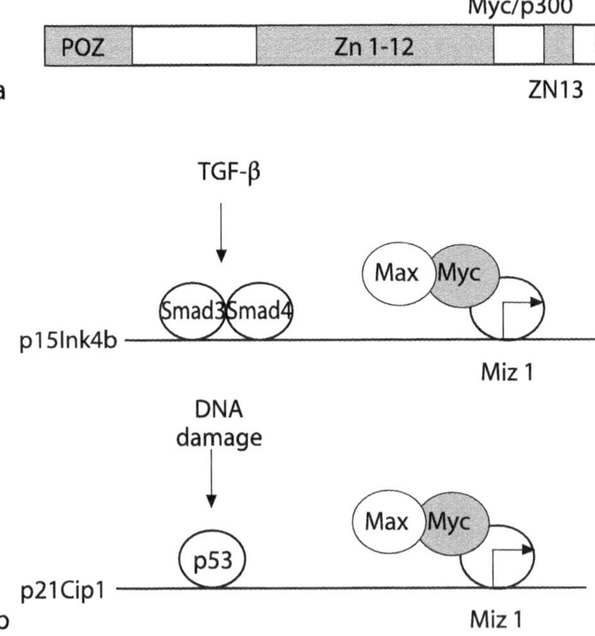

Fig. 1a, b Miz1-dependent repression by Myc. **a** Schematic diagram of Miz1 structure indicating the BTB/POZ-domain, the localization of the 13 Zn fingers, and of a short domain thought to alternatively bind either Myc or the p300 co-activator. **b** The binding sites of the Myc/Miz1 complex in the p15Ink4b and the p21Cip1 promoters immediately adjacent to the transcription start sites

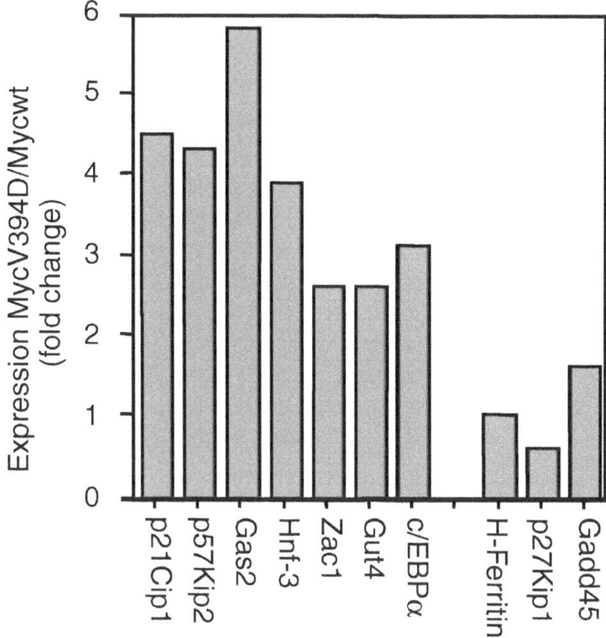

Fig. 2 Miz1-dependent and -independent repression of target gene expression by Myc. Shown is the fold expression of the indicated genes in primary mouse embryo fibroblasts infected with either wild-type Myc (*Mycwt*) or MycV394D, a mutant allele unable to bind to Miz1

through Miz1. Microarray analyses of mouse embryo fibroblasts expressing either wild-type Myc or MycV394D also reveal that there are genes repressed by Myc in a manner that depends on its ability to interact with Miz1 (e.g., p21Cip1, C/EBP-α) and genes that are repressed in a Miz1-independent manner (see Fig. 2). The data clearly support the notion that several pathways of repression exist. Which individual interactions play a role in Miz1-independent processes remains to be established.

5
Mechanism(s) of Repression

Much remains to be learned about the mechanism of transcriptional repression by Myc. One recent suggestion has been that Myc recruits the DNA methyltransferase Dnmt3a to Miz1-bound target sites of repression (Brenner et al. 2005). Since Dnmt3a is complexed with histone deacetylases, its recruitment by Myc might lead to local histone deacetylation and inhibition of transcription (Fuks et al. 2001). Recruitment of Dnmt3a by Myc is an attractive mechanism for repression, since it might provide an explanation of the aberrant DNA methylation of some tumor suppressor genes that is observed in human tumors. A key example is *P15INK4B*, a target for Myc/Miz1-mediated repression, which is inactivated by promoter methylation in multiple lymphomas (where Myc expression is generally high; Esteller 2000).

Several mutant analyses have pointed to the importance of Myc BoxII in both activation and repression by Myc, and this requirement probably does not reflect binding to Dnmt3a. A recent detailed study by Penn and colleagues indicates that two distinct mechanisms of repression with slightly different requirements in the amino-terminus of Myc exist (Oster et al. 2003). Myc BoxII interacts either directly or indirectly with TRRAP, Tip60, GCN5, Tip48, Tip49, p400, and Skp2; of these, Tip48 and Tip49 have been implicated directly in Myc-dependent transcriptional repression (Etard et al. 2005). p400 has been demonstrated to repress the p21Cip1 promoter, but whether this activity is linked to Myc is unknown (Chan et al. 2005).

One of the key open questions is how Myc can function as a transcriptional activator on one set of targets sites and a repressor on another set of binding sites. One possibility is that Myc recruits a distinct set of cofactors to sites of activation and repression, but no systematic study on which of these factors is recruited to which site has been published so far. Perhaps most exiting is the recent demonstration that the ARF tumor suppressor protein may be involved in this process (Datta et al. 2004; Qi et al. 2004). In these studies, ARF was found to bind Myc directly and to inhibit transcriptional activation, but not transcriptional repression by Myc. How ARF exerts these effects remained unclear; it should be noted, however, that ARF has been known to inhibit cell proliferation in a p53-independent manner (Weber et al. 2000).

A further unresolved issue is whether other members of the Myc/Max/Mad network have a functional role in repression by Myc. Neither Max nor Mad proteins bind to Miz1 directly, and the amino acids of Myc involved in contacting Miz1 are not conserved in Max, Mad, or Mnt proteins. Max is present at the core promoters of Myc-repressed genes (Mao et al. 2003; Staller et al. 2001), and we originally suggested it was recruited there through Myc binding to Miz1. However, recent data show that Max is present at core promoters independently of Myc (Mao et al. 2003) and indicate that Mad1 is present at certain core promoters too (E. Sanchez, personal communication). Most likely, therefore, additional interactions recruit Max and Mad proteins to core promoters, raising the possibility that they have a functional role in repression.

6 Summary

Repression of target genes contributes to specific phenotypes of Myc-transformed cells and will almost certainly contribute to Myc-induced tumorigenesis. In contrast to activation, which appears always mediated by binding of Myc/Max complexes to E-box elements, several pathways of repression exist, due to the interaction of Myc with different transcription factors. One such factor, Miz1, links Myc to the TGF-β and p53 signaling pathways and to cellular differentiation. Precisely how Myc switches from being activator to "repressor" is an open question; similarly, the chromatin-modifying events triggered by Myc in gene repression remain to be identified.

Acknowledgements Work in the author's laboratory is supported by the Deutsche Forschungsgemeinschaft, the European Community through the Framework 5 and 6 programs, the Thyssen and the Sander-Stiftung and the Deutsche Krebshilfe.

References

Amundson SA, Zhan Q, Penn LZ, Fornace AJ Jr (1998) Myc suppresses induction of the growth arrest genes gadd34, gadd45, and gadd153 by DNA-damaging agents. Oncogene 17:2149–2154

Berwanger B, Hartmann O, Bergmann E, Nielsen D, Krause M, Kartal A, Flynn D, Wiedemeyer R, Schwab M, Schäfer H, Christiansen H, Eilers M (2002) Loss of a Fyn-regulated differentiation and growth arrest pathway in advanced stage neuroblastoma. Cancer Cell 2:377–386

Brenner C, Deplus R, Didelot C, Loriot A, Vire E, De Smet C, Gutierrez A, Danovi D, Bernard D, Boon T, Pelicci PG, Amati B, Kouzarides T, de Launoit Y, Di Croce L, Fuks F (2005) Myc represses transcription through recruitment of DNA methyltransferase corepressor. EMBO J 24:336–346

Chan HM, Narita M, Lowe SW, Livingston DM (2005) The p400 E1A-associated protein is a novel component of the p53->p21 senescence pathway. Genes Dev 19:196–201

Claassen GF, Hann SR (2000) A role for transcriptional repression of p21CIP1 by c-Myc in overcoming transforming growth factor beta-induced cell-cycle arrest. Proc Natl Acad Sci U S A 97:9498–9503

Coller HA, Grandori C, Tamayo P, Colbert T, Lander ES, Eisenman RN, Golub TR (2000) Expression analysis with oligonucleotide microarrays reveals that MYC regulates genes involved in growth, cell cycle, signaling, and adhesion. Proc Natl Acad Sci U S A 97:3260–3265

Crescenzi M, Crouch DH, Tato F (1994) Transformation by myc prevents fusion but not biochemical differentiation of C2C12 myoblasts: mechanisms of phenotypic correction in mixed culture with normal cells. J Cell Biol 125:1137–1145

Datta A, Nag A, Pan W, Hay N, Gartel AL, Colamonici O, Mori Y, Raychaudhuri P (2004) Myc-ARF (alternate reading frame) interaction inhibits the functions of Myc. J Biol Chem 279:36698–36707

Dauphinot L, De Oliveira C, Melot T, Sevenet N, Thomas V, Weissman BE, Delattre O (2001) Analysis of the expression of cell cycle regulators in Ewing cell lines: EWS-FLI-1 modulates p57KIP2and c-Myc expression. Oncogene 20:3258–3265

Esteller M (2000) Epigenetic lesions causing genetic lesions in human cancer: promoter hypermethylation of DNA repair genes. Eur J Cancer 36:2294–2300

Etard C, Gradl D, Kunz M, Eilers M, Wedlich D (2005) Pontin and Reptin regulate cell proliferation in early Xenopus embryos in collaboration with c-Myc and Miz-1. Mech Dev 122:545–556

Feng XH, Liang YY, Liang M, Zhai W, Lin X (2002) Direct interaction of c-Myc with Smad2 and Smad3 to inhibit TGF-beta-mediated induction of the CDK inhibitor p15(Ink4B). Mol Cell 9:133–143

Freytag SO, Geddes TJ (1992) Reciprocal regulation of adipogenesis by Myc and C/EBP alpha. Science 256:379–382

Frye M, Gardner C, Li ER, Arnold I, Watt FM (2003) Evidence that Myc activation depletes the epidermal stem cell compartment by modulating adhesive interactions with the local microenvironment. Development 130:2793–2808

Fuks F, Burgers WA, Godin N, Kasai M, Kouzarides T (2001) Dnmt3a binds deacetylases and is recruited by a sequence-specific repressor to silence transcription. EMBO J 20:2536–2544

Gartel AL, Ye X, Goufman E, Shianov P, Hay N, Najmabadi F, Tyner AL (2001) Myc represses the p21(WAF1/CIP1) promoter and interacts with Sp1/Sp3. Proc Natl Acad Sci U S A 98:4510–4515

Herold S, Wanzel M, Beuger V, Frohme C, Beul D, Hillukkala T, Syvaoja J, Saluz HP, Hänel F, Eilers M (2002) Negative regulation of the mammalian UV response by Myc through association with Miz-1. Mol Cell 10:509–521

Inghirami G, Grignani F, Sternas L, Lombardi L, Knowles DM, Dalla Favera R (1990) Down-regulation of LFA-1 adhesion receptors by C-myc oncogene in human B lymphoblastoid cells. Science 250:682–686

Izumi H, Molander C, Penn LZ, Ishisaki A, Kohno K, Funa K (2001) Mechanism for the transcriptional repression by c-Myc on PDGF beta-receptor. J Cell Sci 114:1533–1544

Kaufmann J, Ahrens K, Koop R, Smale ST, Muller R (1998) CIF150, a human cofactor for transcription factor IID-dependent initiator function. Mol Cell Biol 18:233–239

Kime L, Wright SC (2003) Mad4 is regulated by a transcriptional repressor complex that contains Miz-1 and c-Myc. Biochem J 370:291–298

Knoepfler PS, Cheng PF, Eisenman RN (2002) N-myc is essential during neurogenesis for the rapid expansion of progenitor cell populations and the inhibition of neuronal differentiation. Genes Dev 16:2699–2712

Lee LA, Dolde C, Barrett J, Wu CS, Dang CV (1996) A link between c-Myc-mediated transcriptional repression and neoplastic transformation. J Clin Invest 97:1687–1695

Lee TC, Li L, Philipson L, Ziff EB (1997) Myc represses transcription of the growth arrest gene gas1. Proc Natl Acad Sci U S A 94:12886–12891

Li L, Nerlov C, Prendergast G, MacGregor D, Ziff EB (1994) c-Myc represses transcription in vivo by a novel mechanism dependent on the initiator element and Myc box II. EMBO J 13:4070–4079

Li Q, Dang CV (1999) c-Myc overexpression uncouples DNA replication from mitosis. Mol Cell Biol 19:5339–5351

Mao DY, Watson JD, Yan PS, Barsyte-Lovejoy D, Khosravi F, Wong WW, Farnham PJ, Huang TH, Penn LZ (2003) Analysis of Myc bound loci identified by CpG island arrays shows that Max is essential for Myc-dependent repression. Curr Biol 13:882–886

O'Connell BC, Cheung AF, Simkevich CP, Tam W, Ren X, Mateyak MK, Sedivy JM (2003) A large-scale genetic analysis of c-Myc-regulated gene expression patterns. J Biol Chem 278:12563–12573

Oster SK, Mao DY, Kennedy J, Penn LZ (2003) Functional analysis of the N-terminal domain of the Myc oncoprotein. Oncogene 22:1998–2010

Pelengaris S, Khan M, Evan GI (2002) Suppression of myc-induced apoptosis in Beta cells exposes multiple oncogenic properties of myc and triggers carcinogenic progression. Cell 109:321–334

Peukert K, Staller P, Schneider A, Carmichael G, Hanel F, Eilers M (1997) An alternative pathway for gene regulation by Myc. EMBO J 16:5672–5686

Qi Y, Gregory MA, Li Z, Brousal JP, West K, Hann SR (2004) p19ARF directly and differentially controls the functions of c-Myc independently of p53. Nature 431:712–717

Lohr K, Moritz C, Contente A, Dobbelstein M (2003) p21/CDKN1A mediates negative regulation of transcription by p53. J Biol Chem 278:32507–32516

Roy AL, Carruthers C, Gutjahr T, Roeder RG (1993) Direct role for Myc in transcription initiation mediated by interactions with TFII-I. Nature 365:359–361

Seoane J, Pouponnot C, Staller P, Schader M, Eilers M, Massague J (2001) TGFbeta influences Myc, Miz-1 and Smad to control the CDK inhibitor p15INK4b. Nat Cell Biol 3:400–408

Seoane J, Le HV, Massague J (2002) Myc suppression of the p21(Cip1) Cdk inhibitor influences the outcome of the p53 response to DNA damage. Nature 419:729–734

Shrivastava A, Saleque S, Kalpana GV, Artandi S, Goff SP, Calame K (1993) Inhibition of transcriptional regulator Yin-Yang-1 by association with c-Myc. Science 262:1889–1891

Staller P, Peukert K, Kiermaier A, Seoane J, Lukas J, Karsunky H, Moroy T, Bartek J, Massague J, Hanel F, Eilers M (2001) Repression of p15INK4b expression by Myc through association with Miz-1. Nat Cell Biol 3:392–399

Tanaka H, Matsumura I, Ezoe S, Satoh Y, Sakamaki T, Albanese C, Machii T, Pestell RG, Kanakura Y (2002) E2F1 and c-Myc potentiate apoptosis through inhibition of NF-kappaB activity that facilitates MnSOD-mediated ROS elimination. Mol Cell 9:1017–1029

Tikhonenko AT, Black DJ, Linial ML (1996) Viral Myc oncoproteins in infected fibroblasts down-modulate thrombospondin-1, a possible tumor suppressor gene. J Biol Chem 271:30741–30747

van de Wetering M, Sancho E, Verweij C, de Lau W, Oving I, Hurlstone A, van der Horn K, Batlle E, Coudreuse D, Haramis AP, Tjon-Pon-Fong M, Moerer P, van den Born M, Soete G, Pals S, Eilers M, Medema R, Clevers H (2002) The beta-catenin/TCF-4 complex imposes a crypt progenitor phenotype on colorectal cancer cells. Cell 111:241–250

Verrijzer CP, Chen JL, Yokomori K, Tjian R (1995) Binding of TAFs to core elements directs promoter selectivity by RNA polymerase II. Cell 81:1115–1125

Waikel RL, Kawachi Y, Waikel PA, Wang XJ, Roop DR (2001) Deregulated expression of c-Myc depletes epidermal stem cells. Nat Genet 28:165–168

Wang XW, Zhan Q, Coursen JD, Khan MA, Kontny HU, Yu L, Hollander MC, O'Connor PM, Fornace AJ Jr, Harris CC (1999) GADD45 induction of a G2/M cell cycle checkpoint. Proc Natl Acad Sci U S A 96:3706–3711

Warner BJ, Blain SW, Seoane J, Massague J (1999) Myc downregulation by transforming growth factor beta required for activation of the p15(Ink4b) G(1) arrest pathway. Mol Cell Biol 19:5913–5922

Watnick RS, Cheng YN, Rangarajan A, Ince TA, Weinberg RA (2003) Ras modulates Myc activity to repress thrombospondin-1 expression and increase tumor angiogenesis. Cancer Cell 3:219–231

Weber JD, Jeffers JR, Rehg JE, Randle DH, Lozano G, Roussel MF, Sherr CJ, Zambetti GP (2000) p53-independent functions of the p19(ARF) tumor suppressor. Genes Dev 14:2358–2365

Wilson A, Murphy MJ, Oskarsson T, Kaloulis K, Bettess MD, Oser GM, Pasche AC, Knabenhans C, Macdonald HR, Trumpp A (2004) c-Myc controls the balance between hematopoietic stem cell self-renewal and differentiation. Genes Dev 18:2747–2763

Wu KJ, Polack A, Dalla-Favera R (1999) Coordinated regulation of iron-controlling genes, H-ferritin and IRP2, by c-MYC. Science 283:676–679

Wu S, Cetinkaya C, Munoz-Alonso MJ, von der Lehr N, Bahram F, Beuger V, Eilers M, Leon J, Larsson LG (2003) Myc represses differentiation-induced p21CIP1 expression via Miz-1-dependent interaction with the p21 core promoter. Oncogene 22:351–360

Yang B-S, Gilbert JD, Freytag SO (1993) Overexpression of Myc suppresses CCAAT transcription factor/nuclear factor 1-dependent promoters in vivo. Mol Cell Biol 13:3093–3102

Yang W, Shen J, Wu M, Arsura M, FitzGerald M, Suldan Z, Kim DW, Hofmann CS, Pianetti S, Romieu-Mourez R, Freedman LP, Sonenshein GE (2001) Repression of transcription of the p27(Kip1) cyclin-dependent kinase inhibitor gene by c-Myc. Oncogene 20:1688–1702

CTMI (2006) 302:63–122

The Mad Side of the Max Network: Antagonizing the Function of Myc and More

S. Rottmann · B. Lüscher (✉)

Abteilung Biochemie und Molekularbiologie, Institut für Biochemie,
Klinikum der RWTH, Pauwelsstrasse 30, 52074 Aachen, Germany
luescher@rwth-aachen.de

Abstract A significant body of evidence has been accumulated that demonstrates decisive roles of members of the Myc/Max/Mad network in the control of various aspects of cell behavior, including proliferation, differentiation, and apoptosis. The components of this network serve as transcriptional regulators. Mad family members, including Mad1, Mxi1, Mad3, Mad4, Mnt, and Mga, function in part as antagonists of Myc oncoproteins. At the molecular level this antagonism is reflected by the different cofactor/chromatin remodeling complexes that are recruited by Myc and Mad family members. One important function of the latter is their ability to repress gene transcription. In this review we summarize the current view of how this repression is achieved and what the consequences of Mad action are for cell behavior. In addition, we point out some of the many aspects that have not been clarified and thus leave us with a rather incomplete picture of the functions, both molecular and at the cellular level, of Mad family members.

1
Introduction: The Myc/Max/Mad Network

The correct regulation of cell behavior in particular within the context of a multicellular organism is a highly complicated and demanding but vitally important process. Cellular homeostasis requires that cells proliferate, differentiate, migrate, or apoptose as a consequence of the needs of an organism. Not surprisingly, therefore, these decisions are controlled at the level of organs or the organism. This is necessary to guarantee that individual cells behave in a manner that maintains the integrity of the whole organism. As a consequence of the activation of oncogenes and the inactivation of tumor suppressor genes, cell behavior is uncoupled from the organismal control. This can lead to neoplastic growth and potentially to tumor formation threatening the survival of the organism. Among the oncogenes that were identified first are the *myc* genes. Their deregulation or overexpression (or both) is strongly associated with neoplastic growth. Although these findings have been rather clear cut, it remained unclear for many years how Myc proteins trigger and support uncontrolled proliferation.

It is in recent years that we have seen a substantial increase in our knowledge regarding the biological and molecular functions of Myc proteins. Important was the realization that Myc acts as a component of a group of proteins referred to as the Myc/Max/Mad network. Significantly it has become evident that the components of this network function as transcriptional regulators controlling the expression of a large number of different genes. To understand the biological consequences of the functions of Myc/Max/Mad network members, the identification of target genes is critical, since deciphering the functions of the encoded proteins will help us unravel the biological consequences of network activities. Indeed, the analysis of Myc target genes supports previously defined roles of Myc proteins and identifies additional, new aspects of cell behavior that are controlled by Myc. These behaviors include proliferation, differentiation, and apoptosis but also different aspects of tumorigenesis (Fig. 1). At the molecular level, several cofactors have been identified that appear to mediate the transcriptional potential of Myc and Mad proteins. These findings provide a first, still-limited view into the molecular complexity of how these proteins regulate gene transcription.

While Myc has been the focus of many reviews in recent years (Amati et al. 2001; Grandori et al. 2000; Lüscher 2001; Lutz et al. 2002; Oster et al. 2002), other components of the Myc/Max/Mad network have not obtained

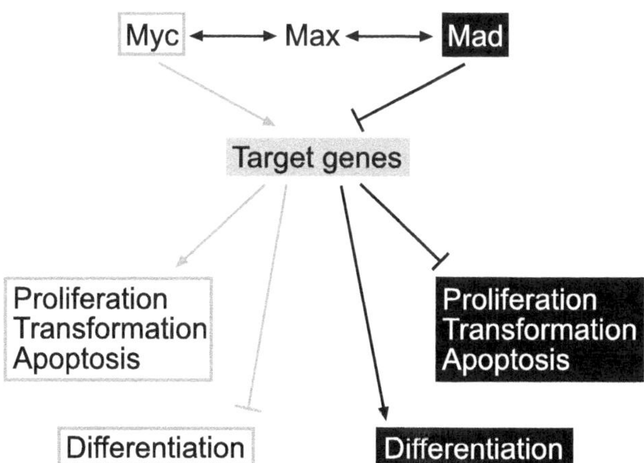

Fig. 1 Summary of the transcriptional regulation of cell behavior by the Myc/Max/Mad network. This represents a simplified view of the role of Myc and Mad proteins in the control of gene expression and cell behavior. Myc and Mad proteins form heterodimers with Max. These bind to E-box DNA elements and regulate gene transcription by recruiting cofactors. Myc proteins bind to coactivators while Mad proteins interact with corepressors. The target genes are at least in part overlapping and include genes that regulate cell growth, proliferation, apoptosis, and differentiation. The consequences of the regulation of many different target genes are summarized. For more details, see the text

comparable attention (Baudino and Cleveland 2001; Nilsson and Cleveland 2004; Zhou and Hurlin 2001). This reflects at least in part the potent role of Myc proteins, particularly in tumor formation (see M. Wade and G.M. Wahl in this volume). With the important identification of Mad1 and Mxi1 as additional interaction partners of Max (Ayer et al. 1993; Zervos et al. 1993), the protein previously defined as Myc heterodimerization partner (Blackwood and Eisenman 1991; Blackwood et al. 1992; Prendergast et al. 1991), the hypothesis quickly developed that these novel proteins might function as tumor suppressors. However, to date little evidence for such a function of Mad family members has been obtained. Also the knockout studies carried out targeting individual *mad* genes have been a little disappointing as will be discussed Sect. 7. These aspects have tended to decrease interest in *mad*, *mnt*, and *mga* genes and proteins. In this chapter we review the function of Mad proteins and of Mnt and Mga. We discuss among other aspects the antagonistic role of these proteins in comparison to the functions of Myc proteins (Fig. 1). In addition we also will point out that while we have accumulated a considerable amount of information on Mad proteins, we know less about Mnt and, in particular, Mga.

2
The Mad, Mnt, and Mga Proteins: Structure-Function Analysis

In the late 1980s, the disrecpancies surrounding the function of Myc began to be disolved by the realization that Myc contained a basic region–helix-loop-helix–leucine zipper motif (bHLHZip) which was also present in several known transcription factors. Within this domain, the basic region had been shown to function in DNA binding subsequent to dimerization mediated by the HLHZip segments. With the realization that Myc proteins possess a bHLHZip domain, a search for dimerization partners was initiated (see Lüscher and Eisenman 1990 for review). While in vitro Myc can be forced into oligomers, homodimerization could not be demonstrated in cells, suggesting that Myc forms heterodimers under physiological conditions (Dang et al. 1991). Thus, Myc was in need of a dimerization partner. Indeed, interaction cloning led to the identification of Max, a novel bHLHZip protein that proved to be the essential heterodimerization factor of Myc proteins (Blackwood and Eisenman 1991; Blackwood et al. 1992; Henriksson and Lüscher 1996; Prendergast et al. 1991).

The prominent role of Myc proteins in the control of cell behavior and in particular the strong selection for deregulated Myc expression in tumors implied that these proteins are tightly regulated under physiological conditions. One model of regulation developed after the identification of Max suggested that the bHLHZip domain of Myc might bind variants of Max or Max-like proteins that lack a basic region. This could lead to the formation of heterodimers incapable of DNA binding. Furthermore, it was suggested that Max might interact with other bHLHZip proteins and thus form alternative heterodimers. Such proteins were imagined to antagonize Myc in several ways including competition for binding to Max, interaction with DNA, or some other functional way. It was the latter model that proved significant since various screening approaches have identified six additional Max heterodimerization partners to date: four different Mad proteins, Mnt, and Mga. At least the Mad proteins and Mnt appear to have opposite activities to Myc. While Myc recruits transcriptional activators, Mad proteins and Mnt bind corepressors providing evidence for the molecular basis of the Myc–Mad antagonism as will be discussed in detail in Sect. 2.1.

Screening of bacterial expression libraries and yeast two-hybrid approaches using Max as bait led to the identification of first Mad1 and Mxi1 (Ayer et al. 1993; Zervos et al. 1993) and later of Mad3, Mad4, Mnt, and Mga (Hurlin et al. 1995a, 1997, 1999; Meroni et al. 1997). While the four Mad proteins share high homology, the two other Max partners, Mnt and Mga, are rather distinct in sequence (Fig. 2a). This suggests significant functional differences, although detailed analyses of these two proteins are still missing. What all six proteins have in common is obviously their ability to interact with Max. HLHZip domains, the regions showing the highest degree of homology among these proteins, mediate this interaction. Whereas the Mad proteins also reveal significant homology throughout the entire sequence, Mnt and Mga are otherwise rather distinct.

2.1
The bHLHZip Domain Defines the Myc/Max/Mad Network

The four members of the Mad family are highly homologous proteins that, in addition, display functional similarities, at least when expressed exogenously. The former is documented by the finding that, e.g., Mad1 and Mxi1 share 43% identity at the amino acid level. The highest identity (66%) is seen between the bHLHZip regions and the adjacent C-terminal 60 amino acids of the two proteins. The other Mad proteins show a similar degree of homology. This identifies the bHLHZip as the most conserved domain in Mad proteins and indeed throughout the Myc/Max/Mad network. This domain is the common structural element of the network and is responsible for specificity and stability of dimer formation. Interaction assays showed that the four Mad proteins specifically interact with Max but not with each other nor with

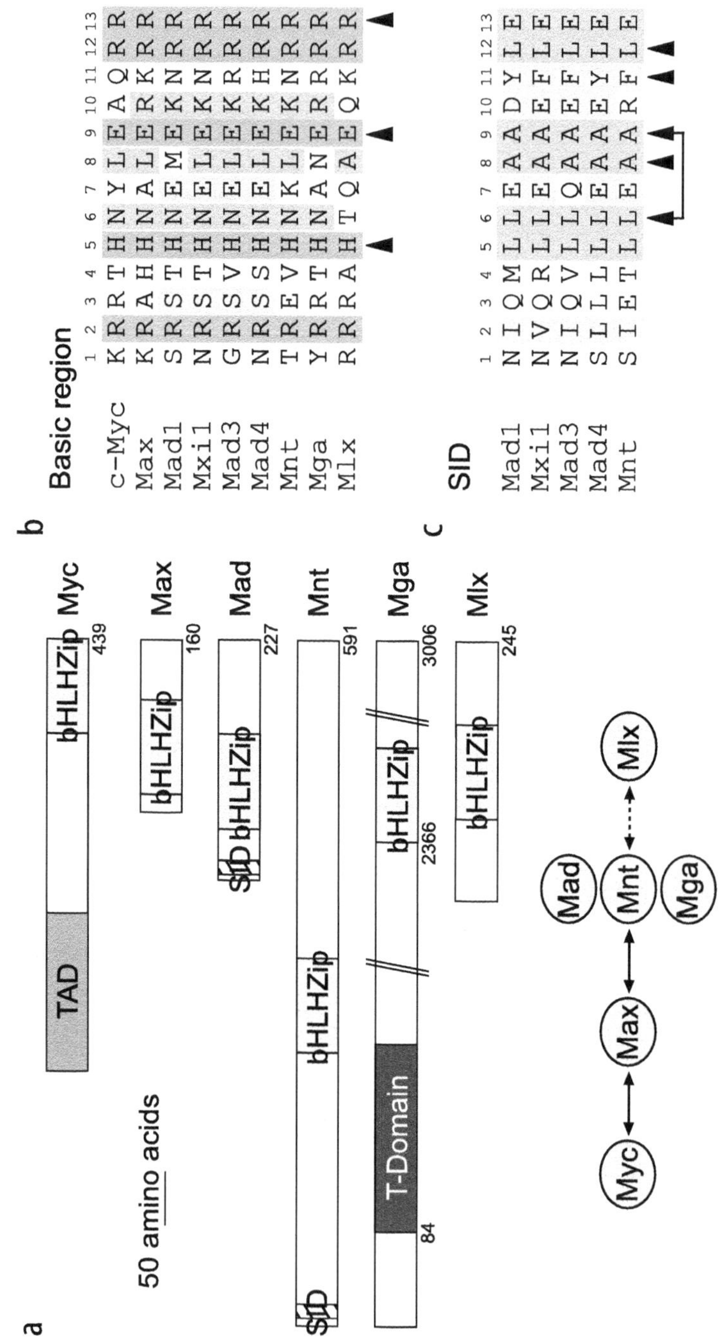
a
50 amino acids
TAD
bHLHZip
439
Myc
160
Max
SID
227
Mad
591
Mnt
T-Domain
84
2366
3006
Mga
245
Mlx
Myc
Max
Mad
Mnt
Mga
Mlx
b
Basic region
1 2 3 4 5 6 7 8 9 10 11 12 13
c-Myc K R R T H N Y L E A Q R R
Max K R A H H N A L E R K R R
Mad1 S R S T H N E M E K N R R
Mxi1 N R S T H N E L E K N R R
Mad3 G R S V H N E L E K R R R
Mad4 N R S S H N E L E K H R R
Mnt T R E V H N K L E K N R R
Mga Y R R T H N A N E R R R R
Mlx R R R A H T Q A E Q K R R
c
SID
1 2 3 4 5 6 7 8 9 10 11 12 13
Mad1 N I Q M L L E A A D Y L E
Mxi1 N V Q R L L E A A E F L E
Mad3 N I Q V L L Q A A E F L E
Mad4 S L L L L L E A A E Y L E
Mnt S I E T L L E A A R F L E

◀

Fig. 2a–c Comparison of structural features of proteins within the Myc/Max/Mad network. **a** The common motif of all members of the Myc/Max/Mad network is the basic region–helix loop helix–leucine zipper domain (bHLHZip). The HLHZip mediates heterodimerization and the basic region enables binding to DNA. As a transcriptional activator Myc possesses a transactivation domain (TAD) through which Myc interacts with coactivators. In contrast, the Mad family members (Mad1, Mxi1/Mad2, Mad3, Mad4, and Mnt, but not Mga) possess a Sin3-interacting domain (SID) that binds to the PAH2 domain of mSin3, thereby recruiting the mSin3–HDAC repressor complex. Mga has an additional DNA binding domain, the T-domain that is common to proteins of the brachyury-T-box-family involved in the regulation of developmental processes. Max and Mlx are the common heterodimerization partners used by the two parallel transcriptional networks, one being centered around Max, the other around Mlx. The different interactions are indicated in the *lower panel*. The *dotted double arrow* indicates that Mlx only interacts with Mad1, Mad4, and Mnt but not with Mxi1, Mad3, and Mga. Thus, Mad1, Mad4, and Mnt interconnect the Max and Mlx networks. **b** An alignment of the basic regions of all family members is shown. Identical or homologous amino acids are shaded in *gray*. Positions 3 and 7 are conserved within the Mad proteins. *Arrowheads* identify the amino acids that make specific base contacts. These three amino acids are identical in all Myc/Max/Mad network members. **c** An alignment of the minimal 13 amino acids of the SID is displayed. Identical or homologous amino acids are indicated. *Arrowheads* mark amino acids that have been shown to be important for binding to mSin3 (Eilers et al. 1999)

Myc proteins (Ayer et al. 1993; Hurlin et al. 1995a; Zervos et al. 1993). The functional similarities in test systems are not surprising since the dominant functional domains, i.e., the bHLHZip and the mSin3-interaction domain (SID; for a detailed description see the following section; also Sects.2.1.2 and 2.2), are conserved. However, other regions of the Mad proteins have not been studied in any great detail, but may possess functions that distinguish the activities of the four Mads.

2.1.1
The bHLHZip and Protein–Protein Interaction

Dimerization through the HLHZip is critical for the functions of all network members. It is required for proper positioning of the basic regions contributed by each subunit of the different dimers. The two basic regions form the contact surface that wraps around both sides of the DNA helix, penetrates the major groove, and makes specific base contacts. Dimerization between network members is largely dependent on helix 2 (H2) and the Zip, which form a single α-helical segment. These interactions have been defined by mutational analysis and by structural work (for a detailed discussion see S. K. Nair and S. K. Burley, this volume). Solving the crystal structure of the isolated bHLHZip domains of a Mad1–Max heterodimer bound to DNA (Nair and Burley 2003) revealed similarities and differences in the interaction between Mad1 and c-Myc with Max as well as between two Max molecules. The specificity of dimer formation is the result of hydrophobic surfaces of the H2Zip α-helices that mediate interaction, with additional support from intermolecular hydrogen bonds. Recently the interaction of different network members has been visualized using bimolecular fluorescence complementation (BiFC) analysis (Grinberg et al. 2004). This method is based on the formation of a fluorescent complex from two fragments of a fluorescent protein. The two fragments are fused to proteins whose interaction is being tested. Both the structural and the BiFC analyses reveal that Myc–Max is favored over Max–Max interaction. Of note is that Mad3 is a rather weak while Mad4 is a strong Max binding partner (Grinberg et al. 2004). The studies summarized above indicate that dimer formation within the network is not only dependent on the relative concentration of individual components but also on the relative affinities of Myc and Mad proteins to Max. Moreover, this offers as-yet-unexplored possibilities for regulation that might further modify dimer formation within the network.

Interestingly, for Myc proteins several other factors have been identified that use the HLHZip domain for binding (Oster et al. 2002). Thus, this part of Myc has functions beyond dimerization with Max. For some of the Mad proteins, Mlx was identified as an interacting protein that appears to replace Max as a heterodimerization partner. Mlx has been suggested to be at the center of a network of bHLHZip proteins that exists in parallel to the Myc/Max/Mad network, the two being interconnected by Mad proteins and Mnt (Fig. 2a) (Billin et al. 1999; Meroni et al. 2000; see A.N. Billin and D.E. Ayer, this volume).

In addition to Mlx, Mmip1 and Mmip2 have been described as Mad HLHZip interacting proteins while other network members do not bind to these factors (Gupta et al. 1998; Yin et al. 1999). Mmip1 is a Zip-only protein whereas Mmip2 is characterized by a RING finger. In both cases the heterodimerization results in DNA binding-deficient complexes, since neither of the Mmip proteins possesses a DNA binding domain. Consequently, both proteins inhibit Mad functions. These findings are reminiscent of the findings concerning the role of Id proteins, which are HLH-only proteins and interfere with bHLH transcription factors (Davis and Turner 2001; Norton et al. 1998).

2.1.2
The bHLHZip and DNA Binding

Defining the DNA binding specificity of different network dimers is important for at least two reasons. First, it will help to identify genes that are regulated by one or the other network member. Knowing the target genes will facilitate understanding of how the network regulates different aspects of cell behavior that have been outlined above. Second, it will aid in defining whether the genes that are regulated by Myc–Max and Mad–Max are identical or not and consequently whether other dimers of the network affect the expression of these genes. In recent years, a substantial number of Myc target genes has been identified (Oster et al. 2002; see L.A. Lee and C.V. Dang, this volume), while much less is known about Mad-regulated genes. Comparing the basic regions of Myc and Mad proteins reveals that the amino acids that mediate DNA binding are identical (Fig. 2b). This suggests that the different dimers will bind similar DNA sequences. However, since other amino acids within the basic region are not conserved and because amino acids of additional domains, e.g. from the loop region, could affect DNA binding, the specificity of Myc–Max and Mad–Max heterodimers might not be identical.

To determine consensus DNA binding sites, originally performed with homodimeric bHLHZip domains of c-Myc, binding site selection studies were undertaken. These led to the identification of the core sequence 5′-CACGTG, which is referred to as Myc E-box (Lüscher and Larsson 1999). Similarly, E-box binding of recombinant Mad–Max heterodimers was demonstrated (Ayer et al. 1993; Hurlin et al. 1995a; Zervos et al. 1993). As previously found for Myc–Max complexes, Mad1–Max heterodimers have a preference for a C and a G, at the 5′ and 3′ end, respectively, extending the core sequence to 5′-CCACGTGG (Brownlie et al. 1997; James and Eisenman 2002; Sommer et al. 1998). The analysis of cell-derived Mad–Max complexes has not been straightforward, mainly due to the low abundance of Mad proteins. In electrophoretic mobility shift assays (EMSA) Mnt–Max heterodimers and to a lesser degree Max–Max homodimers are the predominant complexes, while Myc–Max heterodimers are only seen under specific circumstances (Nilsson et al. 2004; Pulverer et al. 2000; Sommer et al. 1998). Specific, differentiation-associated E-box binding of endogenous Mad1 could only be demonstrated by using a solid-phase DNA binding assay (Larsson et al. 1997). However, when overexpressed, both Mad–Max and Myc–Max complexes are detectable in EMSA, and these complexes possess similar affinities and comparable sequence specificities for E-boxes (Sommer et al. 1998).

It has to be remembered that the DNA binding studies summarized above relied on different in vitro assays. These reflect only in part the situation in cells, where the nucleotide sequence is but one feature that defines the interaction of proteins with DNA. Other characteristics, including packaging of DNA into nucleosomes and modification of core histones, DNA modifications, positively or negatively cooperating transcriptional regulators, and indirect recruitment of network proteins to DNA, are most likely playing decisive roles. With the invention of the chromatin immunoprecipitation (ChIP) technique it became possible to address where in the genome transcription factors bind. Again for the Myc/Max/Mad network, most studies have been performed with Myc (Oster et al. 2002). The analysis of Mad proteins demonstrated that Mad1 is localized to E-boxes of Myc/Max/Mad-responsive genes, including cyclin D2 and human telomerase reverse transcriptase (hTERT), correlating with repression of gene transcription (Bouchard et al. 2001; Lin and Elledge 2003; Xu et al. 2001). Furthermore, Mnt is associated with the promoter of the ornithine decarboxylase (ODC) gene, another previously identified Myc target (Nilsson et al. 2004). Similar to the observations with Mad1, Mnt binding correlates with gene repression.

These findings support the switch model that suggests an exchange of Myc–Max to Mad–Max complexes during transitions in cell behavior, i.e., from cycling to differentiating or resting cells (Ayer and Eisenman 1993). The switch is accompanied by the recruitment of distinct cofactors that affect gene transcription, as discussed in more detail in Sect. 2.2. However, recent work indicates that the switch model is in need of expansion to accommodate novel insights into the distribution of network proteins in chromatin and the correlation of DNA binding with gene expression. Using ChIP on a genomic scale or tagged fusion proteins that result in DNA methylation in the vicinity of the binding site, many promoters have been identified that bind Myc/Max/Mad network proteins (see also Sect. 4; Fernandez et al. 2003; Orian et al. 2003). The analysis of *Drosophila* Mnt, which seems to be the ortholog of mammalian Mad/Mnt (Peyrefitte et al. 2001; M. Haenlin, personal communication), revealed a number of loci that were not detected by Myc and, somewhat surprisingly, also loci that did not come up with Max (Orian et al. 2003). This leaves room, besides the switch model, for alternative aspects of gene regulation by network members.

2.2 The SID: Recruitment of a Repressor Complex

The second region of homology that is found in Mad proteins and in Mnt, but not in the other network members, is the Sin3-interaction domain or SID. This domain was originally identified in Mad1 and in Mxi1 due to its ability to interact with mSin3 (Ayer et al. 1995; Schreiber-Agus et al. 1995). Subsequently a SID was identified in Mad3, Mad4, and in Mnt (Hurlin et al. 1997; Hurlin et al. 1995a). mSin3 proteins are the mammalian orthologs of yeast Sin3, a factor that has been implicated in negative regulation of gene transcription (Vidal et al. 1991; Wang et al. 1990; Wang and Stillman 1993). Indeed, recruitment of mSin3, a component of a repressor complex that contains histone deacetylase (HDAC) activity, through the SID mediates gene repression by Mad proteins and Mnt, as discussed in Sect. 3.1, and several other transcriptional repressors (Ayer 1999; Knoepfler and Eisenman 1999; Schreiber-Agus and DePinho 1998). The identification of the SID was important since previous models suggested that the antagonism between Mad and Myc was based primarily on competition for a potentially limiting pool of Max and/or for DNA binding sites. Defining the role of the SID strongly implied that Mad proteins actively repress genes and thus possess functions beyond competing with Myc.

The SID, localized near the N-terminus of Mad and Mnt proteins, is conserved with respect to its primary amino acid sequence (Fig. 2c; Brubaker et al. 2000; Eilers et al. 1999). Although the N-terminus of Mad proteins shows a high degree of homology over 30 amino acids, the minimal SID contains 13 amino acids (amino acids 8–20 in human Mad1, see Fig. 2c), is rich in hydrophobic residues and forms an amphipathic α-helix (Eilers et al. 1999). However, a recent study revealed the necessity of additional residues outside the minimal SID to modulate binding affinities (van Ingen et al. 2004). The SID interacts with the paired amphipathic helix (PAH) 2 domain of mSin3 (Fig. 3; Ayer et al. 1995; Schreiber-Agus et al. 1995). mSin3 in turn interacts with a number of other proteins including HDACs (see Sect. 3.2). Thus, mSin3 seems to work as a scaffold protein upon which repression complexes can assemble. The presence of several different protein interaction motifs in mSin3 supports this view (Fig. 3; Ayer 1999; Knoepfler and Eisenman 1999).

The nuclear magnetic resonance (NMR) structure of the Mad1–SID/mSin3–PAH2 complex revealed that the α-helix of the SID is embedded into a hydrophobic pocket formed by four α-helical segments of PAH2 (Brubaker et al. 2000; Spronk et al. 2000). The structure of the two interacting domains appears to be the result of a folding transition, since neither domain shows a comparable structure when analyzed in isolation (Brubaker et al. 2000; Spronk et al. 2000). The binding of the two domains results in extensive hydrophobic interactions. Disturbing the hydrophobic or the α-helical nature of the SID by introducing charged amino acids or Pro residues, respectively, results in loss of mSin3 binding and loss of repressing activity (Ayer et al. 1996; Cowley et al. 2004; Eilers et al. 1999). Besides the Mad family members including Mnt, four other transcriptional regulators have also been shown to interact with the PAH2 domain of mSin3. These are Pf1 (Yochum and Ayer 2001), KLFs (Sp1/Krueppel-like Zn-finger transcriptional repressors) (Zhang et al. 2001), Ume6 (Washburn and Esposito 2001), and HBP1, an HMG box-containing transcriptional repressor that binds to the retinoblastoma tumor suppressor protein Rb (Swanson et al. 2004). The SID of Pf1 is related in sequence to the SID of Mad proteins and appears to bind to the PAH2 in a similar manner (Brubaker et al. 2000; Spronk et al. 2000;

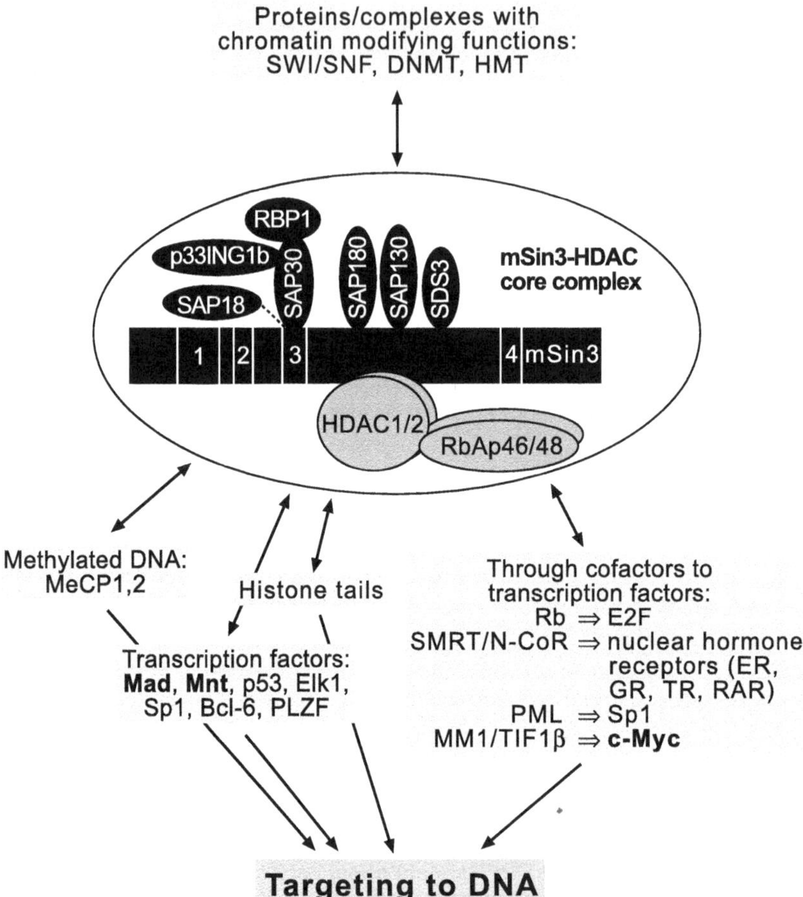

Yochum and Ayer 2001). In contrast, the PAH2-binding motif in HBP1 shares no apparent similarity with the Mad family but seems to be similar to the mSin3 binding motifs found in KLFs and Ume6 (Swanson et al. 2004). In addition, the structural analysis revealed an unexpected reversal in helical orientation of the HBP1 SID in comparison to the Mad1 SID. This correlates with a higher K_d of the mSin3A PAH2–HBP1 SID complex compared to that for PAH2–Mad1. The difference in binding affinity between proteins such as Mad1 and HBP1 might provide one possibility to distinguish interaction of different transcription factors with mSin3 complexes (Swanson et al. 2004).

It is also of interest to note that amphipathic α-helical structural motifs similar to the SID are defining other protein–protein interactions relevant

◄

Fig. 3 The mSin3–HDAC repressor complex. A schematic view is shown of the components of the mSin3–HDAC repressor complex and its interaction with transcriptional regulators and additional components of chromatin remodeling complexes. In general, the mSin3–HDAC core complex can be targeted to DNA by transcriptional repressors that bind to methylated CpG-islands (Jones et al. 1998; Nan et al. 1998) or other transcription factors like Mad1 (Alland et al. 1997; Hassig et al. 1997; Laherty et al. 1997; Sommer et al. 1997), Elk1 (Yang et al. 2002), Sp1 (Won et al. 2002), Bcl-6 (Dhordain et al. 1998), p53 (Kuzmichev et al. 2002; Skowyra et al. 2001), and PLZF (Deltour et al. 1999). Furthermore, the mSin3–HDAC core complex can be recruited indirectly to DNA as in the case of Rb (Lai et al. 2001) or SMRT/N-CoR (Nagy et al. 1997), which are bound by E2F and nuclear hormone receptors, respectively. Sp1 binds both directly and indirectly through PML to the mSin3–HDAC core complex (Wu et al. 2001). The MM1/TIF1β complex bridges the mSin3–HDAC core complex with c-Myc (Satou et al. 2001). DNA methyltransferases (DNMT) (Datta et al. 2003) and histone methyltransferases (HMT) like Suv39H1 (Vaute et al. 2002) have been shown to bind to HDAC1 and presumably to the complete mSin3–HDAC core complex. The *numbers* in mSin3 refer to the four paired amphipathic helix domains. Many additional interactions that are not discussed here have been described involving the mSin3–HDAC repressor complex

for transcriptional regulation, including binding of MDM2 to p53, the herpes simplex virus type 1 VP16 transactivator with $TAF_{II}31$, and the KID domain of the transcription factor cAMP-response element binding protein (CREB) with the coactivator CREB binding protein (CBP) as well as other repressor–corepressor complexes (Han et al. 2003; Kussie et al. 1996; Radhakrishnan et al. 1997; Uesugi et al. 1997; Xu et al. 2002). It has been suggested that these conserved amphipathic α-helical motifs, considering that these are generally short and thus have limited specificity, create the potential for crosstalk between positive and negative transcriptional regulators. This possibility remains to be addressed for Mad proteins.

2.3 The C-Terminal Region of Mad Proteins

While essential functions have been demonstrated for the N-terminal SID and the centrally located bHLHZip, the results regarding the C-terminal region are less clear. In one study the C-terminus of Mad1 was dispensable for the inhibition of c-Myc/Ha-Ras co-transformation (Cerni et al. 1995). In a second study, removal of the C-terminus led to a partial reduction of transcriptional repression by Mad1 (Koskinen et al. 1995). These small differences may depend on the level of overexpression of these proteins in the assay systems used. However, the sequence conservation within the C-terminal region suggests functional relevance. An alignment of C-terminal sequences of Mad proteins revealed five regions of homology (Barrera-Hernandez et al. 2000). Regions I, II, and IV share a high degree of homology across species and among the different Mad family members, while regions III and V are conserved preferentially across species. Removal of the last 18 amino acids of Mad1 (which corresponds to region V) abolishes the ability of Mad1 to inhibit proliferation, a Myc-imposed differentiation block and transformation. In contrast, deletion of the last 42 amino acids (region IV and V) restored all activities, suggesting an interplay of these C-terminal regions in regulating Mad function. Furthermore, the C-terminal region of Mxi1 has been shown to play a role in the repression of the *c-myc* gene (Lee and Ziff 1999). One speculation derived from this work is that the C-terminal region interacts with an as-yet-unidentified co-repressor. However, more detailed understanding of this region of Mad proteins awaits further experimentation and it will be interesting to define the mechanism underlying these observations.

2.4 Additional, Non-conserved Aspects of Mad, Mnt, and Mga Proteins

2.4.1 Mxi1

Unique among the Mad family members is Mxi1 for which two different transcripts have been described that encode Mxi-WR and Mxi-SR for weak and strong repressor, respectively. Mxi-WR lacks the N-terminal 36 amino acids (Schreiber-Agus et al. 1995). This protein is considerably weaker than Mxi-SR in inhibiting cell proliferation and transformation of rat embryo fibroblasts (REF) by c-Myc/Ha-Ras. Subsequent studies demonstrated that Mxi-WR lacks the SID and thus cannot recruit the mSin3 repressor complex but can still bind to Max and to DNA (Rao et al. 1996; Schreiber-Agus et al. 1995). The data suggest that the strength of the Mxi1-dependent antagonism of Myc function can be regulated by altering the ratio between Mxi-WR and Mxi-SR, although differential expression of the two remains to be demonstrated. Comparable alternative splicing of the transcripts of the other *mad* genes has not been reported. Why is this potential regulatory mechanism specific for Mxi1? One possible explanation relates to the observation that Mxi1 is the most broadly expressed Mad family member and may therefore need an additional level of regulation to accommodate all cellular needs.

2.4.2 Mnt

Although Mnt possesses similar activities to Mad proteins when expressed exogenously, Mnt is clearly distinct from Mads. Mnt is much larger and shares beyond the bHLHZip and SID no extensive homologies to other network members (Fig. 2a; Hurlin et al. 1997; Meroni et al. 1997). Although Mnt/Max complexes recognize the E-box consensus 5′-CACGTG, it has been suggested that these heterodimers have higher affinity for the non-canonical 5′-CACGCG binding site (Meroni et al. 1997). The amino acids known to establish base contacts in Myc, Max, and Mad (see Sect. 2.1.2) are conserved in Mnt (Fig. 2b), thus it remains to be determined what the molecular basis is for this proposed shift in specificity. The N-terminus of Mnt contains a SID that interacts in vitro with mSin3. Like Mad proteins, Mnt represses transcription of E-box-driven reporter gene constructs presumably through the mSin3 repressor complex (Hurlin et al. 1997; Meroni et al. 1997). Recent findings support this notion, since Mnt isolated from cells is associated with mSin3. Interestingly, its interaction with Mnt is cell-cycle regulated, suggesting an additional level of control previously not described (N. Popov et al. 2005). It will be important to determine the regulatory circuitry that targets the interaction of Mnt, but possibly also Mad proteins, with mSin3. In this regard, two recent reports are relevant since they indicate that growth factor signaling interferes with the binding of mSin3 to two transcriptional regulators, TIEG2 and AML1 (Ellenrieder et al. 2002; Imai et al. 2004). While basic aspects of the bHLHZip and the SID of Mnt have been analyzed, the role of other regions of this protein have not been studied, and therefore the full spectrum of Mnt functions remains to be determined. In this respect it is important to note that Mnt△SID cooperates with Ha-Ras in REF transformation, indicating that domains in Mnt possess Myc–TAD-like activity (Hurlin et al. 1997).

2.4.3 Mga

Mga, the largest member of the family (300 kDa; Fig. 2a), has been identified in a yeast two-hybrid screen as a Max-interacting protein (Hurlin et al. 1999). Mga is the least-studied member of the Mad/Mnt/Mga family. The four Mad proteins in mammals and Mnt are related at least by having a bHLHZip and a SID. Mga shares with the other family members only the bHLHZip, which specifies interaction with Max and E-box DNA binding. Interestingly, Mga possesses a second DNA binding domain that is homologues to T-box elements of the Tbx/Brachyury family (Kispert and Hermann 1993; Muller and Herrmann 1997; Wilson and Conlon 2002). The presence of two DNA binding domains in the same protein is quite unusual, although other examples have been described (Stuart et al. 1994). In transient systems, Mga can affect transcription through both DNA elements and repress c-Myc/Ha-Ras-dependent transformation of REF cells in a manner dependent on the bHLHZip but not the T-box domain (Hurlin et al. 1999). However, because Mga does not contain a SID, other regions of this large protein may confer repression activity. Despite the lack of detailed knowledge about other functional domains in Mga (see also Sect. 4.4), recent evidence suggests that Mga/Max are indeed part of a repressor complex with histone methyltransferase (HMT) activity (Ogawa et al. 2002). Furthermore, Mga interacts through two PXLXP motifs with BS69, a potential tumor suppressor with a Mynd domain, previously identified as an E1A binding protein (Ansieau and Leutz 2002; Hateboer et al. 1995). Interestingly E1A can compete with Mga for binding to BS69, supporting the notion that the Mga-BS69 interaction is relevant in cell proliferation control and possibly transformation. This interaction is most likely unique to Mga, since the PXLXP motif is found neither in Mad proteins nor in Mnt. While some specific functions for Mga have been uncovered, altogether we know little about biological activities of this protein and its role within as well as outside of the network.

Perspective We have learned some of the basics of Mad, Mnt, and Mga in recent years, but we are still missing the aspects of regulation. While the expression patterns—in particular of *mad* genes and proteins—are complex, we know very little about signaling and regulation of Mad proteins. The evidence from overexpression studies indicates that Mads are constitutive repressors. Is this correct, or can situations be identified in which the interactions of Mad proteins with their partner proteins are modulated? Another important aspect for future studies is the detailed analysis of regions beyond the bHLHZip and the SID in Mad, Mnt, and Mga. Likely some surprises are awaiting us once we obtain a more complete picture of the interaction partners and the functional consequences of domains that are plain white areas on the protein map today.

3 The mSin3–HDAC Repressor Complex, Histone Modification, and Gene Transcription

Transcription in eukaryotic cells is critically dependent on the accessibility of DNA for transcriptional regulators. At the DNA level this is controlled by assembly of DNA into chromatin and by modifications of DNA. The fundamental subunit of chromatin is a highly organized and dynamic protein–DNA complex, the nucleosome. It is composed of an octamer of four core histones, an H3–H4 tetramer, and two H2A–H2B dimers, surrounded by 146 bp of DNA (Luger and Richmond 1998). Furthermore, the quality and the degree of core histone modifications affect the chromatin status (i.e., compaction and position of nucleosomes) and are central determining factors in regulation of gene transcription (Jenuwein and Allis 2001; Strahl and Allis 2000). Several posttranslational modifications of histones, including phosphorylation, acetylation, methylation, and ubiquitination, have been identified. In addition, other proteins associated with the transcription process are subject to similar modifications. Although we are far from understanding in detail the effects of these posttranslational modifications, it is reasonable to suggest that these are vitally important to not only define basic aspects of gene transcription but to fine-tune the expression of each individual gene whenever necessary.

At present, acetylation of core histones is probably the best-understood posttranslational modification of the above-mentioned examples. Acetylation occurs at the ε-amino groups of particular lysine residues. In core histones these lysines are evolutionarily conserved and located predominantly within the N-terminal first 30 amino acids, the so-called histone tails (Jenuwein and Allis 2001; Strahl and Allis 2000). Multiple lysine residues can be acetylated, and it has been suggested that acetylation weakens the interaction between nucleosomes and/or between histones and DNA resulting in a more transcriptionally permissive state. In addition, acetylation affects the interaction between other proteins and the N-terminal histone tails and thus orchestrates, together with other modifications, the recruitment of factors that regulate transcription.

3.1
Repression of Transcription by Mad and Mnt Through an mSin3-Repressor Complex

The last several years have brought to light evidence that the Myc/Max/Mad network controls gene transcription at least in part by recruiting multisubunit protein complexes that contain enzymes capable of modifying histones and possibly other factors. While in most analyses Myc has paved the way to molecular understanding, studies involving Mad proteins were the first to suggest a role for histone-modifying enzymes in Myc/Max/Mad-dependent gene regulation. The antagonism of Myc and Mad proteins addressed above was also apparent in transient reporter gene assays demonstrating that Mad proteins can function as repressors (Knoepfler and Eisenman 1999; Schreiber-Agus and DePinho 1998). This function is mediated through the SID, as discussed above, by an mSin3 corepressor complex that contains HDAC activity (Alland et al. 1997; Hassig et al. 1997; Laherty et al. 1997; Sommer et al. 1997). Similarly, Mnt also recruits HDAC activity most likely through mSin3 (N. Popov et al. 2005). The recruitment of this complex is essential for the biological activities that have been attributed to Mad proteins (Baudino and Cleveland 2001; Grandori et al. 2000). Mutation or deletion of the SID impairs repression in transient assays (Ayer et al. 1996; Hurlin et al. 1997). More recent findings describe the association of histone acetyltransferases (HAT) with Myc (Bouchard et al. 2001; Fernandez et al. 2003; Frank et al. 2001, 2003; Liu et al. 2003; Nikiforov et al. 2002; Park et al. 2001; Vervoorts et al. 2003), suggesting that the switch from Myc/Max to Mad/Max is associated with a change from HATs to HDACs at responsive DNA elements and, in more general terms, from open, accessible to closed, inaccessible chromatin. Indeed, it could be demonstrated that activation of Myc results in an increase in acetylation of core histones, in particular histone H3 and H4, at responsive promoters (Bouchard et al. 2001; Fernandez et al. 2003; Frank et al. 2001, 2003). In contrast, association of Mad proteins with promoters correlates with recruitment of HDAC1 to the promoter and with a decrease in histone acetylation and polymerase loading (Bouchard et al. 2001). Thus, these findings are consistent with the recruitment of opposing acetylation-modifying enzymes by Myc and Mad proteins.

3.2
The mSin3–HDAC Repressor Complex: Subunits and Targeting to DNA

Several HDAC-dependent corepressor complexes have been described in mammalian cells (Narlikar et al. 2002). Of particular relevance for the discussion here are complexes that contain HDAC class I proteins including HDAC1 and HDAC2, which bind mSin3 (Fig. 3; Hassig et al. 1997; Zhang et al. 1997) and Mi-2/NuRD complexes (Wade et al. 1998; Xue et al. 1998; Zhang et al. 1998a), and the HDAC3-containing SMRT and Co-REST complexes (Guenther et al. 2000; Li et al. 2000). From the available data it appears that the mSin3 complex is the only corepressor complex that is recruited by Mad proteins (Li et al. 2002).

The mSin3 and Mi-2/NuRD complexes contain an HDAC core complex with HDAC1/2 and RbAp48 and RbAp46 (Zhang et al. 1997). The latter two were originally identified as proteins associated with Rb (Qian and Lee 1995). It is this core complex that possesses HDAC activity, while isolated HDACs appear to be catalytically inactive. The HDAC1/2 core complex can be used in several ways (Knoepfler and Eisenman 1999). It has been shown recently that it can bind through RbAp48 and RbAp46 to histone tails and thereby may contribute to more global chromatin deacetylation in the absence of DNA binding factors (Li et al. 2002). Furthermore, the core HDAC1/2 complex may interact directly with transcriptional regulators (Knoepfler and Eisenman 1999). In many instances, however, the core complex associates with other proteins, including mSin3 and Mi-2 but also corepressors such as Rb and Groucho, that direct interaction to sequence-specific transcriptional regulators. For the mSin3 and

Mi-2/NuRD complexes that contain the HDAC1/2 core complex, the presence of several additional subunits has been described. It is important to note that these appear to be largely distinct between these two complexes and are presumably involved in complex-specific functions (Knoepfler and Eisenman 1999; Narlikar et al. 2002).

Early work had suggested that the mSin3–HDAC complex possesses at least 10 subunits (Hassig et al. 1997). Indeed, in recent years several proteins have been identified that are part of the mSin3–HDAC complex (Fig. 3). These components include, besides the HDAC1/2 core complex, RBP1 (Lai et al. 2001), p33ING1b (Kuzmichev et al. 2002; Skowyra et al. 2001), SAP18, SAP30 (Laherty et al. 1998; Zhang et al. 1998b), SAP45, SAP130, and SAP180 (Fleischer et al. 2003). Whether these are stoichiometric or conditional members of the complex is not clear.

For some of these subunits specific functions have been identified. mSin3 functions as an adaptor that interacts with multiple subunits of the mSin3–HDAC repressor complex including HDAC1 and 2 and with transcriptional regulators through multiple protein–protein interaction motifs (Fig. 3; Knoepfler and Eisenman 1999; Wang et al. 1990). SAP30 is thought to be required for binding RBP1 and the HDAC1/2 core complex to mSin3, while RPB1 can tether mSin3 to Rb. In addition, SAP30 is required for Mad-dependent repression (Laherty et al. 1998; Lai et al. 2001; Zhang et al. 1998b). Rb is a corepressor that interacts with E2F transcription factors, turning their activity from activation to repression (Muller and Helin 2000). p33ING1b, a negative regulator of cell proliferation, modulates the function of the tumor suppressor p53 (Cheung and Li 2001). Suppression of proliferation by p33ING1b requires interaction with mSin3 (Kuzmichev et al. 2002). It has been shown previously that p53 interacts with mSin3 (Murphy et al. 1999). This appears to be relevant for p53-dependent transcriptional repression and its antiapoptotic function. Indeed, mSin3 binds to and interferes with a region in p53 that is implicated in proapoptotic activities (Vousden 2002). p33ING1b may assist in recruiting the mSin3–HDAC complex to p53 and thus be involved in shifting the transcriptional function of p53 from activation to repression.

The other subunits may be relevant for assembly, for targeting of the complex to yet other transcriptional regulators, or for catalytic activity. The latter has been suggested from studies on the Mi-2/NuRD-complex, which requires MTA2, p70, and p32 for maximal HDAC activity (Feng and Zhang 2003). Furthermore, some subunits may be at the receiving end of signal transduction cascades, thereby regulating the activities of the mSin3–HDAC complex, an aspect poorly studied. We expect that this complex is modulated and fine-tuned, for example, in response to proliferation stimulatory or repressive signals or during the cell cycle. In this respect the finding that TIS7 can be part of the mSin3/HDAC complex is interesting. TIS7 is induced during differentiation and thus may target and/or regulate the mSin3/HDAC complex in response to cellular differentiation processes (Vietor et al. 2002).

In addition to HDAC activity, the mSin3 complex associates with other enzymatic activities. It interacts with components of the SWI/SNF chromatin-remodeling complex that include the Brg1 or hBrm ATPases (Kuzmichev et al. 2002; Sif et al. 2001). In comparison, the Mi-2/NuRD complex includes CHD3/4 ATPases as chromatin remodeling enzymes (Tong et al. 1998; Xue et al. 1998). More recently it has been demonstrated that HDAC1–3 can also bind to the HMT Suv39H1, with RbAp48 and RbAp46 being also part of the complex (Czermin et al. 2001; Vaute et al. 2002). This HMT methylates K9 of histone H3, a modification that has been linked to heterochromatin and to gene repression (Kouzarides 2002; Lachner and Jenuwein 2002). Although it has not been resolved yet whether Suv39H1 is part of an mSin3–HDAC complex, the combination of HDAC and H3K9-specific HMT activities is suggested to cooperate in gene repression. Another interaction of HDAC1 that may be important for efficient repression of transcription is with the DNA methyltransferase Dnmt3a (Fuks et al. 2001). In regard to Mad and Mnt function, it will be of interest to determine whether any of these additional enzymatic activities are associated with the mSin3 complex that binds to Mad. The recruitment of such activities by Mad proteins might have significant consequences on how these proteins affect gene transcription and therefore affect the functions of Mad proteins.

The findings discussed above indicate that the mSin3–HDAC repressor complex is not a static assembly of factors but rather its composition and probably also its functions are dynamic. While some of the components are part of a core complex, other subunits associate under specific circumstances, most likely reflecting different functional requirements during different states of cell behavior. In addition to subunit composition, the recruitment of the mSin3 complex to specific DNA sites further enhances functional diversity. As already mentioned, Mad proteins and Mnt recruit the complex to E-box DNA elements. Several additional transcription factors target the mSin3 complex to specific DNA sites. Like Mad proteins, some of these factors interact directly with mSin3, including Sp1-like factors, Bcl-6, and Elk1 (Dhordain et al. 1998; Yang et al. 2002; Zhang and Reinberg 2001). Ski/Sno are additional transcriptional regulators that bind to mSin3 and recruit HDAC activity (Cohen et al. 1999; Nicol et al. 1999; Nomura et al. 1999). Ski/Sno also interact with N-CoR/SMRT and PML (Khan et al. 2001). It has been suggested that these factors are part of a large complex that is also relevant for Mad function. Other factors recruit the Sin3 complex through linker proteins. Examples for this type of interaction are N-CoR and SMRT (Huang et al. 2000; Nomura et al. 1999).

Of particular interest for understanding the Myc/Max/Mad network is the finding that the c-Myc interacting protein MM-1 interacts with an mSin3A/HDAC complex resulting in c-Myc-dependent repression (Mori et al. 1998; Satou et al. 2001). Thus, not only Mad and Mnt proteins but also Myc may target mSin3–HDAC complexes to DNA. The functional relevance of Myc's ability to bind both coactivators and corepressors and the signals that presumably regulate these interactions are not understood. An immediate question arising is what the functional difference would be between Mad- and Myc-dependent recruitment of mSin3–HDAC complexes for gene transcription.

The functional diversity of mSin3–HDAC complexes is further highlighted by the finding that methylated DNA binding proteins also connect with this repression complex. Methylation of CpG islands has long been known to correlate with gene repression. MeCP1 binds to methylated CpGs and by recruiting the mSin3–HDAC complex mediates core histone deacetylation (Nan et al. 1998). This in turn is thought to inhibit gene transcription by preventing transcription factor binding and polymerase recruitment. Together these findings demonstrate that the mSin3–HDAC repressor complex is involved in gene repression by many different factors.

Perspective A central function of Mad proteins is the recruitment of an mSin3 repressor complex. It is this complex that mediates Mad-dependent gene repression. Although we have obtained some information about the composition of the mSin3 complex, we do not know which additional components associate with the complex that binds to Mad proteins and Mnt. In particular, knowledge about other enzymatic activities might be enlightening in order to define in more detail the molecular mechanisms used by Mad proteins to regulate gene transcription. Furthermore, the identification of additional targeting subunits associated with the Mad-bound mSin3 complex will be important. Since the mSin3 complex may function as a bridging complex between different DNA binding factors, predictions about transcriptional regulators that cooperate with Mad proteins in repression might be possible. So far, Mad proteins have been viewed exclusively as repressors. Since many factors, including c-Myc, SP1, YY1, and Elk1 mentioned above, can recruit both positive and negative cofactors, it will be interesting to determine whether Mad proteins and Mnt may, under specific circumstances, also bind coactivators.

3.3 Function and Regulation of HDACs

Although there may be Mad functions independent of HDAC activity, it is clear that Mad proteins need to recruit HDACs to exert their full biological effects. Therefore it is useful to summarize a few aspects of HDAC function and regulation. Histone deacetylases are generally divided in two protein families: the classical HDAC family and the SIR2 family of NAD-dependent HDACs (de Ruijter et al. 2003). The former is subdivided into class I (containing HDAC1, 2, 3, and 8, which are closely related to the yeast RPD3) and class II (HDAC4, 5, 6, 7, 9, and 10, which share homology with yeast HDA1) enzymes and the recently discovered HDAC11, which shares homology with both classes. Although it is beyond the scope of this review to discuss HDAC expression patterns, it is worth mentioning that expression is rather ubiquitous (de Ruijter et al. 2003). Many studies, including those on Mad, demonstrate important roles for HDACs in regulation of gene transcription. However, a word of caution is reasonable since the role of HDACs has been mainly addressed under conditions of transient overexpression or of broad inhibition of activity. Indeed, the HDAC-dependency of Rb-mediated repression could be demonstrated for some but not all tested endogenous E2F-responsive promoters (Luo et al. 1998). HDACs function in many different processes that affect cell behavior. Nevertheless, somewhat unexpected is the observation that inhibition of HDACs leads preferentially to inhibition of proliferation and to differentiation (Kelly et al. 2002; Kim et al. 2003a).

Aberrant acetylation is associated with many human cancers. Deregulated expression and function of HATs has been observed in a number of tumors. In particular, loss of heterozygosity (LOH) of p300 is detected in the majority of glioblastomas. Furthermore, individuals with Rubinstein–Taybi syndrome who carry mutations in CBP that block its HAT activity have an increased risk of tumor development (Cress and Seto 2000; Mahlknecht and Hoelzer 2000; Marks et al. 2001). So far, mutations in HDAC genes have not been linked with human cancers. But several oncoproteins and tumor suppressor proteins, including PML-RARα, Rb, and Bcl6, repress transcription through HDACs (Klochendler-Yeivin and Yaniv 2001; Timmermann et al. 2001). These findings suggest that decreased acetylation is linked to cancer, which is consistent with HDACs role in proliferation and differentiation. Although this conclusion is not very focused, inhibition of HDAC activity is a promising strategy in the treatment of cancer (Cress and Seto 2000; Mahlknecht and Hoelzer 2000; Marks et al. 2001).

Beside protein–protein interaction, phosphorylation of HDACs has been described as another important mechanism to regulate specific aspects of

HDAC function, including subcellular localization, enzymatic activity, and complex formation (de Ruijter et al. 2003; Tong 2002). Human HDAC1 and HDAC2 are localized predominantly in the nucleus. In contrast, class II HDAC4 and 5 shuttle between nuclear and cytoplasmic compartments, a process that is controlled by Ca^{2+}-calmodulin-dependent kinases and Ras-dependent signal transduction pathways (Grozinger and Schreiber 2000; Kao et al. 2001; McKinsey et al. 2000; Wang et al. 2000a; Zhou et al. 2000). HDAC1 has been suggested to be substrate of protein kinase CK2 and PKA (Cai et al. 2001; Tsai and Seto 2002). Phosphorylation of HDAC1 or HDAC2 by CK2 reduces enzymatic activity and modulates complex formation with mSin3 and Mi-2 (Pflum et al. 2001; Tsai and Seto 2002). In addition, mitotic hyperphosphorylation of HDAC1 and 2 correlates with a small but significant increase in HDAC activity but at the same time with disruption of HDAC1 and Sin3A binding, suggesting that the HDAC core complex is activated but its targeting is altered (Galasinski et al. 2002).

Ubiquitination of proteins has been implicated in transcriptional control in recent years (Muratani and Tansey 2003). It was found that HDAC2, 5, and 6 are ubiquitinated (Hook et al. 2002; Kramer et al. 2003). Furthermore, HDAC6 interacts specifically with polyubiquitin and two proteins of the ubiquitin signaling pathway, Ccd48p and phospholipase A2-activating protein (Hook et al. 2002; Seigneurin-Berny et al. 2001). The finding that HDACs can be ubiquitinated suggests an additional link between acetylation and ubiquitination of proteins involved in transcriptional regulation.

In summary, the data discussed above stress the point that HDACs are tightly regulated, although we are still far from understanding this regulation in depth. But for the purpose of our discussion of the Mad and Mnt components of the Myc/Max/Mad network, it will be interesting to see whether any of these modifications impinge on the Mad–mSin3–HDAC complex.

Another aspect worth considering is that HDACs not only deacetylate core histones but also other proteins involved in transcription. Indeed, an increasing number of transcriptional regulators are found to be acetylated (Sterner and Berger 2000). Acetylation is, similar to phosphorylation, a highly dynamic posttranslational modification that affects numerous functions, including DNA binding, protein stability, and the modulation of protein–protein interactions. Furthermore, since acetylation, methylation, ubiquitination, and sumoylation can occur on Lys residues, competition between these modifications for specific Lys residues is possible. Therefore, acetylation can affect function at least in part by altering the stoichiometry of these alternative modifications. Thus, acetylation can have both positive and negative effects on transcriptional regulators. Consistent with this idea, HDACs can function both as repressors and activators of transcription, compatible with the original studies applying the HDAC inhibitor trichostatin A that revealed both activated and repressed target genes (Van Lint et al. 1996; Vidal et al. 1991). With c-Myc, a member of the Myc/Max/Mad family has been shown recently to be acetylated. Whether the mSin3–HDAC complex tethered to Myc by MM-1 deacetylates Myc thereby affecting its stability or transcriptional activity remains to be studied (Kim et al. 2003b; Vervoorts et al. 2003; von der Lehr et al. 2003). In this context, it will be important to determine whether Mad–mSin3–HDAC complexes deacetylate factors beyond core histones. This may not necessarily result in repression, but instead lead to gene activation under specific, as yet undefined, circumstances.

4 Target Genes

To obtain more insight into the functional consequences of the action of Mad, Mnt, and Mga proteins, it is important to determine their target genes. In contrast to studies on Myc, for which a large number of potential target genes has been reported in several publications, little is known about the targets of the other network members. Due to the high overlap in DNA binding specificity as discussed above, intuitively one would think that the target genes overlap substantially between Mad and Myc. This may not be the case, as a recent study indicates using *Drosophila* as a model system (Orian et al. 2003). Several criteria have to be met to define direct target genes of a specific transcription factor. These include (1) differential expression of the target in response to distinct levels of factor, preferentially using a system that allows analysis in the presence of protein synthesis inhibitors to minimize secondary effects; (2) presence of the factor at or near the promoter of the target gene; and (3) regulation of corresponding reporter genes dependent on the response element and on the relevant domains of the factor. In many instances these criteria are only partially fulfilled, thus various degrees of certainty exist whether a given transcriptional regulator regulates a given gene.

4.1 hTERT, Cyclin D2, and Others

Of the Mad family members, Mad1 is the protein that has been studied most extensively, and at least two target genes seem to be strong candidates for direct regulation. These are the genes encoding hTERT and cyclin D2. hTERT is the catalytic subunit of telomerase, the enzyme responsible for elongating the repetitive sequences at telomeres (Kelleher et al. 2002). Originally it was thought that this enzyme is not active in most somatic cells. However, more recent findings indicate that in general telomerase is found in many proliferating but not in differentiated tissues (Mathon and Lloyd 2001; Rubin 2002). Reactivation or increased expression of this enzyme is observed in most human cancers. Lack or insufficient levels of telomerase can result in erosion of telomeres that is associated with senescence and genetic instability (Cerni 2000; Mathon and Lloyd 2001).

Because hTERT was identified as a c-Myc target gene (Greenberg et al. 1999; Oh et al. 1999; Takakura et al. 1999; Wang et al. 1998; Wu et al. 1999) it was reasonable to address whether Mad functions as a negative regulator of telomerase expression. Indeed, Mad1 represses a reporter gene construct containing the *hTERT* promoter, and this repression is mediated by HDACs (Cong and Bacchetti 2000; Gunes et al. 2000). In addition, Mad1 was identified in a screen designed to define factors that repress the *hTERT* promoter (Oh et al. 2000). These studies, which relied on transient overexpression of Mad1, are supported by ChIP analysis demonstrating Mad1 binding to the promoter in differentiating HL60 promyelocytes (Xu et al. 2001). This correlates with reduced core histone acetylation and repression of the *hTERT* gene. However, a recent study using a pair of isogenic cell lines that differ in telomerase activity indicates that repression of *hTERT* expression is independent of changes in Mad1 levels (Horikawa et al. 2002). Inhibition of *hTERT* is E-box-dependent and is achieved in the telomerase-positive line by transferring chromosome 3, implying the presence of a telomerase repressor gene on this chromosome. Nevertheless, overexpression of either Mad1 or c-Myc was still able to regulate *hTERT*. How can these findings be reconciled? A possible solution comes from a recent study that identifies several repressors of *hTERT* expression, including Mad1 (Lin and Elledge 2003). The removal of just one repressor was sufficient to activate expression in cell lines that are *hTERT* negative, suggesting that the repressors cooperate in silencing the *hTERT* gene. Since none of the genes identified in this study localizes to chromosome 3, an additional repressor is postulated that cooperates with Mad1.

Like *hTERT*, the *cyclin D2* gene was identified as a c-Myc target that mediates activation of cyclin-dependent kinases (Lüscher 2001). In reporter gene assays the *cyclin D2* promoter is repressed by Mad1, an effect that requires the SID (Bouchard et al. 1999; Yochum and Ayer 2001). Furthermore, ChIP experiments demonstrated Mad1 binding specifically to the *cyclin D2* promoter region that contains two E-boxes in differentiating HL60 and U937 promyelocytic cell lines (Bouchard et al. 2001; Vervoorts et al. 2003). Importantly, Mad1 binding to the *cyclin D2* promoter correlates with HDAC1 recruitment, histone deacetylation, reduced polymerase II binding, and inhibition of *cyclin D2* expression (Bouchard et al. 2001). The repression of *cyclin D2* by Mad1 provides one explanation for the observation that Mad1 potently inhibits the G1–S transition of the cell cycle and consequently cell proliferation.

Mxi1 has been postulated to directly repress the *c-myc* gene in reporter assays and blocks serum-induced c-Myc expression in quiescent cells (Lee and Ziff 1999). This repression is independent of the basic region and the SID domain and is antagonized by Max, suggesting that Mxi1 does not function by binding to an E-box element and by recruiting HDAC activity. Further work is required to define in more detail the mechanistic base for Mxi1 function on the *c-myc* promoter. The proposed feedback mechanism within the Myc/Max/Mad network may be important to modulate and even enhance the switch between proliferation and differentiation. Recently, the *id2* gene has been identified as an Mxi1 and Mad4 repressed target in response to transforming growth factor (TGF)-β signaling (Siegel et al. 2003). This gene is regulated by c-Myc and Id2 functions as a negative regulator of Rb, thereby promoting cell cycle progression (Lasorella et al. 2000). Although Mxi1 and Mad4 bind to the *id2* promoter in cells, no change in histone acetylation is seen. In addition, overexpression of either Mad protein is not sufficient to inhibit *id2* transcription (Siegel et al. 2003). Thus, the mechanism of repression seems distinct compared to other examples discussed above and needs further evaluation.

4.2 Chimeric Proteins and the Search for Target Genes

While the analyses of Mad target genes discussed above were guided by studies on Myc, more open approaches to identify Mad-regulated genes have been undertaken. A first study used the chimeric protein Myc(Mxi1-BR) in which the basic region of c-Myc was exchanged with the basic region of Mxi1. When compared to c-Myc, Myc(Mxi1-BR) was poorly transforming (O'Hagan et al. 2000). Interestingly, within the basic region the non-conserved amino acids Ser3 and Glu7 (Fig. 2b) are responsible for the reduced transforming ability. These amino acids are most likely not involved in specifying DNA binding. Instead they may be relevant for an additional function of the basic region such as protein–protein interaction. In microarray studies it became evident that Myc and Myc(Mxi1-BR) have overlapping but not identical sets of target genes. These findings suggest that the basic regions of c-Myc and Mxi1 are not functionally identical.

In a second study, the complete bHLHZip of c-Myc was swapped with the bHLHZip of Mad1, generating Myc(MadbHZ) (James and Eisenman 2002). This molecule now is substantially different from either c-Myc or Mad1 since it possesses the transactivation domain (TAD) of c-Myc but lacks the HLHZip region that interacts with several proteins, including Miz1. Myc(MadbHZ) may, however, acquire additional activities determined by interactions specific for the Mad1 bHLHZip. While some of the basic functions, i.e., DNA binding and activation of reporter genes in transient assays, are indistinguishable from c-Myc, some biological activities are distinct. Myc(MadbHZ) is unable to efficiently induce apoptosis, whereas its effects on cell proliferation are unaltered. This suggests that genes particularly relevant for apoptosis are differentially regulated by c-Myc and the chimeric protein.

A third chimeric protein, Myc(Mad1-BR), was generated by substituting the basic region of Myc with the corresponding domain of Mad1 (Nikiforov et al. 2003). This protein retains its transforming and proapoptotic activities and occupies Myc target promoters, including *hsp60*, *nm23*, *nucleolin*, and *cad*, in cells. These findings are not easily brought in line with the functions of Myc(Mxi1-BR) summarized above. It is possible that some differences in the swapped regions may have profound effects on the functions of these chimeric proteins. Regardless, it should be kept in mind that findings obtained with such chimeric proteins are in general difficult to interpret since the full spectrum of activities associated with individual domains in Myc and Mad proteins has not been characterized.

Together these studies indicate that small conserved alterations in Myc can have profound effects on its function. In addition, the target genes controlled by the basic regions of Mad and Myc are not identical. This conclusion is supported by the recent study on Myc/Max/Mnt family target genes in *Drosophila* (Orian et al. 2003). Each of the three proteins regulates a set of target genes that is only partially overlapping with the targets of the other two network members. What are the rules behind these patterns of transcription factor binding? Chromosomal regions bound by dMnt, but not by dMyc, appear to be associated with DNA replication elements (DREs), a hint for cooperative effects between dMnt and proteins bound to DREs. Such suggested cooperative effects may provide part of the explanation why Myc and Mnt target genes are partially overlapping but not identical. That many dMnt targets are not identified through dMax is less easily understood since Mad, Mnt, and Mga in mammalian systems require Max for most aspects of their function. The findings in *Drosophila* argue for dMnt activities in gene regulation that are independent of dMax. A similar conclusion can be drawn for dMyc. Intuitively, repression by Myc comes to mind. This is mediated by Miz1 and is independent of E-boxes; nevertheless Myc still binds to Max (Staller et al. 2001). For both Mnt and Myc, other as-yet-unidentified modes of gene regulation have to be proposed, but at present we know very little about such alternative regulatory mechanisms.

4.3
Mad1 Target Genes by DNA Microarray Analysis

A more systematic approach to identify Mad1 target genes has been undertaken by comparing the gene expression pattern of murine thymocytes from either wild-type or Mad1 transgenic animals (Iritani et al. 2002). Of the genes identified, the majority falls under the category of cell growth control genes. A substantial portion appears to be directly involved in protein and DNA biosynthesis, in protein degradation, and in ribosome biogenesis. Some of these genes have been identified previously as Myc targets, indicating that Mad proteins also control basic aspects of cell proliferation, similar to the conclusions drawn for Myc (Grandori et al. 2000; Oster et al. 2002).

4.4
The Mga Complex

In mammalian systems, no studies describing Mad3, Mad4, and Mnt target genes have been published. Although evidence for direct regulation of genes by Mga is lacking, a recent report links this protein to E2F-dependent gene transcription. Mga was found as a subunit of a complex that also contains Max, E2F6, DP-1, HP1γ, a histone methyl transferase, and Polycomb group proteins (Ogawa et al. 2002). The HMT is specific for lysine 9 of histone H3, a modification that is thought to be repressive in nature (Kouzarides 2002; Zhang and Reinberg 2001). It is proposed that this complex could potentially regulate promoters with three different DNA elements, i.e., E2F-binding sites, E-boxes, and T-box elements. Indeed, there are a number of promoters of genes involved in cell-cycle control, including *cdc25A*, *cyclin E*, *cyclin D1*, *cyclin D3*, *hTERT*, *mcm7*, *DNA polymerase* α, and *cdc2*, that possess both E2F sites and E-boxes. It will be of interest to determine the contribution of the Mga complex to the regulation of the above-mentioned genes.

Perspective While in vitro DNA binding studies suggested very similar specificity for all Myc/Max/Mad complexes, the situation in cells appears more complex. Thus, one important task is to define DNA binding specificity in cells, i.e., to determine which E-box mediates response to which network members and for what reason. This seems particularly relevant for Mnt, which is rather ubiquitously expressed, and Mad3, which is specifically expressed during S-phase. Besides E-box-dependent gene regulation, accumulating evidence suggests that the network regulates genes also through other elements. Information describing such elements and defining the mechanism by which Myc/Max/Mad network members regulate gene transcription through these elements will be important to acquire.

5
Expression Patterns

As described in the structure-function section (Sect. 2), at least the four Mad proteins exhibit a high degree of sequence homology, as well as similar biochemical and transcriptional repression activities. This suggests functional redundancy of the four Mad proteins. This suggestion was originally based on overexpression studies, but the knockout analyses of *mad* genes also indicate redundancy. What then are the differences between these four molecules? Obviously, functional differences may exist that are currently insufficiently understood at the molecular level. In addition, distinct subcellular distribution might regulate Mad proteins as suggested recently from the finding that Mad4 is cytoplasmic. It bears a CRM1-dependent nuclear export signal and is recruited to the nucleus by Max (Grinberg et al. 2004). The most obvious differences between the Mad proteins and Mnt are their expression patterns. Altogether, *mad* family genes (the endogenous proteins are poorly evaluated due to very low expression levels) are expressed during development and in adulthood in specific patterns. The expression pattern of single *mad* family members is tissue-specific and quite complex. A general observation is that these genes are expressed in differentiating and differentiated cell types, although there are exceptions as discussed in Sects. 2.4.2 and 5.1 (Fig. 4). Frequently the expression of individual mad genes is staggered. We assume that Mad expression is a general phenotype of differentiating cells, as the presence of Myc is a general feature of proliferating cells. Cell types that have been analyzed in more detail include epidermal keratinocytes, chondrocytes, colonic epithelia, motor neurons, and erythroid and myeloid hematopoietic cells (Ayer and Eisenman 1993; Ayer et al. 1993; Cultraro et al. 1997; Foley et al. 1998; Gandarillas and Watt 1995; Hurlin et al. 1995a, b; Kime and Wright 2003; Larsson et al. 1994, 1997; Lymboussaki et al. 1996; Pulverer et al. 2000; Queva et al. 1998; Vastrik et al. 1995; Werner et al. 2001; Zervos et al. 1993). Thus, one aspect of antagonism between Myc and Mad proteins is their distinct expression patterns. This parallels their role in promoting cell-cycle progression in the case of Myc and cell-cycle exit and commitment to differentiation in the case of Mad. In the following sections those *mad* family members are discussed whose expression patterns deviate from the above-discussed general principle.

5.1
mad3

Although the expression pattern of the *mad3* gene has been linked to differentiation, *mad3* transcripts and Mad3 proteins are also found in proliferating cells. In addition, *mad3* messenger (m)RNA is undetectable in most adult

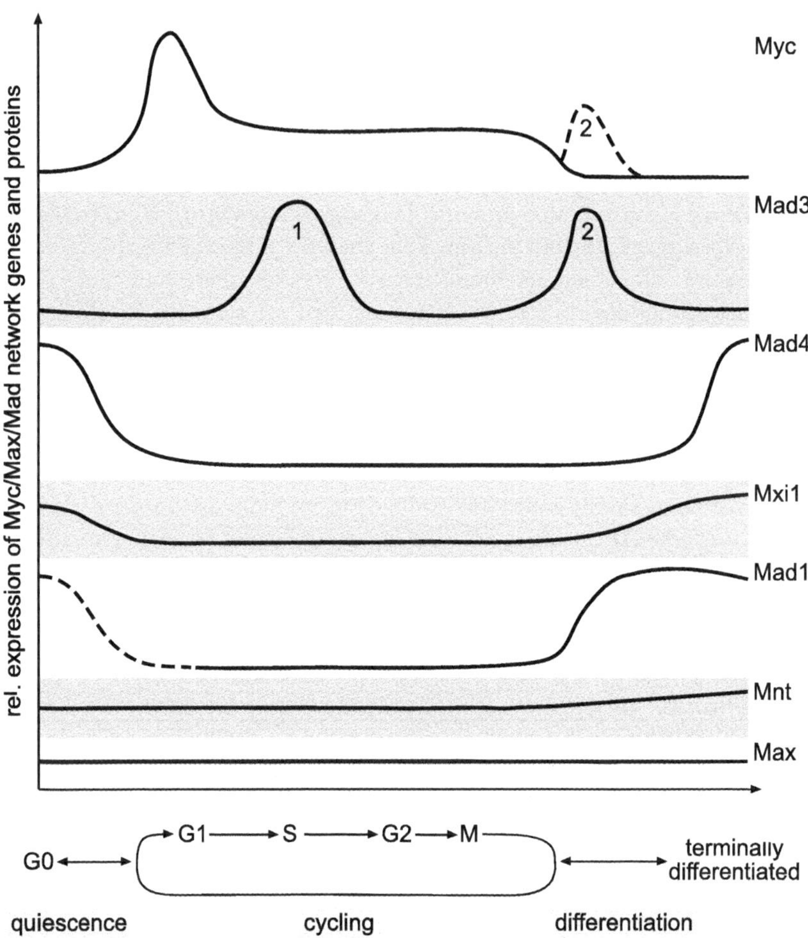

Fig. 4 Expression patterns of the Myc/Max/Mad network genes and proteins during distinct phases of cell behavior. A schematic and simplified expression pattern of the indicated molecules is given. Note that the expression patterns of both mRNA and protein are not known for all of these genes. In general, Max and Mnt show a rather uniform expression while Mad1, Mxi1, and Mad4 are upregulated during differentiation and in quiescence. In contrast, Mad3 expression is associated with S-phase (*1*) and the proliferative burst prior to terminal differentiation (*2*). The available evidence is not sufficient to display an expression pattern for Mga. Myc expression is shown for comparison

tissues except for areas in the testis and the thymus that contain proliferating cells (Hurlin et al. 1995a; Queva et al. 1998). Furthermore, *mad3* expression was observed in neural progenitors of developing mouse embryos that were also positive for BrdU incorporation (Queva et al. 2001). These findings suggest the presence of Mad3 during S-phase of the cell cycle, and more specifically it has been proposed that the expression of this Mad family member is linked to the last S-phase prior to terminal differentiation (Queva et al. 2001). Indeed, during differentiation of cells of an adipoblast line into adipocytes, *mad3* expression occurs during the proliferative burst, a discrete period of mitotic divisions that accompanies the differentiation process. Later in differentiation, *mad3* mRNA is no longer detected, clearly distinguishing it from other *mad* family members (Pulverer et al. 2000). Further analysis revealed a correlation between *mad3* expression and DNA synthesis in adipoblasts and in fibroblasts, indicating that this gene can also be expressed in proliferating cells that are not committed to differentiate (Fox and Wright 2001; Pulverer et al. 2000). These findings were unexpected since expression of other Mad family members is in general associated with differentiation or cell-cycle arrest and since overexpression of Mad3 inhibits proliferation and transformation (Hurlin et al. 1995a). Together the available data suggest that Mad3 plays a role during the cell cycle, but no precise function has been defined. One possibility is that Mad3 antagonizes the proapoptotic function of Myc once cells have entered S-phase. However, a null mutation of mad3 in mice had no effect on viability and only a modest phenotype relating to radiation-induced apoptosis in the nervous system (Queva et al. 2001).

5.2 mnt

While the transcription of *mad* genes is highly regulated, *mnt* expression is more ubiquitous and associated with both proliferating and differentiating tissues (Hurlin et al. 1997; Meroni et al. 1997; Sommer et al. 1999). This is reminiscent of the expression pattern observed for *max* and suggests broader functional activities of Mnt in comparison to Mad proteins (Henriksson and Lüscher 1996). Mnt has a SID and can repress gene transcription as discussed above. However, deletion of the SID results in a Mnt mutant that now functions as an activator in reporter gene assays and cooperates with Ha-Ras in transformation (Hurlin et al., 1997). Thus, Mnt seems more pleiotropic than Mad proteins.

5.3 mga

Little information regarding the expression of *mga* has been published. A high level of expression is observed within specific areas of the developing mouse embryo, leading to the suggestion that Mga may participate in mesoderm development (Hurlin et al. 1999). Indeed, the expression pattern of *mga* appears to overlap with other genes encoding T-box-containing proteins that have been described to regulate mesoderm induction (Papaioannou and Silver 1998).

6 Regulation of *mad, mnt,* and *mga* Genes

The regulation of *mad*, *mnt*, and *mga* gene transcription, despite their rather distinct expression patterns, has not been intensely studied. Although the promoters of the human *mxi1* gene, the murine *mad3*, and *mad4* genes have been published, we have little information regarding regulatory factors and signaling pathways that impinge on the promoters of these genes. One common theme of the three cloned promoters is that they are TATA-less and contain GC-rich sequences with potential Initiator (INR) elements and that they possess multiple transcription initiation sites (Benson et al. 1999; Fox and Wright 2003; Kime and Wright 2003). This theme is repeated also in the human *mad1* promoter (K. Eckert, K. Jiang, and B. Lüscher, unpublished observation). Transcription of the *mxi1* gene appears to be initiated at or near two INR elements in the vicinity of Sp1 sites, while the transcription factor AP2 has been implicated as negative regulator of *mxi1* transcription (Benson et al. 1999). The transcription factor Sp1 has been suggested to play a basal role at many TATA-less promoters, and although it is ubiquitously expressed it functions at many tissue-specific and differentiation-regulated promoters (Suske 1999). Sp1 is also involved in transcription of the *mad1* promoter (K. Jiang and B. Lüscher, unpublished observation).

The *mad3* promoter contains an E2F binding site that mediates cell-cycle-dependent expression of a reporter gene construct (Fox and Wright 2003). This response element binds E2F1, implying that this transcription factor regulates the unique S-phase-specific expression pattern of *mad3*. This classifies this gene into a growing group of E2F-regulated genes that fulfill important functions during cell-cycle progression and in many instances are necessary for the transition from the G1 into the S-phase (Muller and Helin 2000; Trimarchi and Lees 2002).

Within the *mad4* promoter, the critical element for differentiation-specific expression of reporter genes was mapped to the core promoter region that contains an INR consensus sequence (Kime and Wright 2003). In transient transfection assays it has been shown that Miz-1 activates the *mad4* core promoter, an effect that is antagonized by c-Myc. Furthermore, in vitro binding assays imply interaction of Miz-1 and c-Myc with the core promoter. These findings suggest that c-Myc negatively regulates Mad4 expression, one of its own antagonists. It will be interesting to see whether Mad4 represses *c-myc* transcription as has been suggested for Mxi1 (Lee and Ziff 1999). Such a double-negative feedback loop would not only be in support of the switch model but also enhance the potency of the switch.

Expression of individual *mad* genes has been observed under essentially all conditions examined that stimulate differentiation and/or growth arrest of cells as discussed above. Thus, the sum of different signals that can induce transcription of *mad* genes is large and suggests that the promoters are far more complex than the data accumulated thus far imply. It will be important to determine whether substances such as tetradecanoylphorbol acetate (TPA), retinoic acid, certain cytokines, and TGF-β (Ayer and Eisenman 1993; Hurlin et al. 1995b; Hurlin et al. 1995a; Larsson et al. 1994; Werner et al. 2001), to name just a few, regulate *mad* promoters directly and, if so, through which signal transduction pathways.

It is interesting to note that INR elements have been found in genes that are transcriptionally repressed by c-Myc in a Miz-1-dependent mechanism (Herold et al. 2002; Seoane et al. 2001, 2002; Staller et al. 2001; van de Wetering et al. 2002; Wu et al. 2003). In these situations Myc appears to integrate signals from several different pathways, including TGF-β and the phorbol ester TPA. With the observations that the promoters of *mad1* and *mxi1* also contain potential INR elements, it is worth considering whether Myc not only affects *mad4* but also *mad1* and *mxi1* expression. In support of such a model are also the findings that *mad1* is induced in response to both TGF-β and TPA. In addition to Myc-dependent repression through Miz-1, it has also been suggested that Myc can repress through Sp1 binding sites (Gartel et al. 2001). This offers another opportunity for a feedback mechanism, since *mad* promoters are GC-rich and possess potential Sp1 binding sites. Addressing these interactions, i.e., regulation of *mad* gene expression by Myc proteins, will be an important step toward defining in more detail the interplay among components of the network. It is interesting to note that *mxi1*, but neither of the other *mad* genes, has been identified as a gene that is activated by c-Myc (O'Connell et al. 2003; Schuhmacher et al. 2001). Since it is unclear whether the effect is direct and since this observation has not been verified, the relevance remains undefined.

The promoters of *mnt* and *mga* have not been analyzed, and essentially nothing about the regulation of expression of these genes is known. Interestingly, *mnt* was identified as a c-Myc target (Menssen and Hermeking 2002). This appears to be a direct effect since Myc binds to the *mnt* promoter region. Future studies will have to address the control of *mnt* and *mga* gene expression, a relevant task in light of the proposed functions of Mnt and Mga in association with, as well as independent of, the Myc/Max/Mad network.

Perspective The distinct expression patterns of *mad*, *mnt*, and *mga* genes and their sensitivity to many signals suggest the presence of multiple responsive elements within the promoters of these genes. The challenge will be to define these elements and identify the interacting transcription factors and the upstream regulatory pathways that impinge on these factors. Knowledge about these aspects will help us to understand the molecular framework that is in place to control the expression of these genes.

7 Biological Functions of Mad, Mnt, and Mga Proteins

The role of Myc proteins in regulating many aspects of cell behavior including transformation is thoroughly documented (Grandori et al. 2000; Henriksson and Lüscher 1996; Marcu et al. 1992; Oster et al. 2002). The Mad, Mnt, and Mga proteins function at least in part by antagonizing the functions of Myc proteins (Fig. 5). In this section we summarize the biological activities attributed to these proteins as revealed by overexpression analyses and by characterizing knockout and transgenic models (see also M. Pirity et al., this volume).

7.1 Proliferation and Cell Growth

Mad gene and protein expression, with the exception of Mad3, is associated with inhibition of proliferation, cell-cycle exit, and differentiation. Early on it was demonstrated that Mad1 and Mxi1 interfere with proliferation in several cell systems, resulting in the accumulation of cells in the G0/G1 phase. Furthermore, expression of Mad1 reduces the outgrowth of transformed cells in colony formation assays (Cerni et al. 2002; Chen et al. 1995; Gehring et al. 2000; Roussel et al. 1996; Sommer et al. 1997; Wechsler et al. 1997). In all these studies, however, no complete block of proliferation was observed; rather the effects were subtle. A more substantial effect on proliferation was seen when Mad1 was expressed upon microinjection in resting fibroblasts that were subsequently stimulated with serum (Sommer et al. 1997). Under these conditions, a severe reduction in the number of cells that are able to reach S-phase is evident. One possible interpretation is that additional signals that cooperate with Mad1 are required for efficient cell-cycle arrest of cycling cells, whereas Mad1 is sufficient to prevent resting cells from reentering the cell cycle.

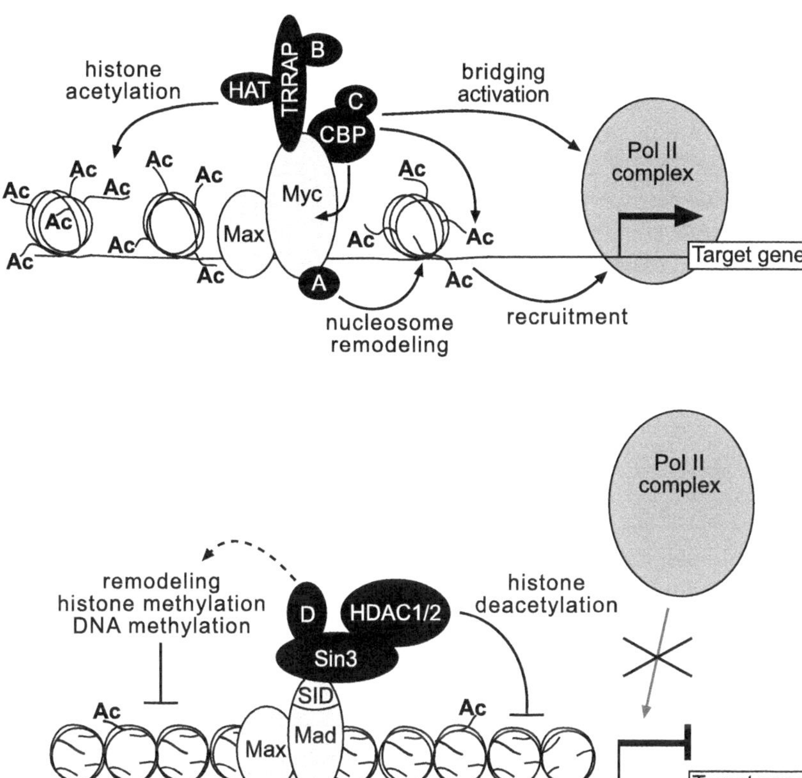

Fig. 5 Mechanistic model of transcriptional regulation by the Myc/Max/Mad network. Myc/Max and Mad/Max heterodimers recognize a specific DNA consensus sequence (CACGTG) belonging to the class of E-box sequences. Binding of Myc/Max and Mad/Max to such elements in promoters of target genes has been demonstrated. For transcriptional activation, Myc recruits cofactors including TRRAP and CBP. TRRAP is a subunit of a histone acetyltransferase-containing complex. Acetylation of core histones correlates with an open chromatin structure that enables binding of the Pol II complex. CBP acetylates components of the Pol II complex as well as c-Myc. In addition, CBP can mediate interaction of transcription factors with the Pol II complex. *A*, *B*, and *C* indicate additional proteins that bind directly Myc or are part of the TRRAP and CBP complexes. Mad represses gene transcription by recruiting an mSin3–HDAC repressor complex. Deacetylation of core histones correlates with gene repression. Interaction of subunits of the mSin3 complex with components of chromatin remodeling complexes and methyltransferases may broaden the functional consequences of recruiting the mSin3 complex. *D* indicates additional proteins interacting with the mSin3–HDAC repressor complex that are described in detail in the text and in Fig. 3

This interpretation is in line with the analysis of *mad1* transgenic mouse models (Iritani et al. 2002; Queva et al. 1999; Rudolph et al. 2001). In one model, broad expression of the transgene was achieved using the β-*actin* promoter. Although animals bearing the transgene were underrepresented, many were born (Queva et al. 1999). All the offspring were considerably smaller, consistent with the interpretation that Mad1 reduces proliferation but does not completely abolish it; otherwise the establishment of *mad1* transgenic mice would not have been possible. Inhibition of proliferation appears to be at least in part the result of an increased sensitivity to high density. Fibroblasts from *mad1* transgenic mice reveal a reduced density at confluence and a delay in their progression from G0 into S-phase in response to mitogens (Queva et al. 1999). In a second model, expression of *mad1* under the control of the *lck* promoter in the T cell compartment of mice reduced the number of thymocytes considerably without affecting the ratio of distinct subpopulations of T cells (Rudolph et al. 2001). In a third model, *mad1* was expressed in T and B cells from a transgene controlled by the *lck* promoter and the immunoglobulin heavy chain enhancer (Iritani et al. 2002). Here a strong decrease in thymocyte cellularity corresponding, unlike in the second model, to a decrease in double-positive and single-positive thymocytes. This appears to be the result of reduced proliferation and decreased maturation of double-negative cells.

In addition, mitogen-stimulated thymocytes and B cells from *mad1* transgenic animals showed a profound reduction in proliferation (Iritani et al. 2002; Rudolph et al. 2001). Furthermore, determination of the intermitotic time in cells that express Mad1 in a tetracycline-regulatable fashion showed only small differences in cell cycle time when Mad1 was turned on (Holzel et al. 2001). However, the cells stopped dividing more readily, a finding that was particularly evident in areas of increased cell density, resulting in a substantial increase in the overall population doubling time. Together these studies demonstrate that Mad1 (the other Mad family members have not been analyzed in detail) interferes with proliferation in subtle ways, i.e., Mad1 does not modulate proliferation in an all-or-nothing process. This suggests that Mad proteins interact with additional signals or regulators to control proliferation. One signal may involve cell–cell contacts; another event may be the downregulation of Myc expression that will further shift the system towards the Mad side. The latter may be relevant in situations where cells reenter the cycle. In these situations, Myc expression is very low and Mad1 shows the strongest effect on proliferation.

Given the many studies that implicate Mad proteins in the control of proliferation in diverse cell systems, it came as a surprise that *mad1*$^{-/-}$ animals did not show any widespread phenotype that could be related to proliferation

(Foley et al. 1998). In particular the animals were neither different in size nor showed altered cellularity. However, distinct phenotypes were noted in granulocytes and in B cells. While the overall number of granulocytes is unaltered, additional rounds of mitotic divisions occur prior to terminal differentiation. It is thought that this is the consequence of a delay in exiting the cell cycle of granulocytic precursor cells. However, the differentiated cells are more sensitive to limiting amounts of growth factors, resulting in increased apoptosis. Together these two processes seem to compensate each other in the animals. In addition, mitogen-induced reentry into the cell cycle of B cells is enhanced, conforming to the observation that B cells from Mad1 transgenic mice divide less upon stimulation (Foley et al. 1998; Iritani et al. 2002). These findings reveal specific but limited effects on cell behavior in *mad1*$^{-/-}$ mice.

Several reports have shown that Myc can stimulate cell growth in the absence of proliferation (Iritani and Eisenman 1999; Johnston et al. 1999; Schuhmacher et al. 1999). This may be connected to Myc's ability to stimulate protein synthesis through modulating a number of relevant target genes, a parameter for the increase in cell size during the G1 phase of the cell cycle. The analysis of B cells in Mad1 transgenic mice showed that the cells are smaller irrespective of their proliferative potential (Iritani et al. 2002). This suggests that this aspect of Myc function can be antagonized by a Mad protein.

An important aspect of Myc-driven cell cycle progression is the activation of cyclin-dependent kinases that regulate the G1 to S-phase transition (Amati et al. 1998; Lüscher 2001). In particular, the CDK inhibitor (CKI) p27^{KIP1} is targeted by Myc through several pathways, thereby resulting in the inhibition of this CKI. Therefore it was of interest to determine the consequences of a *p27/mad1* double knockout (McArthur et al. 2002). These animals reveal a number of synthetic phenotypes. The lack of Mad1 and p27^{KIP1} results in partial embryonic lethality, while the individual knockouts are viable (Fero et al. 1996; Foley et al. 1998; Kiyokawa et al. 1996; Nakayama et al. 1996). In addition, differentiation of double-knockout granulocytes is impaired in response to retinoic acid, a phenotype similar to *mad1*$^{-/-}$ cell but strongly enhanced. This is paralleled by a failure to arrest proliferation and to downregulate cyclin E/CDK2. Blocking this kinase in double-knockout granulocytes is sufficient to render the cells sensitive to retinoic acid. This suggests a cooperative effect of Mad1 and p27^{KIP1} during differentiation of granulocytes. Other phenotypes observed in p27^{KIP1}-null mice, including hyperplasia in different tissues, were not enhanced by the loss of Mad1 (McArthur et al. 2002).

Molecularly, Mad1 expression reduces kinase activity of cyclin D-type complexes, while little effect on cyclin E-dependent complexes was observed (Gehring et al. 2000; Queva et al. 1999). This is surprising considering the above-mentioned role of Myc in cell-cycle control and the multiple Myc-regulated pathways that target cyclin E/CDK2 (Amati et al. 1998; Lüscher 2001). The effect on cyclin D/CDK4,6 suggests that the Mad1-induced inhibition of proliferation should be rescued by this kinase complex. However, microinjection of Mad1 together with cyclin D1/CDK4 into resting fibroblasts was not sufficient to restore serum-induced S-phase progression. Instead cyclin E/CDK2 could efficiently overcome the Mad1 block (S. Rottmann et al. 2005). This indicates that cyclin E/CDK2 can function as a mediator of a feedback mechanism within the Myc/Max/Mad network.

The homozygous deletion of *mxi1* showed a more severe phenotype than seen for the *mad1* knockout (Schreiber-Agus et al. 1998). Hyperplasic changes were observed in several different tissues but—similar to the *mad1* knockout animals—the effects appeared cell-type specific. The broader phenotype is thought to reflect the wider expression pattern of *mxi1* as compared to *mad1*. The tissues affected include the hematopoietic system, albeit not all cell types, kidneys, and the prostate. Despite the very specific expression pattern of *mad3*, it being distinct from the other *mad* family members, *mad3*$^{-/-}$ animals did not reveal any phenotype that could be attributed to altered cell proliferation (Queva et al. 2001).

In comparison to the above-discussed Mad1 transgenic animals, a more severe phenotype is observed when Mnt is expressed under the control of the β-*actin* promoter (Hurlin et al. 1997). Transgenic embryos die in uteri and are considerably smaller. In addition to embryonic size, a number of phenotypes can be detected that appear to be heterogeneous among the transgenic embryos. This indicates that exogenous Mnt expression has deleterious effects on multiple tissues but with altered kinetics from one embryo to the other. The reduced size suggests a strong effect of Mnt on proliferation. This is reminiscent of the *myc* and *max* knockout animals that show an embryonic lethal phenotype (Charron et al. 1992; Davis et al. 1993; Moens et al. 1993; Sawai et al. 1993; Shen-Li et al. 2000; Stanton et al. 1992).

A prominent role for Mnt in the control of cell proliferation is well supported by recent studies employing *mnt*$^{-/-}$ cells and animals and *mnt* knockdown cells (Hurlin et al. 2003; Nilsson et al. 2004; Toyo-oka et al. 2004). *mnt*$^{-/-}$ mouse embryo fibroblasts (MEFs) enter S-phase prematurely and proliferate more rapidly than control cells. This is associated with increased CDK4 and cyclin E expression. In addition, these cells escape senescence (Hurlin et al. 2003). Similarly knockdown of Mnt expression using RNA interference (RNAi) constructs also resulted in enhanced proliferation (Nilsson et al. 2004). This is also true in *myc*$^{-/-}$ fibroblasts, suggesting that Mnt is a dominant-negative regulator of proliferation in the absence of Myc. It is important to note that in both systems, i.e., *mnt* knockout and knockdown, the resulting cells behave similar to Myc-overexpressing cells. Thus, not only proliferation is enhanced, but the *mnt*$^{-/-}$ MEFs can be transformed by oncogenic Ras only (Hurlin et al. 2003). Again similar observations were made in the *mnt* knockdown cells (Nilsson et al. 2004). Having seen the effects of Mnt loss on proliferation and transformation, it is perhaps not surprising that cells without Mnt are highly susceptible to apoptosis (Hurlin et al. 2003; Nilsson et al. 2004). From these studies it appears that Mnt is a very potent antagonist of Myc function. Overcoming Mnt, either by raising Myc levels as seen in many human tumors or by knocking out/down Mnt, is sufficient to induce the changes in cell behavior associated classically with Myc, i.e., enhanced proliferation, tumorigenesis, and increased sensitivity to apoptosis-inducing conditions.

7.2 Differentiation

The association of Mad protein expression with differentiation processes suggested early on that these proteins might function as regulators or even as inducers of differentiation. This suggestion was in line with the overall thinking that Mad proteins antagonize the function of Myc proteins. However, with the exception of one study, exogenous expression of Mad proteins was found insufficient to induce differentiation. This also was unexpected in light of the findings that Myc proteins are potent inhibitors of differentiation in many cellular systems (Henriksson and Lüscher 1996).

Mouse erythroleukemia (MEL) cells respond to dimethylsulfoxide (DMSO), which induces differentiation-accompanied downregulation of *myc* and upregulation of *mad* gene expression comparable to many other differentiation systems (Cultraro et al. 1997; Kime and Wright 2003; Lachman and Skoultchi 1984). Induction of expression of a *mad1* transgene in MEL cells stimulates both spontaneous and DMSO-induced differentiation. By contrast, in 3T3-L1 adipoblasts and in U937 promyelocytes Mad1 blocks differentiation (Pulverer et al. 2000; L.-G. Larsson, personal communication). These cells, similar to MEL cells, also downregulate *myc* and upregulate *mad* gene expression when differentiation is induced by various agents (Freytag 1988; Freytag and Geddes 1992; Larsson et al. 1997, 1988, 1994; Pulverer et al. 2000). How can these apparently contradictory results be reconciled? Once differentiation is induced, adipoblasts go through an additional round of mitotic divisions, referred to as proliferative burst. Blocking this phase by inhibiting DNA synthesis is sufficient to severely reduce the number of terminally differentiated cells (Yeh et al. 1995). The analysis of 3T3-L1–Mad1 cells revealed that DNA synthesis during the proliferative burst is significantly reduced. This led to the suggestion that inhibition of differentiation in adipoblasts by Mad1 is the consequence of a lack of the proliferative burst phase (Pulverer et al. 2000). Retinoic acid-induced, but not TPA-induced, U937 differentiation is accompanied by 1 to 2 mitotic divisions. The former but not the latter is inhibited by Mad1 (L.-G. Larsson, personal communication). This is reminiscent of the situation in adipoblasts and suggests that the effect of Mad1 on differentiation is the result of blocking a proliferative phase required for terminal differentiation. However, in U937 cells, unlike the findings in MEL cells, Mad1 is not sufficient to induce spontaneous differentiation (Cultraro et al. 1997; L.-G. Larsson, personal communication). In addition, the analysis of *mad1* transgenic and *mad1*$^{-/-}$ mice did not reveal clear effects on differentiation that can be separated from proliferation effects (Foley et al. 1998; Iritani et al. 2002; Queva et al. 1999; Rudolph et al. 2001). Together these studies support the notion that Mad1 is a negative regulator of proliferation, whereas its role in differentiation remains ill defined.

Although the effects of Mad1 on differentiation are not very pronounced, two aspects of these studies should be kept in mind. In the transgenic approaches, the regulated expression of the endogenous Mad1 gene and protein is not reproduced, which may disturb aspects of the differentiation process to such a degree that it is blocked. In the knockout studies, the other Mad proteins may compensate for the loss of Mad1. Consequently the full potential of Mad1 may not be revealed until the analyses can be performed in animals and cells that also lack other Mad proteins. Most likely other Mad family members as well as Mnt and Mga are relevant players in the differentiation game. Indeed, loss of Mnt results in craniofacial defects and neonatal lethality, suggesting that this protein is important for developmental processes (Toyooka et al. 2004). Together these findings reveal important insights into the functions of Mad and Mnt proteins associated with differentiation. However, these analyses are still at an early phase and leave many aspects open for future work.

7.3
Apoptosis

The identification of the Myc family members as apoptosis-inducing proteins came as a surprise, since this function of Myc was not, at least at first view, compatible with cellular transformation (Pelengaris et al. 2002; Prendergast 1999). However, it is now thought that apoptosis represents a safeguard mechanism to prevent cells from acquiring an activated Myc or, for that matter, other proteins that possess oncogenic capacity. As for proliferation and differentiation, Mad proteins were suggested early on to also antagonize the proapoptotic function of Myc. Indeed, Mad proteins inhibit apoptosis induced by different stimuli, but the mechanism remains to be determined.

The targeted disruption of *mad1* affects proliferation and apoptosis of granulocytes and their precursors (Foley et al. 1998). In the absence of Mad1 these cells undergo extra rounds of cell divisions prior to terminal differentiation. However, the number of mature granulocytes is not altered, suggesting that the mature cells may be lost at increased frequency. Indeed, these cells are more sensitive to cytokine withdrawal than granulocytes from wild-type animals and respond with increased apoptosis. Consistent with this, myeloid precursor cells of Mad1 transgenic animals are less sensitive to limiting cytokine levels (Queva et al. 1999). These studies were complemented by experiments in tissue culture. In fibroblasts Mad1 reduces apoptosis under conditions of serum starvation (Bejarano et al. 2000) and in response to oncoprotein expression (Gehring et al. 2000). In addition, induction of Mad1 expression in the osteosarcoma cell line U2OS inhibits apoptosis induced by different stimuli including Fas ligand, TNF-related apoptosis inducing ligand (TRAIL), and UV (Gehring et al. 2000). Further analysis revealed a reduction of caspase-8 activation in response to Fas ligand, perhaps providing the first hint of a molecular mechanism of Mad1 function.

In this respect, the association of Myc with caspase-8 activation is of interest. It has been reported that a large proportion of aggressive N-Myc-driven neuroblastomas that are particularly resistant to induction of apoptosis exhibit a methylated and silenced caspase-8 gene (Teitz et al. 2000). Furthermore, in small-cell lung cancer with amplification of *myc*-family members, loss of caspase-8 expression or increased expression of the proteolytically inactive caspase-8 homolog c-FLIP is frequently observed (Shivapurkar et al. 2002). Thus, inactivation of caspase-8 by various means appears to accompany tumors with high Myc expression. In addition, c-Flip expression is negatively regulated by Myc, and overexpression of c-Flip prevents Myc-induced apoptosis (Amanullah et al. 2002). This suggests together with the data discussed above, an interaction between the Myc/Max/Mad network and Caspase-8 function and regulation.

One study reported enhancement of apoptosis in the presence of Mad1. The stimulation of double-negative thymocytes of Mad1 transgenic/rag2$^{-/-}$ animals with anti-CD3ε antibodies results in their maturation to double-positive cells (Iritani et al. 2002). However, a considerable fraction of the cells undergo apoptosis, suggesting that stimulation of proliferation and maturation is not compatible with Mad1 expression, a conflict that appears to be resolved by cell death. Furthermore, the matured double-positive cells have considerably lower levels of Mad1, an indication of selection for thymocytes with low Mad1 expression under these conditions. In addition to Mad1, Mad3 has also been implicated in apoptosis control. Thymocytes of *mad3*$^{-/-}$ animals show a slight increase in sensitivity to γ-irradiation (Queva et al. 2001).

Furthermore, *mad3*$^{-/-}$ embryos subjected to γ-irradiation in utero reveal an increased number of apoptotic neural progenitor cells. These seem to be in S-phase at the time of treatment, indicating a specific anti-apoptotic function of Mad3 in this phase of the cell cycle.

Together these analyses identify Mad proteins as modulators of apoptosis. This interpretation is supported by the finding that Max, the dimerization partner of Mad proteins, is specifically targeted by caspases upon induction of apoptosis (Krippner-Heidenreich et al. 2001). The loss of Max is thought to antagonize the anti-apoptotic activities of Mad proteins. Clearly more work will be required to understand the molecular consequences of Mad activity in cell death.

7.4 Transformation

A prominent function of Myc proteins that can be addressed in tissue culture systems is its ability to transform primary rat embryo fibroblasts in cooperation with Ha-Ras (for reviews see Henriksson and Lüscher 1996; Marcu et al. 1992). This function of Myc is strongly antagonized by Mad, Mnt, and Mga proteins (Cerni et al. 1995; Hurlin et al. 1995a, 1997, 1999; Koskinen et al. 1995; Lahoz et al. 1994; Schreiber-Agus et al. 1995; Vastrik et al. 1995). Furthermore, using an inducible Mad1ER fusion protein, tumor growth of Myc/Ha-Ras-transformed REF cells in syngeneic rats is significantly delayed by Mad1 (Cerni et al. 2002). This Mad1-related repression of transformation is dependent at least in part on the recruitment of the mSin3–HDAC repressor complex, indicating that inhibition of transformation is tightly linked to gene repression. Extrapolating from these studies, it was proposed that these proteins are potential tumor suppressors. Indeed, *mad* genes have distinct chromosomal locations that have been linked to human tumors (Edelhoff et al. 1994; Hurlin et al. 1995a, 1997, 1999; Shapiro et al. 1994). An important question then is whether any of these genes are true tumor suppressors. The short answer is that none of these proteins has been firmly established as a tumor suppressor.

Of the Mad family members, Mxi1 seems to be the best tumor suppressor candidate. Support came from the analysis of *mxi1*$^{-/-}$ mice (Schreiber-Agus et al. 1998). These animals show increased susceptibility to tumor formation when challenged with a carcinogen. In addition, *mxi1*$^{-/-}$/*ink4a*$^{-/-}$ double-knockout animals reveal enhanced tumor development in comparison to single-knockout controls. These findings suggest that the loss of Mxi1, unlike the loss of Mad1 or Mad3, represents a precancerous lesion in mice.

The *mxi1* gene is located on chromosome 10q25. 10q23–25 represents an area that shows alterations in a number of tumors, including prostate cancer

(Edelhoff et al. 1994; Shapiro et al. 1994). Indeed, mutations in the *mxi1* gene associated with prostate cancer were reported in one study (Eagle et al. 1995). However, other studies were unable to demonstrate mutations or LOH of *mxi1* in this tumor type (Gray et al. 1995; Kawamata et al. 1996; Kuczyk et al. 1998). Bladder carcinomas also show LOH at 10q. A recent analysis suggests that the *PTEN* gene is the likely tumor suppressor located on 10q rather than *mxi1* (Wang et al. 2000b). Similarly, no mutations or LOH associated with *mxi1* could be identified in other human tumors (Bartsch et al. 1996; Fults et al. 1998; Kim et al. 1998; Petersen et al. 1998). Nevertheless, the combined data on the role of Mxi1 in transformation suggest that this protein can function as a tumor suppressor. Further studies are required, however, to determine whether it acts as a tumor suppressor in human cancers.

The *mnt* gene is located on chromosome 17p13.3, a region that is frequently deleted in different human malignancies (Meroni et al. 1997). However, as for the analysis of mxi1, no evidence for mutations or LOH of the *mnt* gene has been obtained (Nigro et al. 1998; Sommer et al. 1999; Takahashi et al. 1998). However, recent studies demonstrate that loss of Mnt enhances tumor formation. Fibroblasts that lack Mnt, derived either from *mnt*$^{-/-}$ mice or upon knockdown using RNAi, can be transformed with an oncogenic Ras (Hurlin et al. 2003; Nilsson et al. 2004). Furthermore, disruption of *mnt* in mammary epithelium using a conditional mouse model results in the development of adenocarcinoma (Hurlin et al. 2003). From these findings a function of Mnt in tumor formation is indicated, yet—as is the case for *mad* genes—it remains open whether Mnt functions as a tumor suppressor in humans. It has been suggested that the strong apoptotic response in *mnt* knockdown cells or the reduced expression of *myc* genes in *mnt*$^{-/-}$ cells could be the reason why Mnt does not appear to normally act as a tumor suppressor (Hurlin et al. 2003; Nilsson et al. 2004). Further studies should clarify these issues.

In the context of a function of Mad proteins or Mnt as tumor suppressors, it is worth mentioning that the suggested Mad-interacting complex, which includes Sno/Ski, mSin3, HDAC, and PML, contains tumor suppressor proteins and proto-oncoproteins (Li et al. 1986; Salomoni and Pandolfi 2002; Shinagawa et al. 2000, 2001). The relevance of Mad and Mnt in the context of this complex and its proposed role in transformation remains to be determined but can potentially reveal new leads towards a better understanding of Mad and Mnt in cancer.

Perspective Myc proteins are potent regulators of various aspects of cell behavior. The initial expectation that Mad, Mnt, and/or Mga would antagonize all the functions associated with Myc is true only in part. The most prominent effects of Mad, Mnt, and Mga are on proliferation and apoptosis. The studies so far have not revealed clear effects on differentiation. Furthermore, tumor suppressor function in humans has not been convincingly shown. This may reflect the overlapping expression patterns and functions of these proteins. Consequently, the combination of knockouts of *mad*, *mnt*, and *mga* genes may help to elucidate their function in more detail. Among the activities that have been associated with Myc but have not been linked to the Mad side are genomic instability and vascularization. Overexpression of Myc causes genomic destabilization potentially due to its ability to induce endoreplication (Felsher and Bishop 1999; Felsher et al. 2000; Kuschak et al. 2002; Mai et al. 1999; Taylor and Mai 1998; see M. Wade and G.M. Wahl, this volume). In addition, Myc has been recently shown to be essential for vascularization and angiogenesis in tumor development (Baudino et al. 2002). Both these aspects are important for Myc's role in transformation. Obviously it will be important to define whether Mad, Mnt, and/or Mga regulate any of these aspects. This information will shed more light on their potential role in tumor suppression.

Acknowledgements We thank M. Haenlin, M. Henriksson, and L.-G. Larsson for discussion and sharing unpublished information. We apologize to numerous authors whose work could not be cited in detail. B.L. dedicates this paper to Barbara Schätti-Lüscher. The work carried out in our laboratory relevant to this review was funded by the Deutsche Forschungsgesellschaft.

References

Alland L, Muhle R, Hou H Jr, Potes J, Chin L, Schreiber-Agus N, DePinho RA (1997) Role for N-CoR and histone deacetylase in Sin3-mediated transcriptional repression. Nature 387:49–55

Amanullah A, Liebermann DA, Hoffman B (2002) Deregulated c-Myc prematurely recruits both Type I and II CD95/Fas apoptotic pathways associated with terminal myeloid differentiation. Oncogene 21:1600–1610

Amati B, Alevizopoulos K, Vlach J (1998) Myc and the cell cycle. Front Biosci 3:D250–D268

Amati B, Frank SR, Donjerkovic D, Taubert S (2001) Function of the c-Myc oncoprotein in chromatin remodeling and transcription. Biochim Biophys Acta 1471:M135–145

Ansieau S, Leutz A (2002) The conserved Mynd domain of BS69 binds cellular and oncoviral proteins through a common PXLXP motif. J Biol Chem 277:4906–4910

Ayer DE (1999) Histone deacetylases: transcriptional repression with SINers and NuRDs. Trends Cell Biol 9:193–198

Ayer DE, Eisenman RN (1993) A switch from Myc:Max to Mad:Max heterocomplexes accompanies monocyte/macrophage differentiation. Genes Dev 7:2110–2119

Ayer DE, Kretzner L, Eisenman RN (1993) Mad: a heterodimeric partner for Max that antagonizes Myc transcriptional activity. Cell 72:211–222

Ayer DE, Lawrence QA, Eisenman RN (1995) Mad-Max transcriptional repression is mediated by ternary complex formation with mammalian homologs of yeast repressor Sin3. Cell 80:767–776

Ayer DE, Laherty CD, Lawrence QA, Armstrong AP, Eisenman RN (1996) Mad proteins contain a dominant transcription repression domain. Mol Cell Biol 16:5772–5781

Barrera-Hernandez G, Cultraro CM, Pianetti S, Segal S (2000) Mad1 function is regulated through elements within the carboxy terminus. Mol Cell Biol 20:4253–4264

Bartsch D, Peiffer SL, Kaleem Z, Wells SA Jr, Goodfellow PJ (1996) Mxi1 tumor suppressor gene is not mutated in primary pancreatic adenocarcinoma. Cancer Lett 102:73–76

Baudino TA, Cleveland JL (2001) The Max network gone mad. Mol Cell Biol 21:691–702

Baudino TA, McKay C, Pendeville-Samain H, Nilsson JA, Maclean KH, White EL, Davis AC, Ihle JN, Cleveland JL (2002) c-Myc is essential for vasculogenesis and angiogenesis during development and tumor progression. Genes Dev 16:2530–2543

Bejarano MT, Albihn A, Cornvik T, Brijker SO, Asker C, Osorio LM, Henriksson M (2000) Inhibition of cell growth and apoptosis by inducible expression of the transcriptional repressor Mad1. Exp Cell Res 260:61–72

Benson LQ, Coon MR, Krueger LM, Han GC, Sarnaik AA, Wechsler DS (1999) Expression of MXI1, a Myc antagonist, is regulated by Sp1 and AP2. J Biol Chem 274:28794–28802

Billin AN, Eilers AL, Queva C, Ayer DE (1999) Mlx, a novel Max-like BHLHZip protein that interacts with the Max network of transcription factors. J Biol Chem 274:36344–36350

Blackwood EM, Eisenman RN (1991) Max: a helix-loop-helix zipper protein that forms a sequence-specific DNA-binding complex with Myc. Science 251:1211–1217

Blackwood EM, Lüscher B, Eisenman RN (1992) Myc and Max associate in vivo. Genes Dev 6:71–80

Bouchard C, Thieke K, Maier A, Saffrich R, Hanley-Hyde J, Ansorge W, Reed S, Sicinski P, Bartek J, Eilers M (1999) Direct induction of cyclin D2 by Myc contributes to cell cycle progression and sequestration of p27. EMBO J 18:5321–5333

Bouchard C, Dittrich O, Kiermaier A, Dohmann K, Menkel A, Eilers M, Lüscher B (2001) Regulation of cyclin D2 gene expression by the Myc/Max/Mad network: Myc-dependent TRRAP recruitment and histone acetylation at the cyclin D2 promoter. Genes Dev 15:2042–2047

Brownlie P, Ceska T, Lamers M, Romier C, Stier G, Teo H, Suck D (1997) The crystal structure of an intact human Max-DNA complex: new insights into mechanisms of transcriptional control. Structure 5:509–520

Brubaker K, Cowley SM, Huang K, Loo L, Yochum GS, Ayer DE, Eisenman RN, Radhakrishnan I (2000) Solution structure of the interacting domains of the Mad-Sin3 complex: implications for recruitment of a chromatin-modifying complex. Cell 103:655–665

Cai R, Kwon P, Yan-Neale Y, Sambuccetti L, Fischer D, Cohen D (2001) Mammalian histone deacetylase 1 protein is posttranslationally modified by phosphorylation. Biochem Biophys Res Commun 283:445–453

Cerni C (2000) Telomeres, telomerase, and myc. An update. Mutat Res 462:31–47

Cerni C, Bousset K, Seelos C, Burkhardt H, Henriksson M, Lüscher B (1995) Differential effects by Mad and Max on transformation by cellular and viral oncoproteins. Oncogene 11:587–596

Cerni C, Skrzypek B, Popov N, Sasgary S, Schmidt G, Larsson LG, Lüscher B, Henriksson M (2002) Repression of in vivo growth of Myc/Ras transformed tumor cells by Mad1. Oncogene 21:447–459

Charron J, Malynn BA, Fisher P, Stewart V, Jeannotte L, Goff SP, Robertson EJ, Alt FW (1992) Embryonic lethality in mice homozygous for a targeted disruption of the N-myc gene. Genes Dev 6:2248–2257

Chen J, Willingham T, Margraf LR, Schreiber-Agus N, DePinho RA, Nisen PD (1995) Effects of the MYC oncogene antagonist, MAD, on proliferation, cell cycling and the malignant phenotype of human brain tumour cells. Nat Med 1:638–643

Cheung KJ Jr, Li G (2001) The tumor suppressor ING1: structure and function. Exp Cell Res 268:1–6

Cohen SB, Zheng G, Heyman HC, Stavnezer E (1999) Heterodimers of the SnoN and Ski oncoproteins form preferentially over homodimers and are more potent transforming agents. Nucleic Acids Res 27:1006–1014

Cong YS, Bacchetti S (2000) Histone deacetylation is involved in the transcriptional repression of hTERT in normal human cells. J Biol Chem 275:35665–35668

Cowley SM, Kang RS, Frangioni JV, Yada JJ, DeGrand AM, Radhakrishnan I, Eisenman RN (2004) Functional analysis of the Mad1-mSin3A repressor-corepressor interaction reveals determinants of specificity, affinity, and transcriptional response. Mol Cell Biol 24:2698–2709

Cress WD, Seto E (2000) Histone deacetylases, transcriptional control, and cancer. J Cell Physiol 184:1–16

Cultraro CM, Bino T, Segal S (1997) Regulated expression and function of the c-Myc antagonist, Mad1, during a molecular switch from proliferation to differentiation. Curr Top Microbiol Immunol 224:149–158

Czermin B, Schotta G, Hulsmann BB, Brehm A, Becker PB, Reuter G, Imhof A (2001) Physical and functional association of SU(VAR)3–9 and HDAC1 in Drosophila. EMBO Rep 2:915–919

Dang CV, Barrett J, Villa-Garcia M, Resar LM, Kato GJ, Fearon ER (1991) Intracellular leucine zipper interactions suggest c-Myc hetero-oligomerization. Mol Cell Biol 11:954–962

Datta J, Ghoshal K, Sharma SM, Tajima S, Jacob ST (2003) Biochemical fractionation reveals association of DNA methyltransferase (Dnmt) 3b with Dnmt1 and that of Dnmt 3a with a histone H3 methyltransferase and Hdac1. J Cell Biochem 88:855–864

Davis AC, Wims M, Spotts GD, Hann SR, Bradley A (1993) A null c-myc mutation causes lethality before 10.5 days of gestation in homozygotes and reduced fertility in heterozygous female mice. Genes Dev 7:671–682

Davis RL, Turner DL (2001) Vertebrate hairy and Enhancer of split related proteins: transcriptional repressors regulating cellular differentiation and embryonic patterning. Oncogene 20:8342–8357

de Ruijter AJ, van Gennip AH, Caron HN, Kemp S, van Kuilenburg AB (2003) Histone deacetylases (HDACs): characterization of the classical HDAC family. Biochem J 370:737–749

Deltour S, Guerardel C, Leprince D (1999) Recruitment of SMRT/N-CoR-mSin3A-HDAC-repressing complexes is not a general mechanism for BTB/POZ transcriptional repressors: the case of HIC-1 and gammaFBP-B. Proc Natl Acad Sci U S A 96:14831–14836

Dhordain P, Lin RJ, Quief S, Lantoine D, Kerckaert JP, Evans RM, Albagli O (1998) The LAZ3(BCL-6) oncoprotein recruits a SMRT/mSIN3A/histone deacetylase containing complex to mediate transcriptional repression. Nucleic Acids Res 26:4645–4651

Eagle LR, Yin X, Brothman AR, Williams BJ, Atkin NB, Prochownik EV (1995) Mutation of the MXI1 gene in prostate cancer. Nat Genet 9:249–255

Edelhoff S, Ayer DE, Zervos AS, Steingrimsson E, Jenkins NA, Copeland NG, Eisenman RN, Brent R, Disteche CM (1994) Mapping of two genes encoding members of a distinct subfamily of MAX interacting proteins: MAD to human chromosome 2 and mouse chromosome 6, and MXI1 to human chromosome 10 and mouse chromosome 19. Oncogene 9:665–668

Eilers AL, Billin AN, Liu J, Ayer DE (1999) A 13-amino acid amphipathic alpha-helix is required for the functional interaction between the transcriptional repressor Mad1 and mSin3A. J Biol Chem 274:32750–32756

Ellenrieder V, Zhang JS, Kaczynski J, Urrutia R (2002) Signaling disrupts mSin3A binding to the Mad1-like Sin3-interacting domain of TIEG2, an Sp1-like repressor. EMBO J 21:2451–2460

Felsher DW, Bishop JM (1999) Transient excess of MYC activity can elicit genomic instability and tumorigenesis. Proc Natl Acad Sci U S A 96:3940–3944

Felsher DW, Zetterberg A, Zhu J, Tlsty T, Bishop JM (2000) Overexpression of MYC causes p53-dependent G2 arrest of normal fibroblasts. Proc Natl Acad Sci U S A 97:10544–10548

Feng Q, Zhang Y (2003) The NuRD complex: linking histone modification to nucleosome remodeling. Curr Top Microbiol Immunol 274:269–290

Fernandez PC, Frank SR, Wang L, Schroeder M, Liu S, Greene J, Cocito A, Amati B (2003) Genomic targets of the human c-Myc protein. Genes Dev 17:1115–1129

Fero ML, Rivkin M, Tasch M, Porter P, Carow CE, Firpo E, Polyak K, Tsai LH, Broudy V, Perlmutter RM, Kaushansky K, Roberts JM (1996) A syndrome of multiorgan hyperplasia with features of gigantism, tumorigenesis, and female sterility in p27(Kip1)-deficient mice. Cell 85:733–744

Fleischer TC, Yun UJ, Ayer DE (2003) Identification and characterization of three new components of the mSin3A corepressor complex. Mol Cell Biol 23:3456–3467

Foley KP, McArthur GA, Queva C, Hurlin PJ, Soriano P, Eisenman RN (1998) Targeted disruption of the MYC antagonist MAD1 inhibits cell cycle exit during granulocyte differentiation. EMBO J 17:774–785

Fox EJ, Wright SC (2001) S-phase-specific expression of the Mad3 gene in proliferating and differentiating cells. Biochem J 359:361–367

Fox EJ, Wright SC (2003) The transcriptional repressor gene Mad3 is a novel target for regulation by E2F1. Biochem J 370:307–313

Frank SR, Schroeder M, Fernandez P, Taubert S, Amati B (2001) Binding of c-Myc to chromatin mediates mitogen-induced acetylation of histone H4 and gene activation. Genes Dev 15:2069–2082

Frank SR, Parisi T, Taubert S, Fernandez P, Fuchs M, Chan HM, Livingston DM, Amati B (2003) MYC recruits the TIP60 histone acetyltransferase complex to chromatin. EMBO Rep 4:575–580

Freytag SO (1988) Enforced expression of the c-myc oncogene inhibits cell differentiation by precluding entry into a distinct predifferentiation state in G0/G1. Mol Cell Biol 8:1614–1624

Freytag SO, Geddes TJ (1992) Reciprocal regulation of adipogenesis by Myc and C/EBP alpha. Science 256:379–382

Fuks F, Burgers WA, Godin N, Kasai M, Kouzarides T (2001) Dnmt3a binds deacetylases and is recruited by a sequence-specific repressor to silence transcription. EMBO J 20:2536–2544

Fults D, Pedone CA, Thompson GE, Uchiyama CM, Gumpper KL, Iliev D, Vinson VL, Tavtigian SV, Perry WL 3rd (1998) Microsatellite deletion mapping on chromosome 10q and mutation analysis of MMAC1, FAS, and MXI1 in human glioblastoma multiforme. Int J Oncol 12:905–910

Galasinski SC, Resing KA, Goodrich JA, Ahn NG (2002) Phosphatase inhibition leads to histone deacetylases 1 and 2 phosphorylation and disruption of corepressor interactions. J Biol Chem 277:19618–19626

Gandarillas A, Watt FM (1995) Changes in expression of members of the fos and jun families and myc network during terminal differentiation of human keratinocytes. Oncogene 11:1403–1407

Gartel AL, Ye X, Goufman E, Shianov P, Hay N, Najmabadi F, Tyner AL (2001) Myc represses the p21(WAF1/CIP1) promoter and interacts with Sp1/Sp3. Proc Natl Acad Sci U S A 98:4510–4515

Gehring S, Rottmann S, Menkel AR, Mertsching J, Krippner-Heidenreich A, Lüscher B (2000) Inhibition of proliferation and apoptosis by the transcriptional repressor Mad1. Repression of Fas-induced caspase-8 activation. J Biol Chem 275:10413–10420

Grandori C, Cowley SM, James LP, Eisenman RN (2000) The Myc/Max/Mad network and the transcriptional control of cell behavior. Annu Rev Cell Dev Biol 16:653–699

Gray IC, Phillips SM, Lee SJ, Neoptolemos JP, Weissenbach J, Spurr NK (1995) Loss of the chromosomal region 10q23–25 in prostate cancer. Cancer Res 55:4800–4803

Greenberg RA, O'Hagan RC, Deng H, Xiao Q, Hann SR, Adams RR, Lichtsteiner S, Chin L, Morin GB, DePinho RA (1999) Telomerase reverse transcriptase gene is a direct target of c-Myc but is not functionally equivalent in cellular transformation. Oncogene 18:1219–1226

Grinberg AV, Hu CD, Kerppola TK (2004) Visualization of Myc/Max/Mad family dimers and the competition for dimerization in living cells. Mol Cell Biol 24:4294–4308

Grozinger CM, Schreiber SL (2000) Regulation of histone deacetylase 4 and 5 and transcriptional activity by 14-3-3-dependent cellular localization. Proc Natl Acad Sci U S A 97:7835–7840

Guenther MG, Lane WS, Fischle W, Verdin E, Lazar MA, Shiekhattar R (2000) A core SMRT corepressor complex containing HDAC3 and TBL1, a WD40-repeat protein linked to deafness. Genes Dev 14:1048–1057

Gunes C, Lichtsteiner S, Vasserot AP, Englert C (2000) Expression of the hTERT gene is regulated at the level of transcriptional initiation and repressed by Mad1. Cancer Res 60:2116–2121

Gupta K, Anand G, Yin X, Grove L, Prochownik EV (1998) Mmip1: a novel leucine zipper protein that reverses the suppressive effects of Mad family members on c-myc. Oncogene 16:1149–1159

Han A, Pan F, Stroud JC, Youn HD, Liu JO, Chen L (2003) Sequence-specific recruitment of transcriptional co-repressor Cabin1 by myocyte enhancer factor-2. Nature 422:730–734

Hassig CA, Fleischer TC, Billin AN, Schreiber SL, Ayer DE (1997) Histone deacetylase activity is required for full transcriptional repression by mSin3A. Cell 89:341–347

Hateboer G, Gennissen A, Ramos YF, Kerkhoven RM, Sonntag-Buck V, Stunnenberg HG, Bernards R (1995) BS69, a novel adenovirus E1A-associated protein that inhibits E1A transactivation. EMBO J 14:3159–3169

Henriksson M, Lüscher B (1996) Proteins of the Myc network: essential regulators of cell growth and differentiation. Adv Cancer Res 68:109–182

Herold S, Wanzel M, Beuger V, Frohme C, Beul D, Hillukkala T, Syvaoja J, Saluz HP, Haenel F, Eilers M (2002) Negative regulation of the mammalian UV response by Myc through association with Miz-1. Mol Cell 10:509–521

Holzel M, Kohlhuber F, Schlosser I, Holzel D, Lüscher B, Eick D (2001) Myc/Max/Mad regulate the frequency but not the duration of productive cell cycles. EMBO Rep 2:1125–1132

Hook SS, Orian A, Cowley SM, Eisenman RN (2002) Histone deacetylase 6 binds polyubiquitin through its zinc finger (PAZ domain) and copurifies with deubiquitinating enzymes. Proc Natl Acad Sci U S A 99:13425–13430

Horikawa I, Cable PL, Mazur SJ, Appella E, Afshari CA, Barrett JC (2002) Downstream E-box-mediated regulation of the human telomerase reverse transcriptase (hTERT) gene transcription: evidence for an endogenous mechanism of transcriptional repression. Mol Biol Cell 13:2585–2597

Huang EY, Zhang J, Miska EA, Guenther MG, Kouzarides T, Lazar MA (2000) Nuclear receptor corepressors partner with class II histone deacetylases in a Sin3-independent repression pathway. Genes Dev 14:45–54

Hurlin PJ, Queva C, Koskinen PJ, Steingrimsson E, Ayer DE, Copeland NG, Jenkins NA, Eisenman RN (1995a) Mad3 and Mad4: novel Max-interacting transcriptional repressors that suppress c-myc dependent transformation and are expressed during neural and epidermal differentiation. EMBO J 14:5646–5659

Hurlin PJ, Foley KP, Ayer DE, Eisenman RN, Hanahan D, Arbeit JM (1995b) Regulation of Myc and Mad during epidermal differentiation and HPV-associated tumorigenesis. Oncogene 11:2487–2501

Hurlin PJ, Queva C, Eisenman RN (1997) Mnt, a novel Max-interacting protein is co-expressed with Myc in proliferating cells and mediates repression at Myc binding sites. Genes Dev 11:44–58

Hurlin PJ, Steingrimsson E, Copeland NG, Jenkins NA, Eisenman RN (1999) Mga, a dual-specificity transcription factor that interacts with Max and contains a T-domain DNA-binding motif. EMBO J 18:7019–7028

Hurlin PJ, Zhou ZQ, Toyo-oka K, Ota S, Walker WL, Hirotsune S, Wynshaw-Boris A (2003) Deletion of Mnt leads to disrupted cell cycle control and tumorigenesis. EMBO J 22:4584–4596

Imai Y, Kurokawa M, Yamaguchi Y, Izutsu K, Nitta E, Mitani K, Satake M, Noda T, Ito Y, Hirai H (2004) The corepressor mSin3A regulates phosphorylation-induced activation, intranuclear location, and stability of AML1. Mol Cell Biol 24:1033–1043

Iritani BM, Eisenman RN (1999) c-Myc enhances protein synthesis and cell size during B lymphocyte development. Proc Natl Acad Sci U S A 96:13180–13185

Iritani BM, Delrow J, Grandori C, Gomez I, Klacking M, Carlos LS, Eisenman RN (2002) Modulation of T-lymphocyte development, growth and cell size by the Myc antagonist and transcriptional repressor Mad1. EMBO J 21:4820–4830

James L, Eisenman RN (2002) Myc and Mad bHLHZ domains possess identical DNA-binding specificities but only partially overlapping functions in vivo. Proc Natl Acad Sci U S A 99:10429–10434

Jenuwein T, Allis CD (2001) Translating the histone code. Science 293:1074–1080

Johnston LA, Prober DA, Edgar BA, Eisenman RN, Gallant P (1999) Drosophila myc regulates cellular growth during development. Cell 98:779–790

Jones PL, Veenstra GJ, Wade PA, Vermaak D, Kass SU, Landsberger N, Strouboulis J, Wolffe AP (1998) Methylated DNA and MeCP2 recruit histone deacetylase to repress transcription. Nat Genet 19:187–191

Kao HY, Verdel A, Tsai CC, Simon C, Juguilon H, Khochbin S (2001) Mechanism for nucleocytoplasmic shuttling of histone deacetylase 7. J Biol Chem 276:47496–47507

Kawamata N, Park D, Wilczynski S, Yokota J, Koeffler HP (1996) Point mutations of the Mxil gene are rare in prostate cancers. Prostate 29:191–193

Kelleher C, Teixeira MT, Forstemann K, Lingner J (2002) Telomerase: biochemical considerations for enzyme and substrate. Trends Biochem Sci 27:572–579

Kelly WK, O'Connor OA, Marks PA (2002) Histone deacetylase inhibitors: from target to clinical trials. Expert Opin Investig Drugs 11:1695–1713

Khan MM, Nomura T, Kim H, Kaul SC, Wadhwa R, Shinagawa T, Ichikawa-Iwata E, Zhong S, Pandolfi PP, Ishii S (2001) Role of PML and PML-RARalpha in Mad-mediated transcriptional repression. Mol Cell 7:1233–1243

Kim DH, Kim M, Kwon HJ (2003a) Histone deacetylase in carcinogenesis and its inhibitors as anti-cancer agents. J Biochem Mol Biol 36:110–119

Kim SK, Ro JY, Kemp BL, Lee JS, Kwon TJ, Hong WK, Mao L (1998) Identification of two distinct tumor-suppressor loci on the long arm of chromosome 10 in small cell lung cancer. Oncogene 17:1749–1753

Kim SY, Herbst A, Tworkowski KA, Salghetti SE, Tansey WP (2003b) Skp2 regulates myc protein stability and activity. Mol Cell 11:1177–1188

Kime L, Wright SC (2003) Mad4 is regulated by a transcriptional repressor complex that contains Miz-1 and c-Myc. Biochem J 370:291–298

Kispert A, Hermann BG (1993) The Brachyury gene encodes a novel DNA binding protein. EMBO J 12:4898–4899

Kiyokawa H, Kineman RD, Manova-Todorova KO, Soares VC, Hoffman ES, Ono M, Khanam D, Hayday AC, Frohman LA, Koff A (1996) Enhanced growth of mice lacking the cyclin-dependent kinase inhibitor function of p27(Kip1). Cell 85:721–732

Klochendler-Yeivin A, Yaniv M (2001) Chromatin modifiers and tumor suppression. Biochim Biophys Acta 1551:M1–10

Knoepfler PS, Eisenman RN (1999) Sin meets NuRD and other tails of repression. Cell 99:447–450

Koskinen PJ, Ayer DE, Eisenman RN (1995) Repression of Myc-Ras cotransformation by Mad is mediated by multiple protein-protein interactions. Cell Growth Differ 6:623–629

Kouzarides T (2002) Histone methylation in transcriptional control. Curr Opin Genet Dev 12:198–209

Kramer OH, Zhu P, Ostendorff HP, Golebiewski M, Tiefenbach J, Peters MA, Brill B, Groner B, Bach I, Heinzel T, Gottlicher M (2003) The histone deacetylase inhibitor valproic acid selectively induces proteasomal degradation of HDAC2. EMBO J 22:3411–3420

Krippner-Heidenreich A, Talanian RV, Sekul R, Kraft R, Thole H, Ottleben H, Lüscher B (2001) Targeting of the transcription factor Max during apoptosis: phosphorylation-regulated cleavage by caspase-5 at an unusual glutamic acid residue in position P1. Biochem J 358:705–715

Kuczyk MA, Serth J, Bokemeyer C, Schwede J, Herrmann R, Machtens S, Grunewald V, Hofner K, Jonas U (1998) The MXI1 tumor suppressor gene is not mutated in primary prostate cancer. Oncol Rep 5:213–216

Kuschak TI, Kuschak BC, Taylor CL, Wright JA, Wiener F, Mai S (2002) c-Myc initiates illegitimate replication of the ribonucleotide reductase R2 gene. Oncogene 21:909–920

Kussie PH, Gorina S, Marechal V, Elenbaas B, Moreau J, Levine AJ, Pavletich NP (1996) Structure of the MDM 2 oncoprotein bound to the p53 tumor suppressor transactivation domain. Science 274:948–953

Kuzmichev A, Zhang Y, Erdjument-Bromage H, Tempst P, Reinberg D (2002) Role of the Sin3-histone deacetylase complex in growth regulation by the candidate tumor suppressor p33(ING1). Mol Cell Biol 22:835–848

Lachman HM, Skoultchi AI (1984) Expression of c-myc changes during differentiation of mouse erythroleukaemia cells. Nature 310:592–594

Lachner M, Jenuwein T (2002) The many faces of histone lysine methylation. Curr Opin Cell Biol 14:286–298

Laherty CD, Yang WM, Sun JM, Davie JR, Seto E, Eisenman RN (1997) Histone deacetylases associated with the mSin3 corepressor mediate mad transcriptional repression. Cell 89:349–356

Laherty CD, Billin AN, Lavinsky RM, Yochum GS, Bush AC, Sun JM, Mullen TM, Davie JR, Rose DW, Glass CK, Rosenfeld MG, Ayer DE, Eisenman RN (1998) SAP30, a component of the mSin3 corepressor complex involved in N-CoR-mediated repression by specific transcription factors. Mol Cell 2:33–42

Lahoz EG, Xu L, Schreiber-Agus N, DePinho RA (1994) Suppression of Myc, but not E1a, transformation activity by Max-associated proteins, Mad and Mxi1. Proc Natl Acad Sci U S A 91:5503–5507

Lai A, Kennedy BK, Barbie DA, Bertos NR, Yang XJ, Theberge MC, Tsai SC, Seto E, Zhang Y, Kuzmichev A, Lane WS, Reinberg D, Harlow E, Branton PE (2001) RBP1 recruits the mSIN3-histone deacetylase complex to the pocket of retinoblastoma tumor suppressor family proteins found in limited discrete regions of the nucleus at growth arrest. Mol Cell Biol 21:2918–2932

Larsson LG, Ivhed I, Gidlund M, Pettersson U, Vennstrom B, Nilsson K (1988) Phorbol ester-induced terminal differentiation is inhibited in human U-937 monoblastic cells expressing a v-myc oncogene. Proc Natl Acad Sci U S A 85:2638–2642

Larsson LG, Pettersson M, Oberg F, Nilsson K, Lüscher B (1994) Expression of mad, mxi1, max and c-myc during induced differentiation of hematopoietic cells: opposite regulation of mad and c-myc. Oncogene 9:1247–1252

Larsson LG, Bahram F, Burkhardt H, Lüscher B (1997) Analysis of the DNA-binding activities of Myc/Max/Mad network complexes during induced differentiation of U-937 monoblasts and F9 teratocarcinoma cells. Oncogene 15:737–748

Lasorella A, Noseda M, Beyna M, Yokota Y, Iavarone A (2000) Id2 is a retinoblastoma protein target and mediates signalling by Myc oncoproteins. Nature 407:592–598

Lee TC, Ziff EB (1999) Mxi1 is a repressor of the c-Myc promoter and reverses activation by USF. J Biol Chem 274:595–606

Li J, Wang J, Nawaz Z, Liu JM, Qin J, Wong J (2000) Both corepressor proteins SMRT and N-CoR exist in large protein complexes containing HDAC3. EMBO J 19:4342–4350

Li J, Lin Q, Wang W, Wade P, Wong J (2002) Specific targeting and constitutive association of histone deacetylase complexes during transcriptional repression. Genes Dev 16:687–692

Li Y, Turck CM, Teumer JK, Stavnezer E (1986) Unique sequence, ski, in Sloan-Kettering avian retroviruses with properties of a new cell-derived oncogene. J Virol 57:1065–1072

Lin SY, Elledge SJ (2003) Multiple tumor suppressor pathways negatively regulate telomerase. Cell 113:881–889

Liu X, Tesfai J, Evrard YA, Dent SY, Martinez E (2003) c-Myc transformation domain recruits the human STAGA complex and requires TRRAP and GCN5 acetylase activity for transcription activation. J Biol Chem 278:20405–20412

Luger K, Richmond TJ (1998) DNA binding within the nucleosome core. Curr Opin Struct Biol 8:33–40

Luo RX, Postigo AA, Dean DC (1998) Rb interacts with histone deacetylase to repress transcription. Cell 92:463–473

Lüscher B (2001) Function and regulation of the transcription factors of the Myc/Max/Mad network. Gene 277:1–14

Lüscher B, Eisenman RN (1990) New light on Myc and Myb. Part I. Myc. Genes Dev 4:2025–2035

Lüscher B, Larsson LG (1999) The basic region/helix-loop-helix/leucine zipper domain of Myc proto-oncoproteins: function and regulation. Oncogene 18:2955–2966

Lutz W, Leon J, Eilers M (2002) Contributions of Myc to tumorigenesis. Biochim Biophys Acta 1602:61–71

Lymboussaki A, Kaipainen A, Hatva E, Vastrik I, Jeskanen L, Jalkanen M, Werner S, Stenback F, Alitalo R (1996) Expression of Mad, an antagonist of Myc oncoprotein function, in differentiating keratinocytes during tumorigenesis of the skin. Br J Cancer 73:1347–1355

Mahlknecht U, Hoelzer D (2000) Histone acetylation modifiers in the pathogenesis of malignant disease. Mol Med 6:623–644

Mai S, Hanley-Hyde J, Rainey GJ, Kuschak TI, Paul JT, Littlewood TD, Mischak H, Stevens LM, Henderson DW, Mushinski JF (1999) Chromosomal and extrachromosomal instability of the cyclin D2 gene is induced by Myc overexpression. Neoplasia 1:241–252

Marcu KB, Bossone SA, Patel AJ (1992) myc function and regulation. Annu Rev Biochem 61:809–860

Marks P, Rifkind RA, Richon VM, Breslow R, Miller T, Kelly WK (2001) Histone deacetylases and cancer: causes and therapies. Nat Rev Cancer 1:194–202

Mathon NF, Lloyd AC (2001) Cell senescence and cancer. Nat Rev Cancer 1:203–213

McArthur GA, Foley KP, Fero ML, Walkley CR, Deans AJ, Roberts JM, Eisenman RN (2002) MAD1 and p27(KIP1) cooperate to promote terminal differentiation of granulocytes and to inhibit Myc expression and cyclin E-CDK2 activity. Mol Cell Biol 22:3014–3023

McKinsey TA, Zhang CL, Lu J, Olson EN (2000) Signal-dependent nuclear export of a histone deacetylase regulates muscle differentiation. Nature 408:106–111

Menssen A, Hermeking H (2002) Characterization of the c-MYC-regulated transcriptome by SAGE: identification and analysis of c-MYC target genes. Proc Natl Acad Sci U S A 99:6274–6279

Meroni G, Reymond A, Alcalay M, Borsani G, Tanigami A, Tonlorenzi R, Nigro CL, Messali S, Zollo M, Ledbetter DH, Brent R, Ballabio A, Carrozzo R (1997) Rox, a novel bHLHZip protein expressed in quiescent cells that heterodimerizes with Max, binds a non-canonical E box and acts as a transcriptional repressor. EMBO J 16:2892–2906

Meroni G, Cairo S, Merla G, Messali S, Brent R, Ballabio A, Reymond A (2000) Mlx, a new Max-like bHLHZip family member: the center stage of a novel transcription factors regulatory pathway? Oncogene 19:3266–3277

Moens CB, Stanton BR, Parada LF, Rossant J (1993) Defects in heart and lung development in compound heterozygotes for two different targeted mutations at the N-myc locus. Development 119:485–499

Mori K, Maeda Y, Kitaura H, Taira T, Iguchi-Ariga SM, Ariga H (1998) MM-1, a novel c-Myc-associating protein that represses transcriptional activity of c-Myc. J Biol Chem 273:29794–29800

Muller CW, Herrmann BG (1997) Crystallographic structure of the T domain-DNA complex of the Brachyury transcription factor. Nature 389:884–888

Muller H, Helin K (2000) The E2F transcription factors: key regulators of cell proliferation. Biochim Biophys Acta 1470:M1–12

Muratani M, Tansey WP (2003) How the ubiquitin-proteasome system controls transcription. Nat Rev Mol Cell Biol 4:192–201

Murphy M, Ahn J, Walker KK, Hoffman WH, Evans RM, Levine AJ, George DL (1999) Transcriptional repression by wild-type p53 utilizes histone deacetylases, mediated by interaction with mSin3a. Genes Dev 13:2490–2501

Nagy L, Kao HY, Chakravarti D, Lin RJ, Hassig CA, Ayer DE, Schreiber SL, Evans RM (1997) Nuclear receptor repression mediated by a complex containing SMRT, mSin3A, and histone deacetylase. Cell 89:373–380

Nair SK, Burley SK (2003) X-ray structures of Myc-Max and Mad-Max recognizing DNA. Molecular bases of regulation by proto-oncogenic transcription factors. Cell 112:193–205

Nakayama K, Ishida N, Shirane M, Inomata A, Inoue T, Shishido N, Horii I, Loh DY (1996) Mice lacking p27(Kip1) display increased body size, multiple organ hyperplasia, retinal dysplasia, and pituitary tumors. Cell 85:707–720

Nan X, Ng HH, Johnson CA, Laherty CD, Turner BM, Eisenman RN, Bird A (1998) Transcriptional repression by the methyl-CpG-binding protein MeCP2 involves a histone deacetylase complex. Nature 393:386–389

Narlikar GJ, Fan HY, Kingston RE (2002) Cooperation between complexes that regulate chromatin structure and transcription. Cell 108:475–487

Nicol R, Zheng G, Sutrave P, Foster DN, Stavnezer E (1999) Association of specific DNA binding and transcriptional repression with the transforming and myogenic activities of c-Ski. Cell Growth Differ 10:243–254

Nigro CL, Venesio T, Reymond A, Meroni G, Alberici P, Cainarca S, Enrico F, Stack M, Ledbetter DH, Liscia DS, Ballabio A, Carrozzo R (1998) The human ROX gene: genomic structure and mutation analysis in human breast tumors. Genomics 49:275–282

Nikiforov MA, Chandriani S, Park J, Kotenko I, Matheos D, Johnsson A, McMahon SB, Cole MD (2002) TRRAP-dependent and TRRAP-independent transcriptional activation by Myc family oncoproteins. Mol Cell Biol 22:5054–5063

Nikiforov MA, Popov N, Kotenko I, Henriksson M, Cole MD (2003) The Mad and Myc basic domains are functionally equivalent. J Biol Chem 278:11094–11099

Nilsson JA, Cleveland JL (2004) Mnt: master regulator of the max network. Cell Cycle 3:588–590

Nilsson JA, Maclean KH, Keller UB, Pendeville H, Baudino TA, Cleveland JL (2004) Mnt loss triggers Myc transcription targets, proliferation, apoptosis, and transformation. Mol Cell Biol 24:1560–1569

Nomura T, Khan MM, Kaul SC, Dong HD, Wadhwa R, Colmenares C, Kohno I, Ishii S (1999) Ski is a component of the histone deacetylase complex required for transcriptional repression by Mad and thyroid hormone receptor. Genes Dev 13:412–423

Norton JD, Deed RW, Craggs G, Sablitzky F (1998) Id helix-loop-helix proteins in cell growth and differentiation. Trends Cell Biol 8:58–65

O'Connell BC, Cheung AF, Simkevich CP, Tam W, Ren X, Mateyak MK, Sedivy JM (2003) A large-scale genetic analysis of c-Myc-regulated gene expression patterns. J Biol Chem 278:12563–12573

O'Hagan RC, Schreiber-Agus N, Chen K, David G, Engelman JA, Schwab R, Alland L, Thomson C, Ronning DR, Sacchettini JC, Meltzer P, DePinho RA (2000) Gene-target recognition among members of the myc superfamily and implications for oncogenesis. Nat Genet 24:113–119

Ogawa H, Ishiguro K, Gaubatz S, Livingston DM, Nakatani Y (2002) A complex with chromatin modifiers that occupies E2F- and Myc-responsive genes in G0 cells. Science 296:1132–1136

Oh S, Song YH, Kim UJ, Yim J, Kim TK (1999) In vivo and in vitro analyses of Myc for differential promoter activities of the human telomerase (hTERT) gene in normal and tumor cells. Biochem Biophys Res Commun 263:361–365

Oh S, Song YH, Yim J, Kim TK (2000) Identification of Mad as a repressor of the human telomerase (hTERT) gene. Oncogene 19:1485–1490

Orian A, van Steensel B, Delrow J, Bussemaker HJ, Li L, Sawado T, Williams E, Loo LW, Cowley SM, Yost C, Pierce S, Edgar BA, Parkhurst SM, Eisenman RN (2003) Genomic binding by the Drosophila Myc, Max, Mad/Mnt transcription factor network. Genes Dev 17:1101–1114

Oster SK, Ho CS, Soucie EL, Penn LZ (2002) The myc oncogene: MarvelouslY complex. Adv Cancer Res 84:81–154

Papaioannou VE, Silver LM (1998) The T-box gene family. Bioessays 20:9–19

Park J, Kunjibettu S, McMahon SB, Cole MD (2001) The ATM-related domain of TRRAP is required for histone acetyltransferase recruitment and Myc-dependent oncogenesis. Genes Dev 15:1619–1624

Pelengaris S, Khan M, Evan G (2002) c-MYC: more than just a matter of life and death. Nat Rev Cancer 2:764–776

Petersen S, Wolf G, Bockmuhl U, Gellert K, Dietel M, Petersen I (1998) Allelic loss on chromosome 10q in human lung cancer: association with tumour progression and metastatic phenotype. Br J Cancer 77:270–276

Peyrefitte S, Kahn D, Haenlin M (2001) New members of the Drosophila Myc transcription factor subfamily revealed by a genome-wide examination for basic helix-loop-helix genes. Mech Dev 104:99–104

Pflum MK, Tong JK, Lane WS, Schreiber SL (2001) Histone deacetylase 1 phosphorylation promotes enzymatic activity and complex formation. J Biol Chem 276:47733–47741

Popov N, Wahlström T, Hurlin PJ, Henriksson M (2005) Mnt transcriptional repressor is functionally regulated during cell cycle progression. Oncogen (in press)

Prendergast GC (1999) Mechanisms of apoptosis by c-Myc. Oncogene 18:2967–2987

Prendergast GC, Lawe D, Ziff EB (1991) Association of Myn, the murine homolog of max, with c-Myc stimulates methylation-sensitive DNA binding and ras cotransformation. Cell 65:395–407

Pulverer B, Sommer A, McArthur GA, Eisenman RN, Lüscher B (2000) Analysis of Myc/Max/Mad network members in adipogenesis: inhibition of the proliferative burst and differentiation by ectopically expressed Mad1. J Cell Physiol 183:399–410

Rottmann S, Menkel AR, Bouchard C, Mertsching J, Loidl P, Kremmer E, Eilers M, Lüscher-Firzlaff J, Lilischis R, Lüscher B (2005) Mad1 function in cell proliferation and transcriptional repression is antagonized by Cyclin E/CDK2. J Biol Chem 280:15489–15492

Qian YW, Lee EY (1995) Dual retinoblastoma-binding proteins with properties related to a negative regulator of ras in yeast. J Biol Chem 270:25507–25513

Queva C, Hurlin PJ, Foley KP, Eisenman RN (1998) Sequential expression of the MAD family of transcriptional repressors during differentiation and development. Oncogene 16:967–977

Queva C, McArthur GA, Ramos LS, Eisenman RN (1999) Dwarfism and dysregulated proliferation in mice overexpressing the MYC antagonist MAD1. Cell Growth Differ 10:785–796

Queva C, McArthur GA, Iritani BM, Eisenman RN (2001) Targeted deletion of the S-phase-specific Myc antagonist Mad3 sensitizes neuronal and lymphoid cells to radiation-induced apoptosis. Mol Cell Biol 21:703–712

Radhakrishnan I, Perez-Alvarado GC, Parker D, Dyson HJ, Montminy MR, Wright PE (1997) Solution structure of the KIX domain of CBP bound to the transactivation domain of CREB: a model for activator:coactivator interactions. Cell 91:741–752

Rao G, Alland L, Guida P, Schreiber-Agus N, Chen K, Chin L, Rochelle JM, Seldin MF, Skoultchi AI, DePinho RA (1996) Mouse Sin3A interacts with and can functionally substitute for the amino-terminal repression of the Myc antagonist Mxi1. Oncogene 12:1165–1172

Roussel MF, Ashmun RA, Sherr CJ, Eisenman RN, Ayer DE (1996) Inhibition of cell proliferation by the Mad1 transcriptional repressor. Mol Cell Biol 16:2796–2801

Rubin H (2002) The disparity between human cell senescence in vitro and lifelong replication in vivo. Nat Biotechnol 20:675–681

Rudolph B, Hueber AO, Evan GI (2001) Expression of Mad1 in T cells leads to reduced thymic cellularity and impaired mitogen-induced proliferation. Oncogene 20:1164–1175

Salomoni P, Pandolfi PP (2002) The role of PML in tumor suppression. Cell 108:165–170

Satou A, Taira T, Iguchi-Ariga SM, Ariga H (2001) A novel transrepression pathway of c-Myc. Recruitment of a transcriptional corepressor complex to c-Myc by MM-1, a c-Myc-binding protein. J Biol Chem 276:46562–46567

Sawai S, Shimono A, Wakamatsu Y, Palmes C, Hanaoka K, Kondoh H (1993) Defects of embryonic organogenesis resulting from targeted disruption of the N-myc gene in the mouse. Development 117:1445–1455

Schreiber-Agus N, DePinho RA (1998) Repression by the Mad(Mxi1)-Sin3 complex. Bioessays 20:808–818

Schreiber-Agus N, Chin L, Chen K, Torres R, Rao G, Guida P, Skoultchi AI, DePinho RA (1995) An amino-terminal domain of Mxi1 mediates anti-Myc oncogenic activity and interacts with a homolog of the yeast transcriptional repressor SIN3. Cell 80:777–786

Schreiber-Agus N, Meng Y, Hoang T, Hou H Jr, Chen K, Greenberg R, Cordon-Cardo C, Lee HW, DePinho RA (1998) Role of Mxi1 in ageing organ systems and the regulation of normal and neoplastic growth. Nature 393:483–487

Schuhmacher M, Staege MS, Pajic A, Polack A, Weidle UH, Bornkamm GW, Eick D, Kohlhuber F (1999) Control of cell growth by c-Myc in the absence of cell division. Curr Biol 9:1255–1258

Schuhmacher M, Kohlhuber F, Holzel M, Kaiser C, Burtscher H, Jarsch M, Bornkamm GW, Laux G, Polack A, Weidle UH, Eick D (2001) The transcriptional program of a human B cell line in response to Myc. Nucleic Acids Res 29:397–406

Seigneurin-Berny D, Verdel A, Curtet S, Lemercier C, Garin J, Rousseaux S, Khochbin S (2001) Identification of components of the murine histone deacetylase 6 complex: link between acetylation and ubiquitination signaling pathways. Mol Cell Biol 21:8035–8044

Seoane J, Pouponnot C, Staller P, Schader M, Eilers M, Massague J (2001) TGFbeta influences Myc, Miz-1 and Smad to control the CDK inhibitor p15INK4b. Nat Cell Biol 3:400–408

Seoane J, Le HV, Massague J (2002) Myc suppression of the p21(Cip1) Cdk inhibitor influences the outcome of the p53 response to DNA damage. Nature 419:729–734

Shapiro DN, Valentine V, Eagle L, Yin X, Morris SW, Prochownik EV (1994) Assignment of the human MAD and MXI1 genes to chromosomes 2p12-p13 and 10q24-q25. Genomics 23:282–285

Shen-Li H, O'Hagan RC, Hou H Jr, Horner JW 2nd, Lee HW, DePinho RA (2000) Essential role for Max in early embryonic growth and development. Genes Dev 14:17–22

Shinagawa T, Dong HD, Xu M, Maekawa T, Ishii S (2000) The sno gene, which encodes a component of the histone deacetylase complex, acts as a tumor suppressor in mice. EMBO J 19:2280–2291

Shinagawa T, Nomura T, Colmenares C, Ohira M, Nakagawara A, Ishii S (2001) Increased susceptibility to tumorigenesis of ski-deficient heterozygous mice. Oncogene 20:8100–8108

Shivapurkar N, Reddy J, Matta H, Sathyanarayana UG, Huang CX, Toyooka S, Minna JD, Chaudhary PM, Gazdar AF (2002) Loss of expression of death-inducing signaling complex (DISC) components in lung cancer cell lines and the influence of MYC amplification. Oncogene 21:8510–8514

Siegel PM, Shu W, Massague J (2003) Mad upregulation and Id2 repression accompany TGF-beta mediated epithelial cell growth suppression. J Biol Chem 278:35444–35540

Sif S, Saurin AJ, Imbalzano AN, Kingston RE (2001) Purification and characterization of mSin3A-containing Brg1 and hBrm chromatin remodeling complexes. Genes Dev 15:603–618

Skowyra D, Zeremski M, Neznanov N, Li M, Choi Y, Uesugi M, Hauser CA, Gu W, Gudkov AV, Qin J (2001) Differential association of products of alternative transcripts of the candidate tumor suppressor ING1 with the mSin3/HDAC1 transcriptional corepressor complex. J Biol Chem 276:8734–8739

Sommer A, Hilfenhaus S, Menkel A, Kremmer E, Seiser C, Loidl P, Lüscher B (1997) Cell growth inhibition by the Mad/Max complex through recruitment of histone deacetylase activity. Curr Biol 7:357–365

Sommer A, Bousset K, Kremmer E, Austen M, Lüscher B (1998) Identification and characterization of specific DNA-binding complexes containing members of the Myc/Max/Mad network of transcriptional regulators. J Biol Chem 273:6632–6642

Sommer A, Waha A, Tonn J, Sorensen N, Hurlin PJ, Eisenman RN, Lüscher B, Pietsch T (1999) Analysis of the Max-binding protein MNT in human medulloblastomas. Int J Cancer 82:810–816

Spronk CA, Tessari M, Kaan AM, Jansen JF, Vermeulen M, Stunnenberg HG, Vuister GW (2000) The Mad1-Sin3B interaction involves a novel helical fold. Nat Struct Biol 7:1100–1104

Staller P, Peukert K, Kiermaier A, Seoane J, Lukas J, Karsunky H, Moroy T, Bartek J, Massague J, Hanel F, Eilers M (2001) Repression of p15INK4b expression by Myc through association with Miz-1. Nat Cell Biol 3:392–399

Stanton BR, Perkins AS, Tessarollo L, Sassoon DA, Parada LF (1992) Loss of N-myc function results in embryonic lethality and failure of the epithelial component of the embryo to develop. Genes Dev 6:2235–2247

Sterner DE, Berger SL (2000) Acetylation of histones and transcription-related factors. Microbiol Mol Biol Rev 64:435–459

Strahl BD, Allis CD (2000) The language of covalent histone modifications. Nature 403:41–45

Stuart ET, Kioussi C, Gruss P (1994) Mammalian Pax genes. Annu Rev Genet 28:219–236

Suske G (1999) The Sp-family of transcription factors. Gene 238:291–300

Swanson KA, Knoepfler PS, Huang K, Kang RS, Cowley SM, Laherty CD, Eisenman RN, Radhakrishnan I (2004) HBP1 and Mad1 repressors bind the Sin3 corepressor PAH2 domain with opposite helical orientations. Nat Struct Mol Biol 11:738–746

Takahashi T, Konishi H, Kozaki K, Osada H, Saji S (1998) Molecular analysis of a Myc antagonist, ROX/Mnt, at 17p13.3 in human lung cancers. Jpn J Cancer Res 89:347–351

Takakura M, Kyo S, Kanaya T, Hirano H, Takeda J, Yutsudo M, Inoue M (1999) Cloning of human telomerase catalytic subunit (hTERT) gene promoter and identification of proximal core promoter sequences essential for transcriptional activation in immortalized and cancer cells. Cancer Res 59:551–557

Taylor C, Mai S (1998) c-Myc-associated genomic instability of the dihydrofolate reductase locus in vivo. Cancer Detect Prev 22:350–356

Teitz T, Wei T, Valentine MB, Vanin EF, Grenet J, Valentine VA, Behm FG, Look AT, Lahti JM, Kidd VJ (2000) Caspase 8 is deleted or silenced preferentially in childhood neuroblastomas with amplification of MYCN. Nat Med 6:529–535

Timmermann S, Lehrmann H, Polesskaya A, Harel-Bellan A (2001) Histone acetylation and disease. Cell Mol Life Sci 58:728–736

Tong JK (2002) Dissecting histone deacetylase function. Chem Biol 9:668–670

Tong JK, Hassig CA, Schnitzler GR, Kingston RE, Schreiber SL (1998) Chromatin deacetylation by an ATP-dependent nucleosome remodelling complex. Nature 395:917–921

Toyo-oka K, Hirotsune S, Gambello MJ, Zhou ZQ, Olson L, Rosenfeld MG, Eisenman R, Hurlin P, Wynshaw-Boris A (2004) Loss of the Max-interacting protein Mnt in mice results in decreased viability, defective embryonic growth and craniofacial defects: relevance to Miller-Dieker syndrome. Hum Mol Genet 13:1057–1067

Trimarchi JM, Lees JA (2002) Sibling rivalry in the E2F family. Nat Rev Mol Cell Biol 3:11–20

Tsai SC, Seto E (2002) Regulation of histone deacetylase 2 by protein kinase CK2. J Biol Chem 277:31826–31833

Uesugi M, Nyanguile O, Lu H, Levine AJ, Verdine GL (1997) Induced alpha helix in the VP16 activation domain upon binding to a human TAF. Science 277:1310–1313

van de Wetering M, Sancho E, Verweij C, de Lau W, Oving I, Hurlstone A, van der Horn K, Batlle E, Coudreuse D, Haramis AP, Tjon-Pon-Fong M, Moerer P, van den Born M, Soete G, Pals S, Eilers M, Medema R, Clevers H (2002) The beta-catenin/TCF-4 complex imposes a crypt progenitor phenotype on colorectal cancer cells. Cell 111:241–250

van Ingen H, Lasonder E, Jansen JF, Kaan AM, Spronk CA, Stunnenberg HG, Vuister GW (2004) Extension of the binding motif of the Sin3 interacting domain of the Mad family proteins. Biochemistry 43:46–54

Van Lint C, Emiliani S, Verdin E (1996) The expression of a small fraction of cellular genes is changed in response to histone hyperacetylation. Gene Expr 5:245–253

Vastrik I, Kaipainen A, Penttila TL, Lymboussakis A, Alitalo R, Parvinen M, Alitalo K (1995) Expression of the mad gene during cell differentiation in vivo and its inhibition of cell growth in vitro. J Cell Biol 128:1197–1208

Vaute O, Nicolas E, Vandel L, Trouche D (2002) Functional and physical interaction between the histone methyl transferase Suv39H1 and histone deacetylases. Nucleic Acids Res 30:475–481

Vervoorts J, Lüscher-Firzlaff JM, Rottmann S, Lilischkis R, Walsemann G, Dohmann K, Austen M, Lüscher B (2003) Stimulation of c-MYC transcriptional activity and acetylation by recruitment of the cofactor CBP. EMBO Rep 4:484–490

Vidal M, Strich R, Esposito RE, Gaber RF (1991) RPD1 (SIN3/UME4) is required for maximal activation and repression of diverse yeast genes. Mol Cell Biol 11:6306–6316

Vietor I, Vadivelu SK, Wick N, Hoffman R, Cotten M, Seiser C, Fialka I, Wunderlich W, Haase A, Korinkova G, Brosch G, Huber LA (2002) TIS7 interacts with the mammalian SIN3 histone deacetylase complex in epithelial cells. EMBO J 21:4621–4631

von der Lehr N, Johansson S, Wu S, Bahram F, Castell A, Cetinkaya C, Hydbring P, Weidung I, Nakayama K, Nakayama KI, Soderberg O, Kerppola TK, Larsson LG (2003) The F-box protein Skp2 participates in c-Myc proteosomal degradation and acts as a cofactor for c-Myc-regulated transcription. Mol Cell 11:1189–1200

Vousden KH (2002) Activation of the p53 tumor suppressor protein. Biochim Biophys Acta 1602:47–59

Wade PA, Jones PL, Vermaak D, Wolffe AP (1998) A multiple subunit Mi-2 histone deacetylase from Xenopus laevis cofractionates with an associated Snf2 superfamily ATPase. Curr Biol 8:843–846

Wang AH, Kruhlak MJ, Wu J, Bertos NR, Vezmar M, Posner BI, Bazett-Jones DP, Yang XJ (2000a) Regulation of histone deacetylase 4 by binding of 14-3-3 proteins. Mol Cell Biol 20:6904–6912

Wang DS, Rieger-Christ K, Latini JM, Moinzadeh A, Stoffel J, Pezza JA, Saini K, Libertino JA, Summerhayes IC (2000b) Molecular analysis of PTEN and MXI1 in primary bladder carcinoma. Int J Cancer 88:620–625

Wang H, Stillman DJ (1993) Transcriptional repression in Saccharomyces cerevisiae by a SIN3-LexA fusion protein. Mol Cell Biol 13:1805–1814

Wang H, Clark I, Nicholson PR, Herskowitz I, Stillman DJ (1990) The Saccharomyces cerevisiae SIN3 gene, a negative regulator of HO, contains four paired amphipathic helix motifs. Mol Cell Biol 10:5927–5936

Wang J, Xie LY, Allan S, Beach D, Hannon GJ (1998) Myc activates telomerase. Genes Dev 12:1769–1774

Washburn BK, Esposito RE (2001) Identification of the Sin3-binding site in Ume6 defines a two-step process for conversion of Ume6 from a transcriptional repressor to an activator in yeast. Mol Cell Biol 21:2057–2069

Wechsler DS, Shelly CA, Petroff CA, Dang CV (1997) MXI1, a putative tumor suppressor gene, suppresses growth of human glioblastoma cells. Cancer Res 57:4905–4912

Werner S, Beer HD, Mauch C, Lüscher B (2001) The Mad1 transcription factor is a novel target of activin and TGF-beta action in keratinocytes: possible role of Mad1 in wound repair and psoriasis. Oncogene 20:7494–7504

Wilson V, Conlon FL (2002) The T-box family. Genome Biol 3:REVIEWS3008

Won J, Yim J, Kim TK (2002) Sp1 and Sp3 recruit histone deacetylase to repress transcription of human telomerase reverse transcriptase (hTERT) promoter in normal human somatic cells. J Biol Chem 277:38230–38238

Wu KJ, Grandori C, Amacker M, Simon-Vermot N, Polack A, Lingner J, Dalla-Favera R (1999) Direct activation of TERT transcription by c-MYC. Nat Genet 21:220–224

Wu S, Cetinkaya C, Munoz-Alonso MJ, von der Lehr N, Bahram F, Beuger V, Eilers M, Leon J, Larsson LG (2003) Myc represses differentiation-induced p21CIP1 expression via Miz-1-dependent interaction with the p21 core promoter. Oncogene 22:351–360

Wu WS, Vallian S, Seto E, Yang WM, Edmondson D, Roth S, Chang KS (2001) The growth suppressor PML represses transcription by functionally and physically interacting with histone deacetylases. Mol Cell Biol 21:2259–2268

Xu D, Popov N, Hou M, Wang Q, Bjorkholm M, Gruber A, Menkel AR, Henriksson M (2001) Switch from Myc/Max to Mad1/Max binding and decrease in histone acetylation at the telomerase reverse transcriptase promoter during differentiation of HL60 cells. Proc Natl Acad Sci U S A 98:3826–3831

Xu HE, Stanley TB, Montana VG, Lambert MH, Shearer BG, Cobb JE, McKee DD, Galardi CM, Plunket KD, Nolte RT, Parks DJ, Moore JT, Kliewer SA, Willson TM, Stimmel JB (2002) Structural basis for antagonist-mediated recruitment of nuclear co-repressors by PPARalpha. Nature 415:813–817

Xue Y, Wong J, Moreno GT, Young MK, Cote J, Wang W (1998) NURD, a novel complex with both ATP-dependent chromatin-remodeling and histone deacetylase activities. Mol Cell 2:851–861

Yang SH, Bumpass DC, Perkins ND, Sharrocks AD (2002) The ETS domain transcription factor Elk-1 contains a novel class of repression domain. Mol Cell Biol 22:5036–5046

Yeh WC, Bierer BE, McKnight SL (1995) Rapamycin inhibits clonal expansion and adipogenic differentiation of 3T3-L1 cells. Proc Natl Acad Sci U S A 92:11086–11090

Yin XY, Gupta K, Han WP, Levitan ES, Prochownik EV (1999) Mmip-2, a novel RING finger protein that interacts with mad members of the Myc oncoprotein network. Oncogene 18:6621–6634

Yochum GS, Ayer DE (2001) Pf1, a novel PHD zinc finger protein that links the TLE corepressor to the mSin3A-histone deacetylase complex. Mol Cell Biol 21:4110–4118

Zervos AS, Gyuris J, Brent R (1993) Mxi1, a protein that specifically interacts with Max to bind Myc-Max recognition sites. Cell 72:223–232

Zhang JS, Moncrieffe MC, Kaczynski J, Ellenrieder V, Prendergast FG, Urrutia R (2001) A conserved alpha-helical motif mediates the interaction of Sp1-like transcriptional repressors with the corepressor mSin3A. Mol Cell Biol 21:5041–5049

Zhang Y, Reinberg D (2001) Transcription regulation by histone methylation: interplay between different covalent modifications of the core histone tails. Genes Dev 15:2343–2360

Zhang Y, Iratni R, Erdjument-Bromage H, Tempst P, Reinberg D (1997) Histone deacetylases and SAP18, a novel polypeptide, are components of a human Sin3 complex. Cell 89:357–364

Zhang Y, LeRoy G, Seelig HP, Lane WS, Reinberg D (1998a) The dermatomyositis-specific autoantigen Mi2 is a component of a complex containing histone deacetylase and nucleosome remodeling activities. Cell 95:279–289

Zhang Y, Sun ZW, Iratni R, Erdjument-Bromage H, Tempst P, Hampsey M, Reinberg D (1998b) SAP30, a novel protein conserved between human and yeast, is a component of a histone deacetylase complex. Mol Cell 1:1021–1031

Zhou X, Richon VM, Rifkind RA, Marks PA (2000) Identification of a transcriptional repressor related to the noncatalytic domain of histone deacetylases 4 and 5. Proc Natl Acad Sci U S A 97:1056–1061

Zhou ZQ, Hurlin PJ (2001) The interplay between Mad and Myc in proliferation and differentiation. Trends Cell Biol 11:S10–14

CTMI (2006) 302:123–143

Structural Aspects of Interactions Within the Myc/Max/Mad Network

S. K. Nair[1] (✉) · S. K. Burley[2]

[1]Department of Biochemistry and Center for Biophysics & Computational Biology, University of Illinois at Urbana-Champaign, 600 S. Mathews Avenue, Urbana, IL 61801, USA
snair@uiuc.edu

[2]Structural GenomiX, Inc., 10505 Roselle Street, San Diego, CA 92121, USA

Abstract Recently determined structures of a number of Myc family proteins have provided significant insights into the molecular nature of complex assembly and DNA binding. These structures illuminate the details of specific interactions that govern the assembly of nucleoprotein complexes and, in doing so, raise more questions regarding Myc biology. In this review, we focus on the lessons provided by these structures toward understanding (1) interactions that govern transcriptional repression by Mad via the Sin3 pathway, (2) homodimerization of Max, (3) heterodimerization of Myc–Max and Mad–Max, and (4) DNA recognition by each of the Max–Max, Myc–Max, and Mad–Max dimers.

1 Introduction

Mutations of genes of the *myc* family have been shown to be among the most frequently affected in the majority of human malignancies (Nesbit et al. 1999). *Myc* genes were first identified as the transforming agents within chicken retroviruses (Sheiness et al. 1978). Over the last 25 years, compelling evidence has accumulated for the role of *myc* homologs in tumor formation, both in experimental systems and in human cancers (Cole and McMahon 1999; Dang et al. 1999; Eilers 1999; Liao and Dickson 2000; Nesbit et al. 1999).

The Myc gene products are transacting transcriptional regulators containing two independently functioning polypeptide regions: N-terminal transactivating residues and a C-terminal DNA binding segment (for a review see Grandori et al. 2000; Fig. 1). The DNA binding segment tethers Myc family gene products to sequences upstream of the core promoter, thereby enabling activation domains to modulate the efficiency of messenger RNA synthesis (Kato et al. 1990). The initial identification of a DNA binding segment within Myc family genes was based on sequence similarities with other transcription factors possessing a modular DNA binding/dimerization motif consisting of a two amphipathic α-helices (helix H1 and H2) separated by a loop (Murre et al. 1989). Myc family members also contain a basic region preceding the first α-helix and a leucine zipper region carboxy-terminal to the second α-helix. In general, the basic-helix-loop-helix-leucine zipper (bHLHZ) domain specifies dimerization through the helix-loop-helix-leucine zipper (HLHZ) region and DNA recognition through interactions between the basic region (b) and the major groove. However, Myc cannot form homodimers at physiological concentrations in vivo, and is incapable of sequence-specific DNA binding in isolation (Dang et al. 1991).

A better understanding of Myc biology emerged with the identification of a closely related bHLHZ protein Max that serves as an obligate, physiological heterodimerization partner for c-Myc (Blackwood and Eisenman 1991; Prendergast et al. 1991; Fig. 1). While c-Myc is incapable of forming homodimers or interacting specifically with DNA in isolation, the bHLHZ regions of Myc and Max form strong heterodimers, recognize DNA in a sequence-specific manner, and support Myc function in transcriptional activation, cellular transformation, and apoptosis (Amati et al. 1992; Amati et al. 1993). Myc–Max heterodimers recognize a core hexanucleotide element (5′-CACGTG-3′), termed the E-box (Blackwood and Eisenman 1991; Prendergast et al. 1991) and activate transcription at promoters containing E-boxes (Benvenisty et al. 1992; Eilers et al. 1991).

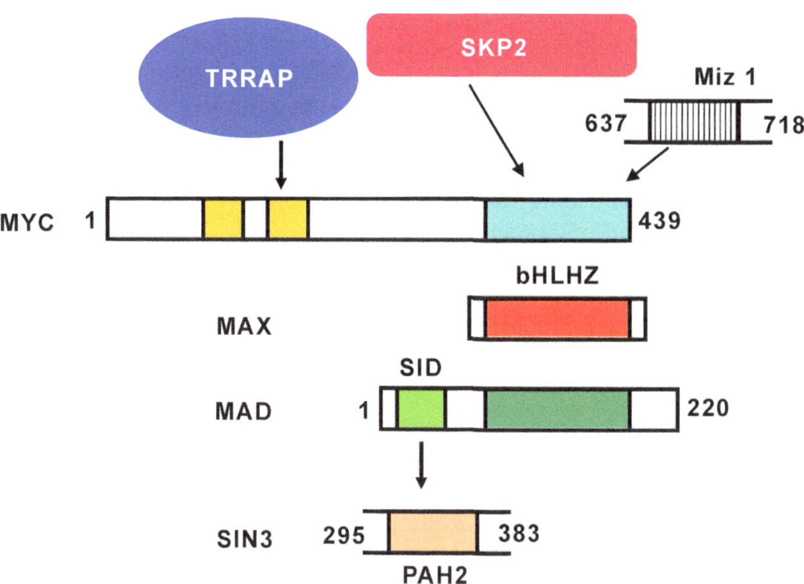

Fig. 1 Domain organization of c-Myc, Max, and Mad, and schematic organization of proteins involved in transcriptional activation and repression within the Myc/Mad/Max network. The basic-helix-loop-helix-leucine zipper domains of the individual proteins are indicated relative to the full-length (Myc, *cyan*; Max, *red*; Mad, *green*). Transcriptional activation by Myc–Max heterodimers is dependent, in part, on recruitment of TRAAP (*purple*) by Myc transactivating residues (*yellow*). Conversely, transcriptional repression by Mad–Max heterodimers requires an interaction between the Sin3 interacting domain (SID) residues of Mad (*light green*) and a paired amphipathic helix (PAH2) domain of Sin3 (*tan*). The bHLHZ domain of Myc can also recruit the Miz-1 transcriptional repressor (*vertical lines*) and the E2 ubiquitin ligase Skp2 (*pink*)

In addition to acting as a heterodimerization partner for Myc, Max can also form homodimers and bind E-box containing DNA sequences. At present, the biological role or roles of the Max homodimer remain unknown, although there are suggestions that Max can function as a transcriptional repressor (Kretzner et al. 1992). While Max homodimers and Myc–Max heterodimers both recognize the same hexanucleotide element, sequence analyses of putative Myc target genes and the results of in vitro binding assays suggest that nucleotides flanking the E-box can confer binding preferences for Myc–Max heterodimers versus Max homodimers (Grandori et al. 1996; Grandori and Eisenman 1997). In addition, Myc–Max heterodimers recognize a number of noncanonical E-boxes containing variant nitrogenous bases at one or more sites in the E box hexanucleotide (e.g., 5′-CATGCG-3′, 5′-CAACGTG-3′, etc.; Blackwell et al. 1991, 1993; Haggerty et al. 2003).

Shortly after the discovery of Max, a second class of bHLHZ proteins, including Mad1 (Ayer et al. 1993) and Mxi1 (Zervos et al. 1993) were independently identified as additional heterodimerization partners of Max. Mad1, Mxi1, and other Mad family members (Hurlin et al. 1995) inhibit cell growth. High levels of *mad* mRNA and Mad protein are found in growth-arrested, differentiated cells in which c-Myc is not expressed. Each of the Mad family member proteins can recognize the E-box as heterodimers with Max and interfere with the transforming function of Myc (Fig. 1). Hence, Mad1, Mxi1, and related members constitute a family of transcriptional repressors (Hurlin et al. 1994; Larsson et al. 1994, 1997; McArthur et al. 1998). Competition between Myc–Max and Mad–Max heterodimers for a common DNA target appears to control cell fate, determining the choice between proliferation/transformation and differentiation/quiescence.

Myc can also act as a transcriptional repressor at a distinct subset of genes (Li et al. 1994; see chapter in this volume by D. Kleine-Kohlbrecher et al.). At least one pathway of Myc repression has been elucidated through the identification of an association of Myc–Max heterodimers with the BTB-POZ domain protein Miz-1 (Peukert et al. 1997). Association of Myc–Max bHLHZ domains with Miz-1 appears to block the ability of Miz-1 to recruit the p300 co-activator, thereby leading to repression of genes normally activated by Miz-1 (Staller et al. 2001; Fig. 1). There is also some evidence that Myc repression can occur through binding of Myc–Max to core promoter elements (Kwon et al. 1996; Yang et al. 2001); however, the physiological significance of this effect has not been established.

2
Topology of the Amino Terminal Domains

Myc and Mad family members have bipartite structures with separable, independently folded domains. The carboxyl terminal bHLHZ domain dictates sequence-specific DNA recognition, while the amino terminal residues dictate transactivation (Myc) or transrepression (Mad). These amino terminal residues mediate specific biological functions via recruitment of different multiprotein complexes.

Transcriptional repression by Mad–Max heterodimers is mediated by interactions between amino terminal Mad residues and the mSin3 co-repressor (Ayer et al. 1995; Schreiber-Agus et al. 1995), a component of the multiprotein histone deacetylase complex. Mad–Max heterodimers recruit the mSin3 co-repressor to promoter DNA, leading to recruitment of histone deacetylases, condensation of chromatin structure, and subsequent transcriptional repression (Hassig et al. 1997; Laherty et al. 1997). Conversely, Myc–Max heterodimers activate gene expression by recruitment of multiprotein complexes bearing histone acetyltransferase activity. Myc interacts with TRRAP, a component of the Gcn5 and Tip60 histone acetyltransferase complexes, and this Myc-mediated recruitment of histone acetyltransferase activity results in upregulation of gene expression (McMahon et al. 2000; Saleh et al. 1998).

While detailed structural analysis of protein recruitment by the amino terminal co-activator domain of Myc has not yet been carried out, structures of the interacting domains of Mad and the mSin3 co-repressor have been determined by nuclear magnetic resonance (NMR) spectroscopy (Brubaker et al. 2000; Spronk et al. 2000). All four Mad paralogs contain a 30-residue amino terminal segment, the Sin3 interaction domain (SID), which is both necessary and sufficient for Sin3 association and transrepression (Ayer et al. 1995; Schreiber-Agus et al. 1995). Deletion mapping studies identified a 13-residue peptide within Mad1 that interacts with mSin3A (Eilers et al. 1999). Sin3 contains four repeats of a 100-residue segment, the paired amphipathic helix (PAH) domain; and the second of these repeats (PAH2) serves as the Mad interaction domain (Ayer et al. 1995; Schreiber-Agus et al. 1995).

Heteronuclear NMR spectroscopic studies of the Sin3 PAH2–Mad1 SID peptide complex demonstrate that the Sin3 PAH2 domain forms a left-handed, four-helix bundle containing an extensive, well-defined hydrophobic core (Brubaker et al. 2000; Spronk et al. 2000; Fig. 2a). α-Helices 1 and 2 form a hydrophobic pocket, defining the interaction surface for the Mad1 SID peptide. The Mad1 SID peptide forms an amphipathic α-helix, and interactions with the Sin3 PAH2 domain engage the nonpolar face of this peptide (Brubaker et al. 2000; Fig. 2a). More recently, the HMG box transcriptional repressor HBP1 has also been shown to interact with the PAH2 of Sin3. The solution structure of the HBP1 SID–Sin3 PAH2 complex demonstrates that the HBP1 SID peptide binds to the PAH2 domain with a reverse orientation relative to that of the Mad1 SID peptide (Fig. 2b). Detailed comparisons of the PAH2–Mad1 SID and PAH2–HBP1 SID structures reveal that both peptides are engaged by the PAH2 domain through similar interactions despite binding in opposite relative helical orientations.

Another intriguing observation that emerges from these structural studies is that both the Mad1 SID peptide and the Sin3 PAH2 domains are partially unfolded in the absence of their respective interaction partners. These mutually induced structural transitions may be representative of a general mechanism for facilitating interactions within multiprotein transcriptional complexes (Dyson and Wright 2002).

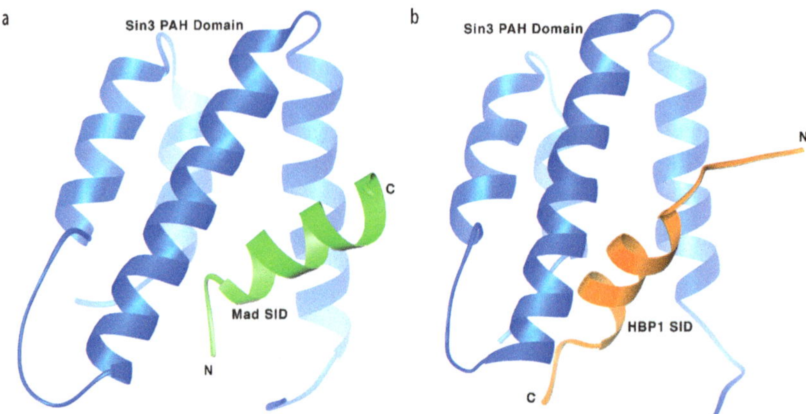

Fig. 2a, b Ribbon diagram showing a representative conformer of the second Sin3 PAH2 domain complexed with the Sin3 interacting domain (SID) peptide. **a** The Sin 3 PAH2 domain is colored *tan* and the SID peptide from Mad1 is shown in *green*. **b** The Sin3 PAH2 domain is colored *tan* and the SID peptide from HBP1 is colored in *blue*. Interactions between the two molecules are mediated by the packing of hydrophobic residues from the SID peptide into a hydrophobic pocket created by the α-helices of the PAH2 domain. Note that the Sin3 PAH engages both Mad1 SID and HBP1 SID1 in similar fashions. However, the helical orientations of the SID peptides are completely reversed relative to each other

3 Topology of the bHLHZ Domain

The co-crystal structure of the bHLHZ domain of the Max homodimer bound to DNA revealed the overall topology of this domain and established the structural bases for DNA recognition by bHLHZ domain proteins (Ferre-D'Amare et al. 1993; Fig. 3a). Co-crystal structures of the Myc–Max and Mad–Max heterodimers recapitulate the disposition of secondary structure elements observed within the Max homodimer structure (Nair and Burley 2003). The bHLHZ domains of Myc, Max, and Mad consist of two lengthy α-helices separated by a random coil loop. Residues from the basic region and helix H1 constitute the first continuous α-helical secondary structure element. A conserved proline residue terminates the first α-helix (bH1) resulting in

Fig. 3a, b Equivalent views of the Max homodimer and the Myc–Max heterodimer bound to oligonucleotides bearing the E-box (Max, *red*; Myc, *cyan*). The tighter packing within the Myc–Max heterodimer structure is mediated by charge complementarity at residues near the c-terminus of the leucine zipper domain

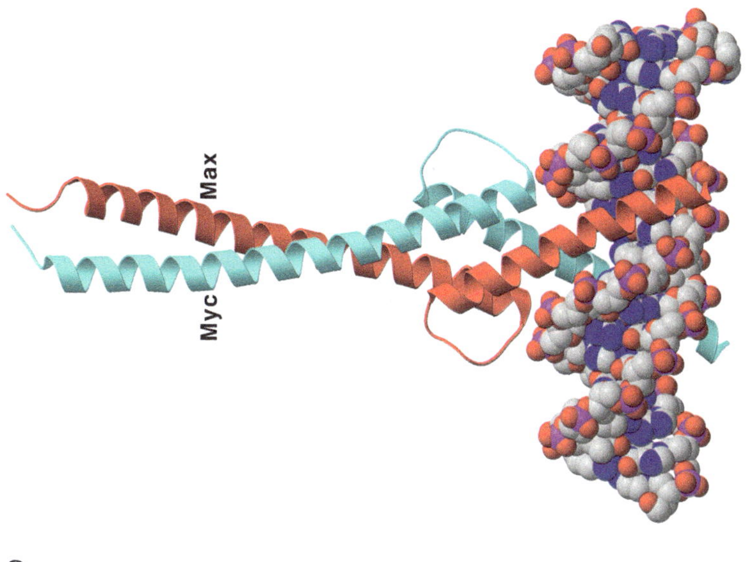
b
Myc
Max

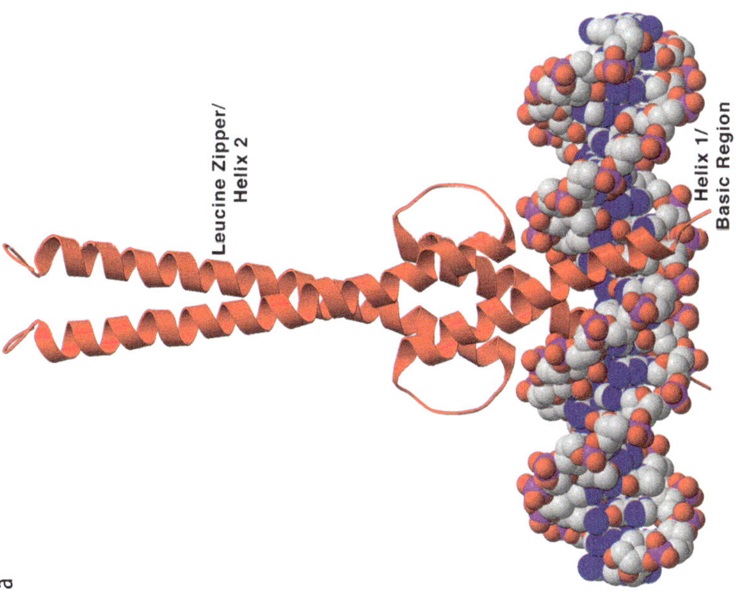
a
Leucine Zipper/
Helix 2
Helix 1/
Basic Region

a turn in the backbone structure at the start of the variable loop region (L) that connects the two α-helical segments. The second α-helix is composed of the H2 and leucine zipper regions (Ferre-D'Amare et al. 1993; Nair and Burley 2003).

The Max homodimer and the Myc–Max and Mad–Max heterodimers all consist of two bHLHZ monomers that fold into a globular, parallel, left-handed, four-helix bundle (Fig. 3). Two pairs of α-helices project in opposite directions from the bundle. Two basic regions project from the amino termini of the four-helix bundle and make sequence-specific contacts with cognate DNA. The carboxy-terminal extensions of the four-helix bundle consist of two α-helical segments that form a parallel, left-handed, coiled coil or leucine zipper, similar in structure to the GCN4 homodimer (O'Shea et al. 1991).

The topology of the bHLHZ domain is distinguished from that of purely coiled-coil leucine zipper proteins, such as GCN4, by the presence of a well-defined globular core formed by α-helices H1 and H2 of the four-helix bundle. Hydrophobic residues conserved within the bHLHZ domain form this globular core, which stabilizes the structure of the Max homodimer. Mutagenesis of Myc–Max heterodimers demonstrates that all of the conserved hydrophobic amino acids within H1 and H2 are required for stable association of the dimer (Davis and Halazonetis 1993).

4 Structural Basis for DNA Recognition

In both the Myc–Max and Mad–Max heterodimer co-crystal structures (Nair and Burley 2003), the DNA adopts a modified B-form conformation, characterized by a narrowed major groove and a widened minor groove. Each monomeric component of the heterodimer interacts with half of the 5′-CACGTG-3′ recognition site. The co-crystal structures revealed three portions of the polypeptide chain responsible for DNA contacts: residues from the basic and loop regions, and the first residue of α-helix H2 (Ferre-D'Amare et al. 1993; Nair and Burley 2003).

4.1 Myc–Max Interactions with DNA

Three invariant residues within the basic region make sequence-specific contacts with selected DNA nucleotides within the 5′-Cyt(1)-Ade(2)-Cyt(3)-Gua(4)-Thy(5)-Gua(6)-3′ recognition sequence. In each half of the homo- or heterodimer co-crystal structures, Max residue His-28 participates in a hydrogen bond with the N7 of Gua(3′) (where ′ denotes opposite strand), residue

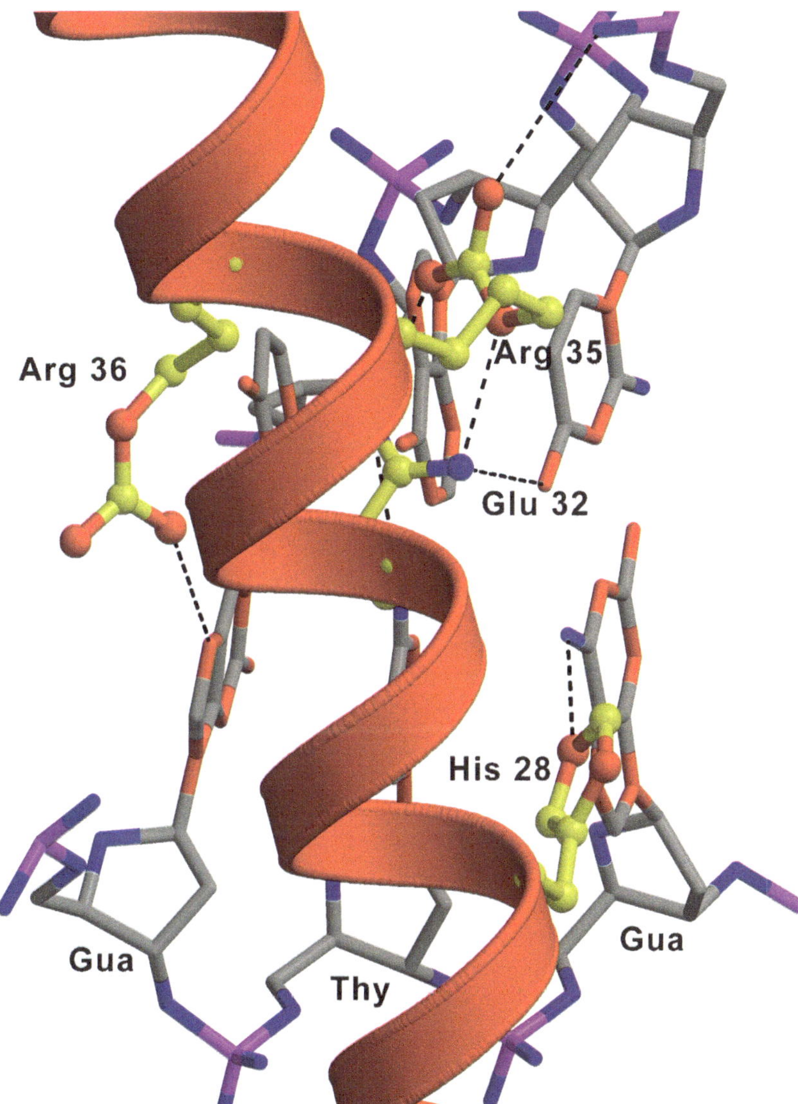

Fig. 4 Ribbon diagram summarizing the DNA contacts made by the basic region of Max. Equivalent contacts are observed with the basic regions of both Myc and Mad. For clarity, numbering derived from the Max bHLHZ domain has been used. The view is perpendicular to the α-helical axis of the basic region and towards the DNA major groove

Glu-32 participates in hydrogen bonds with N4 of Cyt(3) and N6 of Ade(2), and Arg-36 interacts with N7 of Gua(1′). The hydrogen bond between His-28 and N7 of Gua(3′) dictates specificity for a purine base at this position. An additional interaction between Glu-32 and N4 of Cyt(3) further dictates that His-28 and Glu-32 recognize a G:C base pair (Ferre-D'Amare et al. 1993). The sidechain of Glu-32 is oriented relative to the DNA by a hydrogen bond with Arg-35 (Fig. 4). The corresponding Arg→Lys mutation in the mouse bHLHZ transcription factor *mi* results in small eyes and osteoporosis in the heterozygote, thus underscoring the importance of this Arg residue in bHLHZ-DNA interactions (Steingrimsson et al. 1994).

4.2
Class A Vs Class B bHLHZ Proteins

Proteins of the bHLH (similar in structure but lacking the leucine zipper) or bHLHZ families have historically been divided into two classes according to their DNA binding preferences (Blackwell et al. 1993). Class B bHLHZ proteins recognize the central 5′-CG-3′ dinucleotide of the 5′-CA**CG**TG-3′ hexanucleotide. The specific interaction between Arg-36 (Max numbering) and the purine N7, as seen in the co-crystal structures of the Max homodimer (Ferre-D'Amare et al. 1993) and Myc–Max heterodimer (Nair and Burley 2003) structures, almost certainly dictates the sequence preference for class B bHLHZ proteins. Class A bHLHZ proteins recognize 5′-CA**GC**TG-3′ E-boxes. Sequence comparisons between class A and class B bHLHZ proteins suggest that the preference of class A proteins is due to a hydrophobic residue in place of the conserved arginine at position 36. For example, a single amino acid substitution Arg36→Met suffices to convert some class B proteins into class A (Dang et al. 1992). However, co-crystal structures of the class A proteins E47 (Ellenberger et al. 1994) and MyoD (Ma et al. 1994) show that the corresponding valine or leucine are far from the innermost base pair and do not interact directly with DNA. Thus, the binding specificity of class A proteins cannot be explained in terms of direct sidechain-base contacts in the major groove. It is likely that sequence preference differences between class A and class B bHLHZ proteins involve sequence-dependent DNA deformations and/or solvent-mediated effects. Regrettably, the moderate resolution limits (2.8 Å–2.9 Å) of both the E47 (Ellenberger et al. 1994) and MyoD (Ma et al. 1994) co-crystal structures preclude more rigorous examination of this phenomenon.

4.3
Mad–Max Interactions with DNA

Protein–DNA contacts supported by the basic regions of Myc and Mad are essentially identical to those observed for Max with specificity dictated by residues His-359, Glu-363, and Arg-367 in Myc and His-61, Glu-65, and Arg-69 of Mad. In the Myc–Max heterodimer co-crystal structure, several additional contacts are observed between residues specific to Myc and the phosphate backbone of DNA (Nair and Burley 2003). It is possible that these Myc-specific contacts result in differing affinities between the Myc–Max heterodimer and the Max homodimer for the same 5′-CACGTG-3′ element, but this assertion has not been experimentally confirmed.

4.4
The Loop Region Interacts with DNA

The loop regions connecting helices H1 and H2 vary in sequence, amino acid composition, and length among various members of the Myc family. The loop regions lack sequence conservation, with the notable exception of a lysine residue at position 57 in Max (Lys-389 in Myc; Arg-91 in Mad). In the Max homodimer co-crystal structure, Lys-57 interacts with the DNA phosphate backbone. This interaction is conserved in the structure of the Myc–Max heterodimer in which Lys-389 of Myc also makes similar, presumably nonspecific, contacts with the DNA backbone. Loop-deletion studies of MyoD and DNA affinity studies with synthetic bHLH peptides showed that loop residues contribute to DNA binding. Winston and Gottesfeld estimated a roughly 1.3-kcal/mol contribution to DNA binding by an equivalent lysine residue (Lys-80) of the bHLH protein Deadpan (Winston and Gottesfeld 2000). Binding studies of wild-type and mutant Deadpan bHLH with the major groove binding pyrrole-imidazole polyamides further established that Lys-80 contributes to DNA recognition, via interactions with nucleotides outside the core binding element. Contacts between loop residues and the DNA backbone may represent a mechanism for extending DNA binding selectivity to bases that flank the 5′-CACGTG-3′ core element (Nair and Burley 2000).

5
The Bivalent Myc–Max Heterotetramer

In the Myc–Max co-crystal structure, two Myc–Max/DNA complexes constituting the crystallographic asymmetric unit align in a head-to-tail assembly of

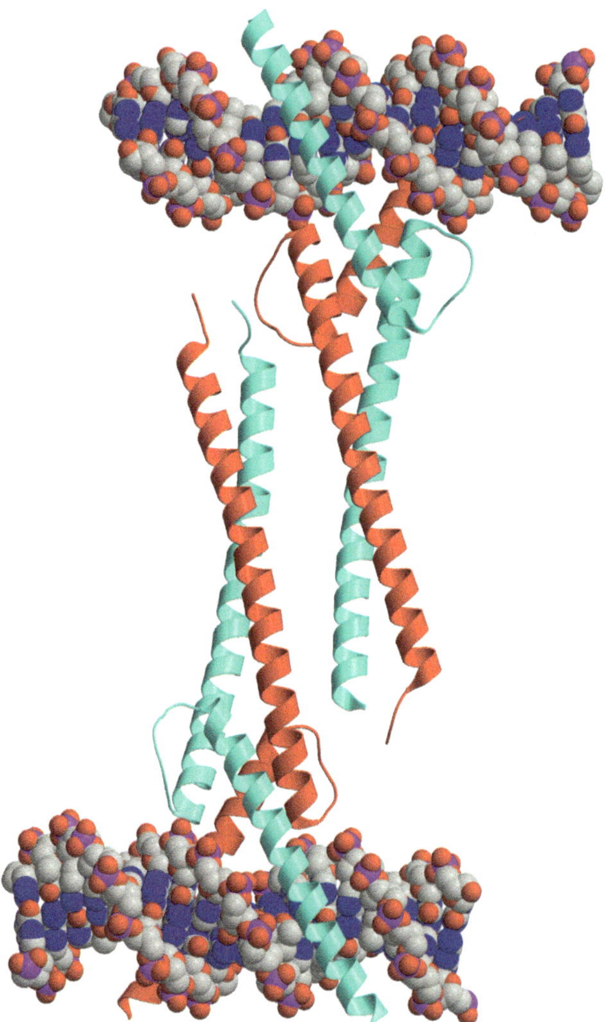

Fig. 5 Ribbon diagram of the bivalent Myc–Max heterotetramer observed in the Myc–Max/DNA co-crystals (Max, *red*; Myc, *cyan*). This head-to-tail assembly of individual leucine zippers of each heterodimer results in the formation of an anti-parallel four-helix bundle

the leucine zippers of each heterodimer, generating an antiparallel four-helix bundle (Nair and Burley 2003; Fig. 5). This four-helix bundle is topologically similar to α-helical bundles observed in members of the cytokine family and in leukemia inhibitory protein (Hill et al. 1993; Somers et al. 1997).

Previously published in vivo and in vitro studies have shown that Myc–Max heterodimers can form higher order oligomers. Solution studies by Dang and co-workers demonstrated that Myc–Max is capable of forming bivalent heterotetramers, and that tetramerization depends on Myc leucine zipper region (Dang et al. 1989). The physiological relevance of the bivalent heterotetramer observed in the Myc–Max co-crystals is supported by solution experiments that demonstrated Myc–Max tetramerization at submicromolar concentrations and analytical ultracentrifugation studies which yielded a tetramer–dimer equilibrium dissociation constant of approximately 90 nM (Nair and Burley 2003). Given that the measured dissociation constant of the Myc–Max tetramer is lower than estimates of physiologic c-Myc concentrations (Moore et al. 1987; Rudolph et al. 1999), these findings document that c-Myc–Max almost certainly exists as a bivalent heterotetramer in cell nuclei.

The biological relevance of the bivalent Myc–Max heterotetramer is borne out by a wealth of genetic and biochemical data. Genetic characterization of the promoters of putative *myc*-regulated genes has provided further evidence for a physiological role for Myc–Max heterotetramerization. Oligonucleotide microarray analysis has identified several Myc target genes that contain multiple E-boxes within promoters, typically separated by at least 100 nucleotides (Coller et al. 2000; see also Grandori and Eisenman 1997). Given the persistence length of DNA, this separation of Myc–Max binding sites is compatible with DNA looping stabilized by bivalent Myc–Max heterotetramers simultaneously bound to two cognate sequences.

An extensive network of hydrogen bonds and salt bridges mediates the protein–protein interface stabilizing the Myc–Max heterotetramer. Residues that are part of this polar interaction network are unique to the Myc–Max heterodimer. It is remarkable that the polarity of many of the residues that make up the interaction network in the Myc–Max heterotetramer in Myc is altered in Mad. This alteration in polarity of residues that stabilize the interaction of the Myc–Max heterotetramer may explain the lack of tetramer formation by Mad–Max heterodimers both in solution and in the co-crystal structure (Nair and Burley 2003).

It is possible that assembly of Myc–Max into bivalent heterotetramers allows for cooperative regulation at promoters and enhancers containing multiple E-boxes. In vitro site selection experiments and chromatin immunoprecipitation studies have documented that Myc–Max heterodimers can bind to sequences that differ from the canonical E-box (5′-CACGTG-3′) hexanucleotide (Blackwell et al. 1993; Grandori et al. 1996). These sequences are not bound with equal affinities. For example, the noncanonical sequences 5′-CACGCG-3′ and 5′-CATGCG-3′ represent low-affinity Myc–Max binding sites (nucleotides that differ from the E-box hexanucleotide are shown in bold).

Given the conservation of amino acids within Myc family proteins, that make direct DNA contacts, this difference in binding affinities of noncanonical sequences is somewhat unexpected. The bivalent heterotetramer observed in the Myc–Max co-crystal structure suggests that cooperative binding may increase the affinity of Myc–Max heterodimers for such noncanonical sites (Walhout et al. 1997). However, this assertion has yet to be validated experimentally.

6 Determinants of Homodimerization Vs Heterodimerization

The bHLHZ segments of Myc, Max, and Mad contain two different dimerization interfaces: the bHLH domain and the leucine zipper domain. Extensive hydrophobic and polar interactions between both of these interfaces stabilize the Max homodimer structure (Ferre-D'Amare et al. 1993) and the quasi-symmetric Myc–Max and Mad–Max heterodimer structures (Nair and Burley 2003). Much of the left-handed coiled-coil that the leucine zipper comprises resembles the structure of canonical leucine zippers, such as the GCN4 homodimer (O'Shea et al. 1991). However, within the Max homodimer structure, a Gln-91–Asn-92–Gln-91–Asn-92 tetrad occurs at the carboxy-terminal end of the zipper region (Ferre-D'Amare et al. 1993). This non-ideal packing scheme results in a flaring of the leucine zipper in the vicinity of the Gln–Asn tetrad.

In contrast, the leucine zipper regions of both the Myc–Max and Mad–Max heterodimers closely resemble the coiled coils found in GCN4 homodimers. The co-crystal structures of both bHLHZ heterodimers demonstrate that the packing defects introduced by the Gln–Asn pairing in Max are compensated for by complementary hydrogen bond interactions with two positively charged Arg–Arg residues located at this position in Myc. Hydrogen bonding between the Max Gln–Asn pair and a Gln–Glu pair at the equivalent position in Mad also results in close packing within the leucine zipper. Mutational analyses documented that residues at these two positions mediate the specificity and avidity for homo- verses heterodimerization within the Myc/Max/Mad network of proteins (Nair and Burley 2003). The packing defects observed in the Max homodimer have been compensated in both Myc–Max and Mad–Max heterodimers. Hence, energetic considerations would suggest that the likely in vivo state for Max polypeptides would be as an obligate *hetero*dimeric species with Myc/Mad.

7
The bHLHZ Domain as an Architectural Scaffold

Work from a number of laboratories has shown that Myc–Max can recruit various cellular factors, such as the zinc-finger protein Miz-1 (Peukert et al. 1997) and the F-box E3 ubiquitin ligase Skp2 (Kim et al. 2003; von der Lehr et al. 2003). Each of these higher order complexes forms as a result of specific interactions with the bHLHZ region of Myc. Given that these proteins are recruited to specific regions of the promoter only in the context of Myc–Max heterodimers, it seems reasonable to suggest that the bHLHZ regions of the Myc–Max heterodimer play an architectural role. Formation of the bivalent heterotetramer observed in the Myc–Max co-crystal structure would provide a substantial platform for recruitment of additional protein factors.

Miz-1 (see chapter by D. Kleine-Kohlbrecher et. al.) encodes a protein of 803 amino acids, bearing 13 putative zinc-finger motifs, which recruits Myc bHLHZ to the core promoter elements of the *P21CIP1* and *P15INK4B* genes (Seoane et al. 2002; Staller et al. 2001; Herold et al. 2002). The interaction between Myc bHLHZ and Miz represses transactivation through competition with the histone acetyltransferase p300 for binding to Miz-1 (Staller et al. 2001). Two-hybrid interaction studies using random mutants of the Myc bHLHZ domain identified several point mutants that retain the ability to heterodimerize Max but do not support interactions with Miz-1. These point mutants of Myc do not repress transcription of *P21CIP1* genes in vivo, thereby demonstrating that residues unique to the bHLHZ domain of Myc support Miz-1-mediated transcriptional repression.

Myc is a target for ubiquitin-mediated proteolysis, and ubiquitination of Myc results in rapid destruction within minutes of Myc synthesis (Salghetti et al. 2001). Thus, turnover plays a fundamental role in the function of Myc and deregulation of this event leads to the onset and development of oncogenic transformations. Recently, two groups independently identified the ubiquitin ligase Skp2 as both a mediator of Myc turnover and a potent stimulator of Myc transcription (Kim et al. 2003; von der Lehr et al 2003). The Skp2 interacting regions have been delimited to two distinct sequences within the Myc polypeptide. The first of these consists of a region within the Myc amino-terminal transactivation domain and the second Skp2 interacting regions consists of the Myc bHLHZ domain. These studies demonstrate that Skp2 is a co-activator of Myc function, and Myc acts to recruit this co-activator activity to target promoters, in part through the bHLHZ domain (Kim et al. 2003; von der Lehr et al 2003).

The assertion that the Myc–Max tetramer is of biological relevance is also borne out by experiments utilizing bHLHZ domain chimeras (Staller et al. 2001; O'Hagan et al. 2000; James and Eisenman 2002). Several laboratories have constructed such chimeric proteins in which the transactivation domain from Myc is attached to the bHLHZ domain from Mad. Given the conservation of protein–DNA contacts observed in the co-crystal structures of both Myc–Max and Mad–Max heterodimers, such chimeric proteins would be expected to have biological activities similar to those of wild-type Myc. While these Myc/Mad–bHLHZ chimeras can activate E-box dependent transcription, clear differences from the behavior of wild-type Myc are observed. Thus, the bHLHZ domain of Myc supports unique aspects of Myc function. It is possible that the ability of Myc–Max bHLHZ heterodimers (and *only* Myc–Max heterodimers) to form higher order tetramers reflects, at least in part, unique properties of Myc.

8
Conclusions

The structures of several Myc family multiprotein and protein–DNA complexes determined over the past few years have offered a number of insights into the biological functions of Myc/Mad/Max. The structure of the Mad SID–Sin3 PAH complex reveals how a small four-helical domain can mediate selective recruitment of a peptide through mutual induction of disorder-to-order structural transitions. Given the unstructured nature of the activation domains in general, this principle may play a role in recognition by the Myc and Mad transactivation domains.

The co-crystal structures of the Myc–Max and Mad–Max heterodimers recognizing their E-box targets demonstrate how bHLHZ heterodimers mediate specific, high-affinity DNA binding. Tetramerization of Myc and Max is mediated by extensive protein–protein interactions between leucine zipper domains, and the resulting antiparallel four-helix bundle could provide a scaffold for recruitment of additional modulators of transcription. Several of the features observed in these structures are consistent with the biology of Myc family proteins and thus serve as a starting point for further directed biochemical and genetic studies to elucidate the roles played by Myc–Max and Mad–Max in cell-fate determination.

Acknowledgements We thank members of the Burley and Nair laboratories for helpful discussions. We apologize to researchers whose papers could not be cited due to space limitations. Work from the authors' laboratory described in this review was supported by the Howard Hughes Medical Institute and the NIH (SKB) and the Leukemia Society of America (SKN).

References

Amati B, Dalton S, Brooks MW, Littlewood TD, Evan GI, Land H (1992) Transcriptional activation by the human c-Myc oncoprotein in yeast requires interaction with Max. Nature 359:423–426

Amati B, Littlewood TD, Evan GI, Land H (1993) The c-Myc protein induces cell cycle progression and apoptosis through dimerization with Max. EMBO J 12:5083–5087

Ayer DE, Kretzner L, Eisenman RN (1993) Mad: a heterodimeric partner for max that antagonizes myc transcriptional activity. Cell 72:211–222

Ayer DE, Lawrence QA, Eisenman RN (1995) Mad–Max transcriptional repression is mediated by ternary complex formation with mammalian homologs of yeast repressor Sin3. Cell 80:767–776

Benvenisty N, Leder A, Kuo A, Leder P (1992) An embryonically expressed gene is a target for c-Myc regulation via the c-Myc binding sequence. Genes Dev 6:2513–2523

Blackwell TK, Huang J, Ma A, Kretzner L, Alt FW, Eisenman RN, Weintraub H (1993) Binding of myc proteins to canonical and noncanonical DNA sequences. Mol Cell Biol 9:5216–5224

Blackwood E, Eisenman RN (1991) Max: A helix-loop-helix zipper protein that forms a sequence-specific DNA-binding complex with Myc. Science 251:1211–1217

Brubaker K, Cowley SM, Huang K, Loo L, Yochum GS, Ayer DE, Eisenman RN, Radhakrishnan I (2000) Solution structure of the interacting domains of the Mad-Sin3 complex: implications for recruitment of a chromatin modifying complex. Cell 103:655–665

Cole MD, McMahon SB (1999) The Myc oncoprotein: a critical evaluation of transactivation and target gene regulation. Oncogene 18:2916–2924

Coller HA, Grandori C, Tamayo P, Colbert T, Lander ES, Eisenman RN, Golub TR (2000) Expression analysis with oligonucleotide microarrays reveals that MYC regulated genes involved in growth, cell cycle, signaling, and adhesions. Proc Natl Acad Sci USA 97:3260–3265

Dalla-Favera R, Bregni M, Erikson J, Patterson D, Gallo RC, Croce CM (1982) Human c-myc onc gene is located on the region of chromosome 8 that is translocated in Burkitt lymphoma cells. Proc Natl Acad Sci U S A 79:7824–7827

Dang CV, McGuire M, Buckmire M, Lee WM (1989) Involvement of the 'leucine zipper' region in the oligomerization and transforming activity of human c myc protein. Nature 337:664–666

Dang CV, Barrett J, Villa-Garcia M, Resar LM, Kato GJ, Fearon ER (1991) Intracellular leucine zipper interactions suggest c-Myc hetero-oligomerization. Mol Cell Biol 11:954–962

Dang CV, Dolde C, Gillison ML, Kato GJ (1992) Discrimination between related DNA sites by a single amino-acid residue of Myc-related basic-helix-loop-helix proteins. Proc Natl Acad Sci USA 89:599–602

Dang CV, Resar LM, Emison E, Kim S, Li Q, Prescott JE, Wonsey D, Zeller K (1999) Function of the c-Myc oncogenic transcription factor. Exp Cell Res 253:63–77

Davis LJ, Halazonetis TD (1993) Both the helix-loop-helix and the leucine zipper motifs of c-Myc contribute to its dimerization specificity with Max. Oncogene 8:125–132

Dyson HJ, Wright PE (2002) Coupling of folding and binding for unstructured proteins. Curr Opin Struct Biol 12:54–60

Eilers AK, Billin AN, Liu J, Ayer DJ (1999) A 13-amino acid ampipathic α-helix is required for the functional interaction between the transcriptional repressor Mad1 and mSin3A. J Biol Chem 274:32750–32756

Eilers M (1999) Control of cell proliferation by Myc family genes. Mol Cells 9:1–6

Eilers M, Schirm S, Bishop JM (1991) The MYC protein activates transcription of the alpha-prothymosin gene. EMBO J 10:133–141

Ellenberger T, Fass D, Arnaud M, Harrison SC (1994) Crystal structure of transcription factor E47: E-box recognition by a basic region helix-loop-helix dimer. Genes Dev 8:970–980

Ferre-D'Amare AR, Prendergast GC, Ziff EB, Burley SK (1993) Recognition by Max of its cognate DNA through a dimeric b/HLH/Z domain. Nature 363:38–45

Freytag SO, Geddes TJ (1992) Reciprocal regulation of adipogenesis by Myc and C/EBP alpha. Science 256:379–382

Grandori C, Eisenman RN (1997) Myc target genes. Trends Biochem Sci 22:177–181

Grandori C, Mac J, Siebelt F, Ayer DE, Eisenman RN (1996) Myc–Max heterodimers activate a DEAD box gene and interact with multiple E box-related sites in vivo. EMBO J 15:4344–4357

Grandori C, Cowley SM, James LP, Eisenman RN (2000) The Myc/Max/Mad network and the transcriptional control of cell behavior. Annu Rev Cell Dev Biol 16:653–699

Haggerty TJ, Zeller KI, Osthus RC, Wonsey DR, Dang CV (2003) A strategy for identifying transcription factor binding sites reveals two classes of genomic c-Myc target sites. Proc Natl Acad Sci USA 100:5313–5318

Hassig CA, Fleischer TC, Billin AN, Schreiber SL, Ayer DE (1997) Histone deacetylase activity is required for full transcriptional repression by mSin3A. Cell 89:341–347

Herold S, Wanzel M, Beuger V, Frohme C, Beul D, Hillukkala T, Syvaoja J, Saluz H-P, Haenel F, Eilers M (2002) Negative regulation of the mammalian UV response by Myc through association with Miz-1. Mol Cell 10:509–521

Hill CP, Osslund TD, Eisenberg D (1993) The structure of granulocyte-colony-stimulating factor and its relationship to other growth factors. Proc Natl Acad Sci USA 90:5167–5171

Hurlin PJ, Ayer DE, Grandori C, Eisenman RN (1994) The Max transcription factor network: involvement of Mad in differentiation and an approach to identification of target genes. Cold Spring Harb Symp Quant Biol 59:109–116

Hurlin PJ, Queva C, Kokinen PJ, Steingrimsson E, Ayer DE, et al (1995) Mad3 and Mad4: novel Max-interacting transcriptional repressors that suppress c-Myc-dependent transformation and are expressed during neural and epidermal differentiation. EMBO J 14:5646–5659

James L, Eisenman RN (2002) Myc and Mad bHLHZ domains possess identical DNA-binding specificities but only partial overlapping functions in vivo. Proc Natl Acad Sci USA 99:10429–10434

Kato GJ, Barrett J, Villa-Garcia M, Dang CV (1990) An amino terminal c-Myc domain required for neoplastic transformation activates transcription. Mol Cell Biol 16:4215–4221

Kim SY, Herbst A, Tworkowski KA, Slghetti SE, Tansey WP (2003) Skp2 regulates Myc protein stability and activity. Mol Cell 11:1177–1188

Kohl NE, Kanda N, Schreck RR, Bruns G, Latt SA, Gilbert F, Alt FW (1983) Transposition and amplification of oncogene-related sequences in human neuroblastomas. Cell 35:359–367

Kretzner L, Blackwood EM, Eisenman RN (1992) Transcriptional activities of the Myc and Max proteins in mammalian cells. Curr Top Microbiol Immunol 182:435–443

Kwon TK, Nagel JE, Buchholz MA, Nordin AA (1996) Characterization of the murine cyclin-dependent kinase inhibitor gene p27Kip1. Gene 180:113–120

Laherty CD, Yang WM, Sun JM, Davie JR, Seto E, Eisenman RN (1997) Histone deacetylases associated with the mSin3 corepressor mediate mad transcriptional repression. Cell 89:349–356

Larsson LG, Pettersson M, Oberg F, Nilsson K, Luscher B (1994) Expression of mad, mxi1, max and c-myc during induced differentiation of hematopoietic cells: opposite regulation of mad and c-myc. Oncogene 9:1247–1252

Larsson LG, Bahram F, Burkhardt H, Luscher B (1997) Analysis of the DNA-binding activities of Myc/Max/Mad network complexes during induced differentiation of U-937 monoblasts and F9 teratocarcinoma cells. Oncogene 15:737–748

Li L, Nerlov K, Prendergast G, MacGregor D, Ziff EB (1994) c-Myc represses transcription in vivo by a novel mechanism dependent on the initiator element and Myc box II. EMBO J 13:4070–4079

Liao DJ, Dickson RB (2000) c-Myc in breast cancer. Endocr Relat Cancer 7:143–164

Lo K, Smale ST (1996) Generality of a functional initiator consensus sequence. Gene 182:13–22

Ma PC, Rould MA, Weintraub H, Pabo CO (1994) Crystal structure of MyoD bHLH domain-DNA complex: perspectives on DNA recognition and implications for transcriptional activation. Cell 77:451–459

McArthur GA, Laherty CD, Queva C, Hurlin PJ, Loo L, James L, Grandori C, Gallant P, Shiio Y, Hokanson WC, et al (1998) The Mad protein family links transcriptional repression to cell differentiation. Cold Spring Harb Symp Quant Biol 63:423–433

McMahon SB, Wood MA, Cole MD (2000) The essential cofactor TRRAP recruits the histone acetyltransferase hGCN5 to c-Myc. Mol Cell Biol 20:556–562

Moore JP, Hancock DC, Littlewood TD, Evan GI (1997) A sensitive and quantitative enzyme-linked immunosorbence assay for the c-Myc and n-myc oncoproteins. Oncogene Res 2:65–80

Murre C, McCaw PS, Baltimore D (1989) A new DNA binding and dimerization motif in immunoglobulin enhancer binding, daughterless, MyoD, and myc proteins. Cell 56:777–783

Nair SK, Burley SK (2000) Recognizing DNA in the library. Nature 404:717–718

Nair SK, Burley SK (2003) X-ray structures of Myc–Max and Mad–Max recognizing DNA: molecular bases of regulation by proto-oncogenic transcription factors. Cell 112:193–205

Nau MM, Brooks BJ, Battey J, Sausville E, Gazdar AF, Kirsch IR, McBride OW, Bertness V, Hollis GF, Minna JD (1985) L-Myc, a new Myc-related gene amplified and expressed in human small cell lung cancer. Nature 318:69–73

Nesbit CE, Tersak JM, Prochownik EV (1999) MYC oncogenes and human neoplastic disease. Oncogene 18:3004–3016

O'Hagan RC, Schreiber-Agus N, Chen K, David G, Engelman JA, Schwab R, Alland L, Thomson C, Ronning DR, Sacchettini JC, et al (2000) Gene-target recognition among members of the myc superfamily and implications for oncogenesis. Nat Genet 24:113–119

O'Shea EK, Klemm JD, Kim PS, Alber T (1991) X-ray structure of the GCN4 leucine zipper, a two-stranded, parallel coiled coil. Science 254:539–544

Peukert K, Staller P, Schneider A, Carmichael G, Hanel F, Eilers M (1997) An alternative pathway for gene regulation by Myc. EMBO J 16:5672–5686

Philipp A, Schneider A, Väsrik I, Finke K, Yiong Y, Beach D, Alitalo K, Eilers M (1994) Repression of cyclin D1: a novel function of MYC. Mol Cell Biol 14:4032–4043

Prendergast GC, Lawe D, Ziff EB (1991) Association of Myn, the murine homolog of Max with c-Myc stiumulates methylation-sensitive DNA binding and Ras cotransformation. Cell 65:395–407

Rudolph C, Adam G, Simm A (1999) Determination of copy number of c-Myc protein per cell by quantitative Western blotting. Anal Biochem 269:66–71

Saleh A, Schieltz D, Ting N, McMahon S, Lichfield DW, Yates JR, Lees-Miller SP, Cole MD, Brandl CJ (1998) Tra1p is a component of the yeast Ada-Spt transcriptional regulatory complexes. J Biol Chem 273:26559–26565

Salghetti SE, Caudy AA, Chenoweth JG, Tansey WP (2003) Regulation of transcriptional activation domain function by ubiquitin. Science 293:1651–1653

Schreiber-Agus N, Chin L, Chen K, Torres R, Rao G, Guida P, Skoultchi AI, DePinho RA (1995) An amino-terminal domain of Mxi1 mediates anti-Myc oncogenic activity and interacts with a homolog of the yeast transcriptional repressor SIN3. Cell 80:777–786

Schwab M, Varmus HE, Bishop JM, Grzeschik KH, Naylor SL, Sakaguchi AY, Brodeur G, Trent J (1984) Chromosome localization in normal human cells and neuroblastomas of a gene related to c-Myc. Nature 308:288–291

Seoane J, Pouponnot C, Staller P, Schader M, Eilers M, Massague J (2001) TGFβ influences Myc, Miz-1 and Smad to control the CDK inhibitor p15INK4b. Nat Cell Biol 3:400–408

Seoane J, Le HV, Massague J (2002) Myc suppression of the p21(Cip1) Cdk inhibitor influences the outcome of the p53 response to DNA damage. Nature 419:729–734

Sheiness D, Fanshier L, Bishop JM (1978) Identification of nucleotide sequences which may encode the oncogenic capacity of avian retrovirus MC29. J Virol 28:600–610

Somers W, Stahl M, Seehra JS (1997) 1.9 Å crystal structure of interleukin 6: implications for a novel mode of receptor dimerization and signaling. EMBO J 16:989–997

Spronk CA, Tessari M, Kaan AM, Jansen JF, Vermeulen M, Stunnenberg HG, Vuister GW (2000) The Mad1-Sin3B interaction involves a novel helical fold. Nat Struct Biol 7:1100–1104

Staller P, Peukert K, Kiermaier A, Seoane J, Lukas J, Karsunky H, Möröy T, Bartek J, Massague J, Hänel Eilers M (2001) Repression of p15INK4b expression by Myc through association with Miz-1. Nat Cell Biol 3:392–399

Steingrimsson E, Moore KJ, Lamoreux ML, Ferre-D'Amare AR, Burley SK, Zimring DC, Skow LC, Hodgkinson CA, Arnheiter H, Copeland NG, et al (1994) Molecular basis of mouse microphthalmia (mi) mutations helps explain their developmental and phenotypic consequences. Nat Genet 8:256–263

Taub R, Kirsch I, Morton C, Lenoir G, Swan D, Tronick S, Aaronson S, Leder P (1982) Translocation of the c-myc gene into the immunoglobulin heavy chain locus in human Burkitt lymphoma and murine plasmacytoma cells. Proc Natl Acad Sci U S A 79:7837–7841

Von der Lehr N, Johansson S, Wu S, Bahram F, Castell A, Cetinkaya C, Hydbring P, Weidung I, Nakayama K, Nakayama KI, Söderberg O, Kerppola T, Larsson L-G (2003) The F-box protein Skp2 participates in c-Myc proteosomal degradation and as a cofactor for c-Myc regulated transcription. Mol Cell 11:1189–1200

Walhout AJM, Gubbels JM, Bernards R, van der Vliet PC, Timmers HTM (1997) c-Myc/Max heterodimers bind cooperatively to the E-box sequences located in the first intron of the rat ornithine decarboxylase (ODC) gene. Nucleic Acids Res 25:1493–1501

Winston RL, Gottesfeld JM (2000) Rapid identification of key amino-acid-DNA contacts through combinatorial peptide synthesis. Chem Biol 7:245–251

Yang W, Shen J, Wu M, Arsura M, FitzGerald M, Suldan Z, Kim DW, Hofmann CS, Pianetti S, Romieu-Mourez R, Freedman LP, Sonenshein GE (2001) Repression of transcription of the p27(Kip1) cyclin-dependent kinase inhibitor gene by c-Myc. Oncogene 20:1688–1702

Zervos AS, Gyuris J, Brent R (1993) Mxi1, a protein that specifically interacts with Max to bind Myc–Max recognition sites. Cell 72:223–232

CTMI (2006) 302:145–167

Myc Target Transcriptomes

L. A. Lee · C. V. Dang (✉)

Department of Medicine, The Johns Hopkins University School of Medicine, Ross 1032, 720 Rutland Avenue, Baltimore, MD 21205, USA
cvdang@jhmi.edu, llee12@jhmi.edu

Abstract The c-Myc oncogenic transcription factor plays a central role in many human cancers through the regulation of gene expression. Although the molecular mechanisms by which c-Myc and its obligate partner, Max, regulate gene expression are becoming better defined, genes or transcriptomes that c-Myc regulate are just emerging from a variety of different experimental approaches. Studies of individual c-Myc target genes and their functional implications are now complemented by large surveys of c-Myc target genes through the use of subtraction cloning, DNA microarray analysis, serial analysis of gene expression (SAGE), chromatin immunoprecipitation, and genome marking methods. To fully appreciate the differences between physiological c-Myc function in normal cells and deregulated c-Myc function in tumors, the challenge now is to determine how the authenticated transcriptomes effect the various phenotypes induced by c-Myc and to define how c-Myc transcriptomes are altered by the Mad family of proteins.

1 Introduction

Despite its initial vague functional definition as an oncoprotein—involved in DNA replication, RNA splicing, or transcription—c-Myc has emerged foremost as a transcription factor. c-Myc dimerizes with Max and binds DNA through its C-terminal basic-helix-loop-helix (bHLH) leucine zipper (LZip) domain and regulates transcription through an N-terminal transcriptional regulatory domain. With transcriptional regulation as its acknowledged function, the search for physiological and pathological c-Myc target genes has intensified over the past decade. Searches for target genes have involved hypothesis-driven, low-throughput studies of candidate c-Myc target genes as well as medium-throughput studies to define a larger repertoire of c-Myc responsive genes through subtraction cloning methods. More recently, the field has rapidly adopted high-throughput technologies for the discovery of c-Myc responsive transcriptomes. Despite impressive advances, major milestones still must be met to achieve a complete understanding of c-Myc responsive transcriptomes and the role of c-Myc in the genesis of human cancers.

2 Approaches to Identify c-Myc Target Genes

To understand fully the network of target genes regulated by c-Myc, it is critical to determine whether c-Myc responsive genes are directly bound by c-Myc or whether the responsive genes are secondary events that require the activities of the direct target genes. Direct target genes are genes that are bound by c-Myc and respond to changes in c-Myc levels or c-Myc activity.

Until factors that alter the activity of c-Myc protein are better defined, most current models to study c-Myc target genes rely on responses to changes in c-Myc protein levels. For example, in the serum starvation and re-stimulation model, rapid activation of *MYC* as an early response gene elevates c-Myc protein levels soon after serum induction. The elevation of c-Myc protein levels causes c-Myc to heterodimerize with Max, which allows the heterodimer to bind specific DNA sequences or E-boxes. Promoter–reporter assays have been used extensively in the past as supporting evidence for direct target genes; however, these artificial constructs are unable to reflect the chromatin context of target genes. The same criticism applies to electromobility gel shift assays for protein–DNA binding.

The inducible Myc estrogen receptor fusion protein (MycER) system has proved to be yet another important model for the study of c-Myc target genes

(Eilers et al. 1989). The expressed chimeric MycER protein is constitutively bound to the chaperone HSP90 in the cytoplasm. Upon exposure to estrogenic compounds, such as 4-OH-tamoxifen, the chimeric protein changes conformation, disengages from the chaperone, and translocates into the nucleus. The MycER protein recognizes c-Myc target sites and initiates transcription of target genes without requiring newly synthesized proteins. Because a direct c-Myc target is one induced by c-Myc alone in the absence of new protein synthesis, genes responding to ligand-stimulated MycER in the presence of cycloheximide are considered direct target genes in this system. While the MycER system has been a powerful one for the validation of direct target genes, the effects of the estrogenic ligands or cycloheximide alone on endogenous gene expression in the absence of MycER has confounded the evaluation of certain genes (O'Connell et al. 2003). Moreover, the MycER system is unable to detect genes that require both Myc and another transcriptional factor that Myc induces directly. Precedent for this type of transcriptional circuitry is provided by the regulation of myeloperoxidase by both C/EBP-α and PU.1, which itself is a target of C/EBP-α (Liu et al. 2003).

A third important technique is chromatin immunoprecipitation (ChIP), which has advanced our knowledge of the association of transcription factors with in vivo cognate genomic sites. To date, ChIP is the only method that provides direct physical evidence of the association of a transcription factor with a specific target gene. With ChIP, target genes in sheared chromatin are crosslinked to specific transcription factors that are subsequently immunoprecipitated with an antibody directed toward that transcription factor. After reversing the chemical crosslinks, the deproteinized DNA may then be assayed through PCR or hybridized to microarrays for specific genomic sequences that are precipitated along with the transcription factor in question.

While immunoprecipitation provides the most direct physical evidence of the association of a transcription factor with target genomic sites, the shearing forces that are usually selected to produce DNA fragments range from several hundred base pairs to about 1 kb, and thus limit the resolution of this technique. More detailed localization of c-Myc binding within these DNA fragments relies on the central hypothesis that c-Myc prefers binding to E-boxes over other non-canonical sites (Mao et al. 2003). Moreover, potential nonspecific binding in ChIP experiments must also be carefully considered when evaluating potential target genes. For example, in a report on a large-scale ChIP study to identify direct c-Myc genomic sites, Amati and co-workers demonstrated a potential for less-specific binding of c-Myc to target genomic sequences when c-Myc protein level is highly elevated (Fernandez et al. 2003). In particular, they showed that nonspecific binding in this assay resulted in ChIP signals that are 0.03% or less of total input DNA for a specific pair of PCR primers. When c-Myc was overexpressed in a human B cell line, they observed that 62% of random (non-promoter) E-boxes and 88% of non-E-box promoters were recovered with c-Myc immunoprecipitation. While it was interpreted from these findings that overexpression of c-Myc leads to widespread association with sites not found in other cells with lower c-Myc expression, it remains unclear as to the signal-to-noise ratio in these experiments when one subjects highly overexpressing cells to the same crosslinking procedure. There are also no appropriate negative controls readily identified for ChIP experiments. The chemical crosslinking by formaldehyde does not take in account the off-rate of c-Myc from its association with chromatin; hence, nonspecific association of c-Myc to DNA via the nonspecific DNA binding domain would not be distinguished from specific binding via the bHLH region. In these same cells, it would be instructive to determine whether the association of another transcription factor to its target site is altered by the overabundance of c-Myc protein. In this system, transcriptional regulation is further complicated by the observation that the binding of E-boxes by c-Myc is not always associated with changes in histone acetylation.

A recent application of ChIP to pinpoint the sites of association of c-Myc among multiple potential genomic sites exploits the ability to scan the specific genes by PCR using the immunoprecipitated DNA product. This approach, termed scanning ChIP (SChIP), demonstrates the resolution of ChIP to be in the neighborhood of about 1 kb due to the distribution of sheared DNA lengths around a specific binding region or site (Zeller et al. 2001). Despite the limited resolution of this method, with SChIP each gene serves as its own control through the use of PCR primer pairs throughout the gene. An instructive observation came from the study of two canonical c-Myc E-boxes in the *NPM1* (B23, nucleophosmin) intron 1 in comparison to a canonical E-box in intron 4. While E-boxes in both intron 1 and intron 4 bound c-Myc–Max equally in gel shift assays, SChIP only detected the tandem intron 1 E-box region as being bound by c-Myc. This emphatically indicates that gel shift assays do not reflect accessibility of chromatin in situ. Furthermore, the tandem intron 1 E-box is phylogenetically highly conserved as compared with the intron 4 E-box that is not conserved. An application of phylogenetic footprinting together with ChIP results suggest that there at least two types of c-Myc genomic E-boxes, those that are highly conserved versus those, such as in cyclin B1, that are not evolutionarily conserved (Haggerty et al. 2003). It is likely that non-canonical c-Myc E-boxes that are evolutionarily conserved will prove to be another category of c-Myc genomic sites that would be missed via searches for the canonical 5′-CACGTG-3′ sequence.

Finally, another approach to identify c-Myc binding sites used Dam-methylase fused to *dMyc* in *Drosophila*. This study of genomic DNA has provided the much-needed evidence for the direct association of *dMyc* with its target genes (Orian et al. 2003). This method tethers the transcription factor with a bacterial DNA methylase, leading to DNA methylation within 1.5 to 2.0 kb of binding and subsequent mapping of DNA binding sites. Full-length dMyc, dMax, and dMnt proteins were fused to the bacterial methylase, and each fusion was expressed separately in *Drosophila* Kc cells. Genomic DNA was then digested at newly methylated sites, labeled, and hybridized to *Drosophila* complementary (c)DNA arrays containing 6,255 cDNA and expressed sequence tags (ESTs). The three proteins bound a total of 968 unique binding sites representing genes of diverse biological function and allowed an assessment of preferential binding sites of each transcription factor, which will be discussed in the following section.

3 Myc Target Transcriptomes

How can we evaluate the functional significance of specific c-Myc target genes or target transcriptomes? The emergence of c-Myc responsive genes from a variety of studies allows for the identification of c-Myc responsive genes that appear recurrently in different cell types, systems, and species (www.myccancergene.org). In addition, the use of ChIP has further identified direct c-Myc targets among the genes that appear to respond to c-Myc regardless of the cell type or species of origin.

Yet there are intriguing differences between the repertoires of c-Myc responsive genes identified from different studies. Some of the variation may be attributed to cell type-specific activation of specific sets of genes in response to c-Myc. For example, while c-Myc represses genes involved in cell–cell interaction or extracellular matrix, the specific type of collagen or integrin affected by c-Myc is dependent on the specific cell type. Despite microarrays featuring oligonucleotides or cDNAs representing thousands of different genes, only a small fraction of genes appear to be regulated by c-Myc independent of cell type or species (Table 1). It stands to reason that the differences among studies may result from differences in the experimental systems as well as the noise inherent in these approaches. Examination of three different studies using the wildtype and myc-null rat fibroblasts reveals that only a few of the targets are common to all three studies (Guo et al. 2000; O'Connell et al. 2003; Watson et al. 2002). Some of these differences may be due to variation in the genes represented on the microarrays that were used by each group. However, some of these differences might also be due to the effects of cell density or the number of culture passages, which are likely to vary among studies. This cur-

Table 1 Microarray studies to identify Myc target genes

	Reference	Approach	Cells	Array	Species	Criteria	Number of genes differentially expressed
1	Orian et al. 2003	Dam ID: DNA methylase fused to dMyc, dMax, or dMnt	*Drosophila* Kc	Custom array 6255 cDNAs	*Drosophila*		dMyc 22 genes (low dMax) dMax 643 genes, dMnt 429 genes dMyc 287 gene (high dMax)
2	Fernandez et al. 2003	Preselected candidate genes	U937 monoblastic leukemia HL60 myeloid leukemia P493 B lymphocytes	Custom array 723 genomic sites with promoter E-boxes 533 loci	Human		257 genes
3	Mao et al. 2003	Anti c-Myc/ anti-Max ChIP	HL60 myeloid leukemia	Custom array 7776 CpG island clones	Human	>3-fold	177 genes
4	Iritani et al. 2002	Mad overexpression	Thymocytes	Affymetrix Mu11 K Sub A & B ~11,000 genes	Mouse	≥2-fold	22 upregulated 55 downregulated
5	O'Hagan et al. 2000	c-Myc and Myc(Mxi-1BR) overexpression	IMR90 cells	Custom array 5272 genes	Human		Myc: 19 genes Myc(Mxi-1BR): 16 genes Overlap: 8 genes
6	Watson et al. 2002	c-Myc reconstitution	HO15.19 Myc null fibroblasts	Synteni 6.3 K 6300 cDNAs	Human		27 upregulated 25 repressed
7	O'Connell et al. 2003	c-Myc reconstitution and overexpression	HO15.19 Myc null fibroblasts	Affymetrix Rat U34 GeneChip 7,000 genes	Rat	>2-fold	45 upregulated (21 direct, 24 indirect) 16 repressed (all indirect)

Table 1 (continued)

	Reference	Approach	Cells	Array	Species	Criteria	Number of genes differentially expressed
8	Yu et al. 2002	DNA damage induced apoptosis	Rat1A fibroblasts myc null fibroblasts	Custom array 6,500 genes	Rat	>2-fold	After exposure to VP-16: rat1-myc 37 genes
9	Guo et al. 2000	c-Myc reconstitution and overexpression	myc null fibroblasts Wild-type fibroblasts	Custom array 4400 genes	Rat	2-fold	147 upregulated 80 repressed
10	Coller et al. 2000	c-Myc overexpression	Primary human Fibroblasts	Custom oligonucleotide array 6,416 genes	Human	>2-fold	27 upregulated 9 repressed
11	Huang et al. 2003	c-Myc overexpression	Mouse embryo fibroblasts	Affymetrix Mu11 K A/B Gene Chip 7706 genes	Mouse		250 genes
12	Menssen and Hermeking 2002	c-Myc overexpression	HVEC	SAGE and IncyteGenomics Human Unigene 1 and Drug Target arrays	Human	>1.7-fold	53 upregulated
13	Boon et al. 2001	N-myc transfected neuroblastoma cells	Neuroblastoma	SAGE	Human		114 upregulated
14	Godfried et al. 2002	N-myc and c-Myc overexpression	SHEP-21N Neuroblastoma cell line	SAGE	Human		No net number given nm23-H1 and nm23-H2
15	Schumacher et al. 2001		P493–6 B lymphocytes	Affymetrix Hu6800 6800 genes	Human	>2-3-fold	82 upregulated **8 downregulated**

Table 1 (continued)

	Reference	Approach	Cells	Array	Species	Criteria	Number of genes differentially expressed
16	Frye et al. 2003	Myc overexpression	Skin/keratinocytes	Affymetrix MGU74A 10,043 genes	Mouse	>2-fold	137 upregulated 81 repressed
17	Schuldiner and Benvenisty 2001	c-MYC and N-MYC tumors and cell lines	Burkitt lymphoma Neuroblastoma	Clontech cDNA Human Atlas 1200 genes	Human		12 genes upregulated in both Burkitt lymphomas and neuroblastomas
18	Kim et al. 2001	c-Myc transgenic mouse	Liver	Clontech cDNA Mouse Atlas 588 genes	Mouse	>2-fold	4 upregulated 2 repressed

Data compiled to 2003

sory inspection illustrates that use of an identical cell line does not guarantee complete reproducibility in identifying target genes. Another difference that may be particularly important when comparing studies are the responses to c-Myc that may be specific to primary cells, as opposed to immortalized cell lines which may have significantly altered cell cycle checkpoints.

With the emergence of many studies using microarray technologies (Table 1), it is necessary to examine the c-Myc responsive genes collectively to appreciate patterns such as cell type-specific c-Myc target genes. Among the 42 genes (Table 2) that appear in three or more studies in the c-Myc target gene database (www.myc-cancer-gene.org), 19 were confirmed in a recent ChIP study (Fernandez et al. 2003) and only two orthologs were found as dMyc targets in *Drosophila* (Orian et al. 2003). These genes are likely to be the core Myc target genes that are independent of cell type, yet they represent only a small fraction of the genes that are predicted to be regulated by c-Myc on the basis of estimates made by several investigators. One group estimates that perhaps 11% of all cellular promoters in the human genome are responsive to c-Myc (Fernandez et al. 2003). A recent ChIP study suggests that Myc globally affects transcription, with only about one quarter of the genes bound to Myc having canonical Myc binding sites within several kilobases of the transcriptional start sites. This study estimates that more than 15% of promoters are affected by Myc (Li et al. 2003). Similar estimates have been made about the rat genome (14%; O'Connell et al. 2003) and Drosophila genome (8%–10%; Orian et al. 2003). In addition, a recent study using ChIP analysis of human chromosomes 21 and 22 has led to an prediction of 24,000 genomic binding sites for Myc (Cawley et al. 2004), highly consistent with the earlier work showing that Myc binding is widespread. Interestingly, many of the Myc binding sites correspond to small non-coding RNAs (see the next section).

3.1 Cell Cycle

The earliest observation about c-Myc function was its ability to promote cell proliferation and inhibit cell differentiation. Thus, it is hardly surprising that many investigators have uncovered target genes that encode molecules that regulate the cell cycle. The genes that consistently emerge and contain E-boxes bound by c-Myc in ChIP assays are cyclins D1 and D2, *CDK4*, and cyclin B1. Cyclin A, cyclin B, and *cdk4* were identified as Myc targets in *Drosophila* as well. c-Myc upregulates the expression of genes that accelerate entry into the cell cycle, as well as downregulates those genes inhibiting cell-cycle progression. c-Myc repression of the CDK inhibitor, p21, appeared on at least one array screen (Coller et al. 2000), consistent with the prior

Table 2 c-Myc target genes commonly identified in at least three separate studies as reported in the c-Myc Target Gene Database*

Gene target	LocusLink ID Number	Description	Regulation	Myc DNA binding
APEX	328	Endonuclease	U	C*
CAD	790	Carbamoyl-phosphate synthetase 2, aspartate transcarbamylase, and dihydroorotase	U	C*
CCNA2	890	Cyclin A2	U	
CCND2	894	CYCLIN D2	U	C*
CCNE1	898	CYCLIN E1	U	
CDK4	1019	CDK4 cyclin-dependent kinase 4	U	G, C*
CDKN1A	1026	Cyclin-dependent kinase inhibitor 1A (p21, Cip1)	D	
CDKN2B	1030	Cyclin-dependent kinase inhibitor 2B (p15, inhibits CDK4)	D	G,C
CHC1	1104	RCC1; chromosome condensation 1	U	G
DDX18	8886	DEAD/H (Asp-Glu-Ala-Asp/His) box polypeptide 18 (Myc-regulated), MrDb	U	
DUSP1	1843	Dual specificity phosphatase 1	D	
EIF4E	1977	Eukaryotic translation initiation factor 4E	U	G, C*
ENO1	2023	Enolase 1 (α)	U	C*
FASN	2194	Fatty acid synthase	U	C*
FKBP4	2288	FK506 binding protein 4, 59 kDa	U	
FN1	2335	Fibronectin 1	D	
GADD45A	1647	Growth arrest and DNA-damage-inducible, α	D	
HSPA4	3308	Heat shock 70-kDa protein 4	U	G
HSPCAL3	3324	Heat shock 90-kDa protein 1, α-like 3	U	C*
HSPD1	3329	Heat shock 60-kDa protein 1 (chaperonin)	U	C*
HSPE1	3336	Heat shock 10-kDa protein 1 (chaperonin 10)	U	C*
LDHA	3939	Lactate dehydrogenase A	U	G, C
MGST1	4257	Glutathione transferase. GST-1	U	C*
MYC	4609	v-myc myelocytomatosis viral oncogene homolog (avian)	D	
NCL	4691	Nucleolin	U	C*

Table 2 (continued)

Gene target	LocusLink ID Number	Description	Regu-lation	Myc DNA binding
NME1	4830	Non-metastatic cells 1, protein NM23A	U	C*
NME2	4831	NM23-H2, non-metastatic cells 2, expressed in protein NM23B	U	C
NPM1	4869	Nucleophosmin, B23	U	C
ODC1	4953	Ornithine decarboxylase 1	U	G, C
PPAT	5471	Phosphoribosyl pyrophosphate amidotransferase	U	C*
PTMA	5757	Prothymosin, α (gene sequence 28)	U	C*, D, G
RPL23	9349	Ribosomal protein L23	U	C*
RPL3	6122	Ribosomal protein L3	U	
RPL6	6128	Ribosomal protein L6	U	
RPS15A	6210	Ribosomal Protein S15A	U	
SRM	6723	Spermidine synthase	U	C*
TERT	7015	Telomerase reverse transcriptase	U	G, C*
TFRC	7037	Transferrin receptor (p90, CD71)	U	C*
THBS1	7057	Thrombospondin 1	D	
TNFSF6	356	Tumor necrosis factor (ligand) superfamily, member 6	U	G
TP53	7157	Tumor protein p53 (Li-Fraumeni syndrome)	U	G
TPM1	7168	Tropomyosin 1 (α)	D	

"Regulation": U, up; D, Down; "Myc DNA Binding": C, ChIP analysis. C*, ChIP (Fernandez et al. 2003); D, DNA footprint analysis; G, gel shift; #, *Drosophila* (Orian et al. 2003). Adapted from http://www.myccancergene.org/site/mycTargetDB.asp

observation that c-Myc represses p21 through an interaction with the Miz-1 protein at its core promoter (Claassen and Hann 2000; Seoane et al. 2002; Wu et al. 2003). c-Myc also represses those genes involved in growth arrest such as Gadd45 (Marhin et al. 1997) and gas1 (Lee et al. 1997). The repression of Gadd45 appears to be universal among studies that have included it on their arrays. Remarkably, Gadd45 was one of only two known target genes whose expression is dysregulated in the serum stimulation model using c-*Myc*-null fibroblasts (Bush et al. 1998). Finally, very recent experiments have demonstrated that Myc induces microRNAs which inhibit E2F expression, perhaps limiting E2F's cell-cycle and apoptotic effects (O'Donnell et al. 2005).

3.2
Protein Synthesis

The protein synthetic machinery consistently appears to be affected at multiple levels by c-Myc. Given that c-Myc accelerates entry into S-phase of the cell cycle, increased cell proliferation would demand a commensurate increase in protein synthesis. The rate of protein synthesis is increased nearly threefold in c-Myc-overexpressing fibroblasts when compared to knockout cells (Mateyak et al. 1997). Target genes repeatedly identified from global screening approaches include those that encode ribosomal (r)RNAs, ribosome biogenesis proteins (e.g., BN51, nucleophosmin, nucleolin), transfer (t)RNAs, RNA helicases, and translation elongation factors (e.g., eIF4E). About 20% of genes upregulated by c-Myc in the Myc-null fibroblast system involve ribosomal biogenesis and protein synthesis (Guo et al. 2000; O'Connell et al. 2003). Similarly, of the 114 genes N-myc was found to upregulate in an array study, over 50% represented ribosomal and protein synthesis genes (Boon et al. 2001). Anti-c-Myc ChIP experiments performed in several labs have demonstrated that the promoters of some of these genes are directly bound by c-Myc (Fernandez et al. 2003; Zeller et al. 2001). These include nucleophosmin, nucleolin, BN51, and several ribosomal protein gene promoters. c-Myc regulation of protein synthesis genes is integral to its control of cell growth and proliferation, and is further bolstered by functional assays in which c-Myc has been either deleted or overexpressed. In particular, several ribosomal protein genes have been identified as direct Myc targets (Fernandez et al. 2003; Mao et al. 2003; Orian et al. 2003), and their importance to the c-Myc-induced phenotypes are underscored by recently identified *Drosophila* mutants. The existence of both the *Drosophila Minute*, whose small body size results from smaller cells as a result of a ribosomal protein gene mutation, and the small-sized *dMyc* hypomorphs (Johnston et al. 1999) links Myc and ribosomal proteins in the control of cell and organismal size. The observation extends to vertebrates as well. An increase in expression of ribosomal protein genes accompanied the increase in cell size when c-Myc was overexpressed in the liver in vivo (Kim et al. 2000) and in B lymphocytes (Iritani and Eisenman 1999). Whether an increase in protein synthesis is necessary for c-Myc-mediated cell transformation remains uncertain, but the recognition that a defect in ribosome biogenesis predisposes to hematopoietic malignancies, such as in Diamond-Blackfan anemia, suggest that regulation of genes involved in some aspects of protein synthesis could be essential to the function of c-Myc in neoplasia. In this respect it is of interest that very recent experiments have demonstrated that c-Myc directly binds to the rDNA promoter and termination regions and stimulates rRNA transcription in human cells (Arabi et al. 2005; Grandori et al. 2005). Myc also stimulates rRNA transcription in *Drosophila* but, in contrast to the case in human cells, this occurs indirectly, through activation of components of the RNA polymerase I transcription complex (Grewal et al. 2005). Thus, Myc appears to activate several components of the protein synthesis machinery.

3.3 Cell Adhesion

Genes that encode cytoskeletal and cell adhesion proteins appear to be coordinately downregulated by c-Myc. A hallmark of c-Myc overexpression in susceptible cell lines is neoplastic transformation, which allows cells to grow in an anchorage-independent manner. First recognized via subtraction cloning approaches and present on global screening arrays, the collagen genes tend to be reproducibly repressed by c-Myc overexpression (Yang et al. 1991). Downregulation of cell adhesion proteins may explain the morphological changes induced by *myc* overexpression. *myc*-null fibroblasts have a characteristic flattened appearance but become more spindle-shaped with *myc* reconstitution. Activation of Myc in the keratinocytes of a transgenic mouse led to the identification of 137 upregulated and 81 downregulated genes (Frye et al. 2003); 30% of those downregulated were genes involved in cellular adhesion and 11% encoded cytoskeletal related proteins. Cell surface proteins that mediate adhesion to the extracellular matrix, including N- and R-cadherins, integrin β_1, fibronectin, and fibrillin 1 and 2, are repressed. Intriguingly, studies in different cell lines suggest an interplay between Myc and integrin expression. Cell adhesion activates the β_1-integrin signaling pathway, which in turn induces c-Myc expression (Benaud and Dickson 2001). Anchorage-dependent growth of normal cells is regulated through adhesion molecules, whose expression is likely suppressed during the normal lifecycle of a cell to allow for mitosis and cytokinesis. Exaggerated responses in this regulatory loop may be induced by constitutive c-Myc expression, which represses β_1-integrin expression and bypasses the need for cell adhesion. For example, c-Myc transforms and enhances the anchorage-independent growth of a human small-cell lung cancer cell line and represses the expression of $\alpha_3\beta_1$ (Barr et al. 1998). Reconstitution of $\alpha_3\beta_1$ expression through ectopic expression of α_3 was shown to suppress Myc-mediated anchorage-independent growth.

The decreased expression of cytoskeletal genes may also relate to functional alterations in cell phenotype that results from c-Myc overexpression. Human keratinocytes with activated c-Myc expression display impaired motility and spread to a lesser extent than control cells (Frye et al. 2003). These changes may explain impaired wound healing in the c-Myc transgenic mice. In contrast, Myc-null fibroblasts possess many actin stress fibers and focal adhesions as compared to null fibroblasts reconstituted with c-Myc (Shiio et al. 2002). A proteomics approach identified the cytoskeletal proteins actin and cdc42 as downregulated in c-Myc-reconstituted fibroblasts. These changes suggest that Myc plays a role in enhancing fibroblast motility, which contrasts with its inhibitory effect in keratinocytes. The potential relationship of these morphological changes to the invasive potential of human tumors that overexpress c-Myc remains elusive.

3.4
Metabolism

Many pathways of cell metabolism are also regulated by c-Myc in vertebrates and *Drosophila*. Several key enzymes of glucose metabolism are found in multiple c-Myc target studies, such as enolase A, hexokinase II, lactate dehydrogenase A, phosphofructokinase, and glucose transporter I (Menssen and Hermeking 2002; O'Connell et al. 2003; Osthus et al. 2000), consistent with the putative role of c-Myc in enhancing glucose uptake and glycolysis. Genes of mitochondrial biogenesis and function constitute another group upregulated in response to c-Myc overexpression in mammalian systems and in *Drosophila*, but how most of these genes facilitate the c-Myc-associated growth phenotypes requires further investigation (Morrish et al. 2003; O'Connell et al. 2003; Orian et al. 2003; Wonsey et al. 2002). Iron metabolism may be affected by c-Myc with ferritin, IRP1, IRP2, and the transferrin receptor gene as potential targets (Bowen et al. 2002; O'Connell et al. 2003; Wu et al. 1999). The transferrin receptor, which is upregulated by c-Myc, is a cell surface glycoprotein that transports iron bound to transferrin into the cell, where iron is then incorporated into enzymes that catalyze energy metabolism and DNA synthesis. For example, ribonucleotide reductase, which is required for the synthesis of deoxyribonucleotide triphosphates (dNTPs) and DNA synthesis, needs iron for its catalytic function. *CAD* and *ODC*, which are repeatedly confirmed as c-Myc targets, are essential for nucleotide synthesis (Bello-Fernandez et al. 1993; Miltenberger et al. 1995). The mitochondrial serine hydroxymethyltransferase, important for folate metabolism and involved in nucleotide metabolism, is a c-Myc target that partially complements the growth defects exhibited by c-Myc-null fibroblasts; its promoter is bound by c-Myc in ChIP assays (Nikiforov et al. 2002). It is intriguing to note that APEX1, an endonuclease involved in DNA repair, is repeatedly identified as a c-Myc target that has also been validated by proteomics analysis of wildtype and Myc-null cells (Shiio et al. 2002). Among its many diverse influences, c-Myc also globally regulates genes involve in DNA synthesis and repair.

4
The Search for Myc Transcriptomes in the Max Network

The emerging protein network to which the c-Myc/Max heterodimer belongs has complicated the search for c-Myc responsive transcriptomes. What is known about c-Myc transcriptional activity relies on the premise that c-Myc heterodimerizes exclusively with Max, but Max is capable of forming dimers with members of the Mad family of proteins (Mad1, Mxi1, Mad3, and Mad4), and Mnt and Mga (see S. Rottmann and B. Lüscher, this volume). Except for Mad3 (Fox and Wright 2001; Quéva et al. 2001), all Mad proteins appear to be most highly expressed in differentiating cells (Ayer and Eisenman 1993; Hurlin et al. 1995b; Larsson et al. 1994; Wechsler et al. 1997; Zervos et al. 1993) in contrast to c-Myc, whose expression is induced in actively proliferating cells. Early studies showed that Mad1 protein accumulates and forms dimers with Max upon induction of differentiation of hematopoietic cell lines (Ayer and Eisenman 1993). This observation has been upheld in other cell lines (Hurlin et al. 1995a; Roussel et al. 1996). Overexpression of the Mad proteins antagonizes c-Myc-mediated transformation and negatively regulates cell growth in part by competitively binding the same canonical E-box. In addition, Max/Mad dimers repress transcription by recruiting mSin3 repressor molecules and histone deacetylase via the N-terminal domains of Mad proteins (Ayer et al. 1995; Schreiber-Agus et al. 1995).

A long-held assumption has been that Mad and c-Myc interchangeably bind the same set of genes to achieve their associated biological phenotypes, depending on their intracellular levels. This paradigm had relied on observations that c-Myc and Mad-like protein levels are inversely proportional in proliferating and differentiating cells. Moreover, both c-Myc and Mad-like proteins heterodimerize with Max and recognize the same canonical E-box. In support of this view is the observation from a microarray experiment involving murine thymocytes expressing Mad1 that 77% of genes repressed by Mad1 are involved in cell growth (Iritani et al. 2002); 80% of the Mad1 repressed genes have been previously shown to be induced by c-Myc. These findings, however, do not exclude the possibility raised by other microarray studies that the cell phenotypes associated with c-Myc and Mad may involve the regulation of unique sets of genes.

Chimeric proteins in which the basic domains responsible for DNA binding were interchanged to determine whether c-Myc and Mxi1 (Mad2) bind the same DNA sites to achieve their biological functions. The chimera, Myc(Mxi-1BR), was not functionally equivalent to wildtype c-Myc in its ability to form foci in transformation assays. Furthermore, a custom cDNA array of 5,272 human genes was hybridized with total RNA derived from IMR90 cells in which the expression of chimeric protein Myc(Mxi-1BR) or wildtype c-Myc was induced (O'Hagan et al. 2000). Only 8 genes changed expression to either c-Myc or Myc(Mxi-1BR) induction. Myc(Mxi-1BR), but not c-Myc, altered the expression of another 8 genes encoding proteins of diverse functions. Conversely, c-Myc altered the expression of 11 additional genes whose expression did not change with Myc(Mxi-1BR). Thus, it appears that the basic regions of c-Myc and Mxi-1, though capable of recognizing the same E-box, are functionally distinct and that some of the c-Myc-associated phenotypes may result from the regulation of specific subsets of genes. Gene specificity may in part be determined by nucleotides that flank the E-box, but the crystal structures of these proteins are unable to resolve these interactions (Nair and Burley 2003). For example, some of the c-Myc-specific targets have E-boxes flanked by 5′C and 3′G nucleotides in contrast to the Myc(Mxi-BR)-specific targets which contain an E-box flanked by a 5′T.

The DNA binding preferences of Mad1, which suppresses proliferation, was determined by using selection and amplification of randomized oligonucleotides and found to be identical with that of c-Myc, which induces growth and apoptosis (James and Eisenman 2002). Interestingly, however, a chimeric c-Myc containing both the Mad basic and HLHLZip dimerization domain could only promote proliferation and not apoptosis. In another study limited only to exchange of the basic region, a chimeric Myc(Mad-BR) rescued the growth phenotype of c-Myc-null cells and restored transformation and apoptosis to the same extent as wildtype Myc. Chromatin immunoprecipitation to detect c-Myc and Mad1 binding at a select group of target genes demonstrated that c-Myc and Mad1 occupy identical E-boxes (Nikiforov et al. 2003). In aggregate, these studies suggest a more complex role for the bHLHLZip domain than previously suspected.

Several important findings about Myc and Mad-like protein binding were made in the Dam-methylase study of *Drosophila* chromatin described above (Orian et al. 2003). By comparing *Drosophila* chromatin expression profiles generated by dMyc, dMax, and dMnt, the authors showed that the number of dMyc targets increased substantially in the presence of high dMax compared to low dMax, implying that for many Myc target genes the amount of endogenous dMax is limiting. Target genes shared by all three proteins were surprisingly few, and consisted of 73 genes from biological pathways affecting cell migration, protein synthesis, mitochondria biogenesis, and transcription factors. Targets recognized by only one or two proteins were also identified. In the case of the transcription factor *bic*, which appears to be a target of dMyc, dMnt, and dMax, the authors showed by chromatin immunoprecipitation that dMyc and dMnt binding is interchangeable at the same chromatin containing an E-box, and that binding correlated with histone acetylation in the case of dMyc and histone deacetylation in the case of dMnt.

The existence of targets recognized exclusively by specific heterodimers containing Max suggests that transcriptional regulation is more intricate than we currently understand. The existence of additional interacting proteins for each of these molecules could explain the unique target genes identified. For example, Mad proteins interact with a RING-finger protein, Mmip2, that causes their translocation to the cytoplasm, thus enhancing c-Myc activity (Yin et al. 2001). Roles for the Myc bHLHLZip, such as the potential for tetramerization (Nair and Burley 2003; see S.K. Nair and S.K. Burley, this volume) or interaction with Miz-1 (Staller et al. 2001; see D. Kleine-Kohlbrecher et al., this volume) and Skp2 (Kim et al. 2003; von der Lehr et al. 2003), may all contribute to the additional specificity of c-Myc in regulating its target genes. It is also possible that the binding specificities of these proteins are further defined by nucleotides that flank the E-box, as indicated by the c-Myc and Mxi-1 chimera studies (O'Hagan et al. 2000), although such extended binding site specificity was not apparent in other studies (James and Eisenman 2002). The other possibility is that the E-box is not the sole DNA binding site through which these proteins operate. The study using Dam-methylase chimeras to identify dMyc, dMax, and dMnt transcriptional targets in *Drosophila* further illustrates the power of the global screening approach by using a bioinformatics algorithm called REDUCE to analyze the binding sites within target genes identified by all three proteins (Orian et al. 2003). When dMax levels were limiting, dMyc tended to associate with AT-rich sites as determined by REDUCE. When dMax was not limiting, association with the E-box was much more prevalent. Moreover, REDUCE demonstrated an additional site, a palindromic DNA region, TATCGATA, that is frequently found in fragments generated by all three proteins. Despite the strength of this statistical method, chromatin immunoprecipitation or electromobility shift assays have not yet been performed to confirm the authenticity of these sites.

5
An Integrated Database of Myc Responsive Genes for the Future

Given the diverse cell types and experimental systems used to study c-Myc target genes, how does the field begin to achieve a comprehensive accounting of c-Myc responsive transcriptomes? To begin to glean a collective view of c-Myc responsive transcriptomes, a publicly accessible c-Myc target gene database has been launched as mentioned above. The Myc Target Gene database (www.myccancergene.org) is designed so that it is searchable and provides the ability to prioritize the putative target genes according to the level of experimental evidence supporting the validity of the gene in question as an authentic target gene. The genes are listed according to official names in the LocusLink database (http://www.ncbi.nlm.nih.gov/), to which each entry is linked, or to the best available name. Each gene is defined, when possible, by the function of its product. Each gene is annotated as to whether it is upregulated (U) or downregulated (D) in response to Myc induction or overexpression. References for the target genes are provided as direct links to PubMed.

The level of experimental evidence for each gene is tabulated according to the technique used [M, microarray; D, differential cloning; S, serial analysis of gene expression (SAGE); and G, guess]. Whether the genes were studied by ChIP, promoter-reporter assays or footprinting is highlighted as evidence for direct regulation by c-Myc. The inducible Myc-estrogen receptor hormone-binding assay is also specifically highlighted in the database.

As more studies are reported, this database will continue to provide a central clearing house for *MYC* responsive genes. Through this collective examination of c-Myc target genes, it is envisioned that a defined core set of c-Myc target genes will become apparent independent of cell types, although there may be species-specific differences (Table 2). As the database is further enriched, patterns will emerge that define the tissue-specific regulation of gene expression by c-Myc. Not only will this database yield insights into the tumorigenic effects of c-Myc, but it also serves as a prototype for other central regulators of transcription such as p53 or the E2F family of proteins.

An anticipated use of the Myc target gene database is in the analysis of gene expression profiles of human cancers. For example, it is noted in a study on altered gene expression associated with the progression of follicular lymphomas that about half of the transformed follicular lymphomas have increased Myc target gene expression (Lossos et al. 2002). In particular, it is important to note that while Myc may initiate tumorigenesis, it is not always required for the progression of the tumor (D'Cruz et al. 2001; Felsher and Bishop 1999). As such, the "hit-and-run" effect of Myc may only be evident through the examination of its target genes that may continue to be expressed through other genetic or epigenetic means. More recently, the patterns of gene expression induced by initiating oncogenic events such as Myc or Ras provide gene expression fingerprints of the oncogenic suspects involved in initiating the tumor (Huang et al. 2003). The use of an integrated Myc responsive/target gene database will facilitate the analysis of altered gene expression profiles in cancers and will lead to the identification of likely suspects that may be targeted for therapy.

Acknowledgements Our work is supported by NIH grants CA51497, CA57341, CA091596, and LM07515. We are grateful for the critical reviews of L. Gardner and C. Eberhart.

References

Arabi A, Wu S, Shiue C, Ridderstrale K, Larsson L-G, Wright APH (2005) c-Myc associates with ribosomal DNA in the nucleolus and activates RNA polymerase I transcription. Nat Cell Biol 7:303–310

Ayer DE, Eisenman RN (1993) A switch from Myc:Max to Mad:Max heterocomplexes accompanies monocyte/macrophage differentiation. Genes Dev 7:2110–2119

Ayer DE, Lawrence QA, Eisenman RN (1995) Mad-Max transcriptional repression is mediated by ternary complex formation with mammalian homologs of yeast repressor Sin3. Cell 80:767–776

Barr LF, Campbell SE, Bochner BS, Dang CV (1998) Association of the decreased expression of alpha3beta1 integrin with the altered cell: environmental interactions and enhanced soft agar cloning ability of c-myc-overexpressing small cell lung cancer cells. Cancer Res 58:5537–5545

Bello-Fernandez C, Packham G, Cleveland JL (1993) The ornithine decarboxylase gene is a transcriptional target of c-Myc. Proc Natl Acad Sci U S A 90:7804–7808

Benaud CM, Dickson RB (2001) Regulation of the expression of c-Myc by beta1 integrins in epithelial cells. Oncogene 20:759–768

Boon K, Caron HN, van Asperen R, Valentijn L, Hermus MC, van Sluis P, Roobeek I, Weis I, Voute PA, Schwab M, Versteeg R (2001) N-myc enhances the expression of a large set of genes functioning in ribosome biogenesis and protein synthesis. EMBO J 20:1383–1393

Bowen H, Biggs TE, Baker ST, Phillips E, Perry VH, Mann DA, Barton CH (2002) c-Myc represses the murine Nramp1 promoter. Biochem Soc Trans 30:774–777

Bush A, Mateyak M, Dugan K, Obaya A, Adachi S, Sedivy J, Cole M (1998) c-myc null cells misregulate cad and gadd45 but not other proposed c-Myc targets. Genes Dev 12:3797–3802

Cawley S, Bekiranov S, Ng HH, Kapranov P, Sekinger EA, Kampa D, Piccolboni A, Sementchenko V Cheng J, Williams AJ, Wheeler R, Wong B, Drenkow J, Yamanaka M, Patel S, Brubaker S, Tammana H, Helt G, Struhl K, Gingeras TR (2004) Unbiased mapping of transcription factor binding sites along human chromosomes 21 and 22 points to widespread regulation of noncoding RNAs. Cell 116:499–509

Claassen GF, Hann SR (2000) A role for transcriptional repression of p21CIP1 by c-Myc in overcoming transforming growth factor beta -induced cell-cycle arrest. Proc Natl Acad Sci U S A 97:9498–9503

Coller HA, Grandori C, Tamayo P, Colbert T, Lander ES, Eisenman RN, Golub TR (2000) Expression analysis with oligonucleotide microarrays reveals that MYC regulates genes involved in growth, cell cycle, signaling, and adhesion. Proc Natl Acad Sci U S A 97:3260–3265

D'Cruz CM, Gunther EJ, Boxer RB, Hartman JL, Sintasath L, Moody SE, Cox JD, Ha SI, Belka GK, Golant A, et al (2001) c-MYC induces mammary tumorigenesis by means of a preferred pathway involving spontaneous Kras2 mutations. Nat Med 7:235–239

Eilers M, Picard D, Yamamoto KR, Bishop JM (1989) Chimaeras of myc oncoprotein and steroid receptors cause hormone-dependent transformation of cells. Nature 340:66–68

Felsher DW, Bishop JM (1999) Transient excess of MYC activity can elicit genomic instability and tumorigenesis. Proc Natl Acad Sci U S A 96:3940–3944

Fernandez PC, Frank SR, Wang L, Schroeder M, Liu S, Greene J, Cocito A, Amati B (2003) Genomic targets of the human c-Myc protein. Genes Dev 17:1115–1129

Fox EJ, Wright SC (2001) S-phase-specific expression of the Mad3 gene in proliferating and differentiating cells. Biochem J 359:361–367

Frye M, Gardner C, Li ER, Arnold I, Watt FM (2003) Evidence that Myc activation depletes the epidermal stem cell compartment by modulating adhesive interactions with the local microenvironment. Development 130:2793–2808

Grandori C, Gomez-Roman N, Felton-Edkins ZANgouenet C, Galloway DA, Eisenman RNand White RJ (2005) c-Myc binds to human ribosomal DNA and stimulates transcription of rRNA genes by RNA polymerase I. Nat Cell Biol 7:311–318

Grewal SS, Li L, Orian A, Eisenman RN, Edgar BA (2005) Myc-dependent regulation of ribosomal RNA synthesis during Drosophila development. Nat Cell Biol 7:295–302

Guo QM, Malek RL, Kim S, Chiao C, He M, Ruffy M, Sanka K, Lee NH, Dang CV, Liu ET (2000) Identification of c-myc responsive genes using rat cDNA microarray. Cancer Res 60:5922–5928

Haggerty TJ, Zeller KI, Osthus RC, Wonsey DR, Dang CV (2003) A strategy for identifying transcription factor binding sites reveals two classes of genomic c-Myc target sites. Proc Natl Acad Sci U S A 100:5313–5318

Huang E, Ishida S, Pittman J, Dressman H, Bild A, Kloos M, D'Amico M, Pestell RG, West M, Nevins JR (2003) Gene expression phenotypic models that predict the activity of oncogenic pathways. Nat Genet 18:18

Hurlin PJ, Foley KP, Ayer DE, Eisenman RN, Hanahan D, Arbeit JM (1995a) Regulation of Myc and Mad during epidermal differentiation and HPV-associated tumorigenesis. Oncogene 11:2487–2501

Hurlin PJ, Quéva C, Koskinen PJ, Steingrimsson E, Ayer DE, Copeland NG, Jenkins NA, Eisenman RN (1995b) Mad3 and Mad4: novel Max-interacting transcriptional repressors that suppress c-myc dependent transformation and are expressed during neural and epidermal differentiation. EMBO J 14:5646–5659

Iritani BM, Eisenman RN (1999) c-Myc enhances protein synthesis and cell size during B lymphocyte development. Proc Natl Acad Sci U S A 96:13180–13185

Iritani BM, Delrow J, Grandori C, Gomez I, Klacking M, Carlos LS, Eisenman RN (2002) Modulation of T-lymphocyte development, growth and cell size by the Myc antagonist and transcriptional repressor Mad1. EMBO J 21:4820–4830

James L, Eisenman RN (2002) Myc and Mad bHLHZ domains possess identical DNA-binding specificities but only partially overlapping functions in vivo. Proc Natl Acad Sci U S A 99:10429–10434

Johnston LA, Prober DA, Edgar BA, Eisenman RN, Gallant P (1999) Drosophila myc regulates cellular growth during development. Cell 98:779–790

Kim S, Li Q, Dang CV, Lee LA (2000) Induction of ribosomal genes and hepatocyte hypertrophy by adenovirus-mediated expression of c-Myc in vivo. Proc Natl Acad Sci U S A 97:11198–11202

Kim SY, Herbst A, Tworkowski KA, Salghetti SE, Tansey WP (2003) Skp2 regulates myc protein stability and activity. Mol Cell 11:1177–1188

Larsson LG, Pettersson M, Oberg F, Nilsson K, Luscher B (1994) Expression of mad, mxi1, max and c-myc during induced differentiation of hematopoietic cells: opposite regulation of mad and c-myc. Oncogene 9:1247–1252

Lee TC, Li L, Philipson L, Ziff EB (1997) Myc represses transcription of the growth arrest gene gas1. Proc Natl Acad Sci U S A 94:12886–12891

Li Z, Van Calcar S, Qu C, Cavenee WK, Zhang MQ, Ren B (2003) A global transcriptional regulatory role for c-Myc in Burkitt's lymphoma cells. Proc Natl Acad Sci U S A 100:8164–8169

Liu H, Keefer JR, Wang QF, Friedman AD (2003) Reciprocal effects of C/EBPalpha and PKCdelta on JunB expression and monocytic differentiation depend upon the C/EBPalpha basic region. Blood 101:3885–3892

Lossos IS, Alizadeh AA, Diehn M, Warnke R, Thorstenson Y, Oefner PJ, Brown PO, Botstein D, Levy R (2002) Transformation of follicular lymphoma to diffuse large-cell lymphoma: alternative patterns with increased or decreased expression of c-myc and its regulated genes. Proc Natl Acad Sci U S A 99:8886–8891

Mao DY, Watson JD, Yan PS, Barsyte-Lovejoy D, Khosravi F, Wong WW, Farnham PJ, Huang TH, Penn LZ (2003) Analysis of Myc bound loci identified by CpG island arrays shows that Max is essential for Myc-dependent repression. Curr Biol 13:882–886

Marhin WW, Chen S, Facchini LM, Fornace AJ Jr, Penn LZ (1997) Myc represses the growth arrest gene gadd45. Oncogene 14:2825–2834

Mateyak MK, Obaya AJ, Adachi S, Sedivy JM (1997) Phenotypes of c-Myc-deficient rat fibroblasts isolated by targeted homologous recombination. Cell Growth Differ 8:1039–1048

Menssen A, Hermeking H (2002) Characterization of the c-MYC-regulated transcriptome by SAGE: identification and analysis of c-MYC target genes. Proc Natl Acad Sci U S A 99:6274–6279

Miltenberger RJ, Sukow KA, Farnham PJ (1995) An E-box-mediated increase in cad transcription at the G1/S-phase boundary is suppressed by inhibitory c-Myc mutants. Mol Cell Biol 15:2527–2535

Morrish F, Giedt C, Hockenbery D (2003) c-MYC apoptotic function is mediated by NRF-1 target genes. Genes Dev 17:240–255

Nair SK, Burley SK (2003) X-ray structures of Myc-Max and Mad-Max recognizing DNA. Molecular bases of regulation by proto-oncogenic transcription factors. Cell 112:193–205

Nikiforov MA, Chandriani S, O'Connell B, Petrenko O, Kotenko I, Beavis A, Sedivy JM, Cole MD (2002) A functional screen for Myc-responsive genes reveals serine hydroxymethyltransferase, a major source of the one-carbon unit for cell metabolism. Mol Cell Biol 22:5793–5800

Nikiforov MA, Popov N, Kotenko I, Henriksson M, Cole MD (2003) The Mad and Myc basic domains are functionally equivalent. J Biol Chem 278:11094–11099

O'Connell BC, Cheung AF, Simkevich CP, Tam W, Ren X, Mateyak MK, Sedivy JM (2003) A large scale genetic analysis of c-Myc-regulated gene expression patterns. J Biol Chem 278:12563–12573

O'Donnell KA, Wentzel EA, Zeller KI, Dang CV, Mendell JT (2005) c-Myc-regulated microRNAs modulate E2F1 expression. Nature 435:839–843

O'Hagan RC, Schreiber-Agus N, Chen K, David G, Engelman JA, Schwab R, Alland L, Thomson C, Ronning DR, Sacchettini JC, et al (2000) Gene-target recognition among members of the myc superfamily and implications for oncogenesis. Nat Genet 24:113–119

Orian A, Van Steensel B, Delrow J, Bussemaker HJ, Li L, Sawado T, Williams E, Loo LW, Cowley SM, Yost C, et al (2003) Genomic binding by the Drosophila Myc, Max, Mad/Mnt transcription factor network. Genes Dev 17:1101–1114

Osthus RC, Shim H, Kim S, Li Q, Reddy R, Mukherjee M, Xu Y, Wonsey D, Lee LA, Dang CV (2000) Deregulation of glucose transporter 1 and glycolytic gene expression by c-Myc. J Biol Chem 275:21797–21800

Quéva C, McArthur GA, Iritani BM, Eisenman RN (2001) Targeted deletion of the S-phase specific Myc antagonist Mad3 sensitizes neural and lymphoid cells to radiation-induced apoptosis. Mol Cell Biol 21:703–712

Roussel MF, Ashmun RA, Sherr CJ, Eisenman RN, Ayer DE (1996) Inhibition of cell proliferation by the Mad1 transcriptional repressor. Mol Cell Biol 16:2796–2801

Schreiber-Agus N, Chin L, Chen K, Torres R, Rao G, Guida P, Skoultchi AI, DePinho RA (1995) An amino-terminal domain of Mxi1 mediates anti-Myc oncogenic activity and interacts with a homolog of the yeast repressor SIN3. Cell 80:777–786

Schuldiner O, Benvenisty NA (2001) DNA microarray screen for genes involved in c-MYC and N-MYC oncogenesis in human tumors. Oncogene 20:4984–4994

Seoane J, Le HV, Massague J (2002) Myc suppression of the p21(Cip1) Cdk inhibitor influences the outcome of the p53 response to DNA damage. Nature 419:729–734

Shiio Y, Donohoe S, Yi EC, Goodlett DR, Aebersold R, Eisenman RN (2002) Quantitative proteomic analysis of Myc oncoprotein function. EMBO J 21:5088–5096

Staller P, Peukert K, Kiermaier A, Seoane J, Lukas J, Karsunky H, Moroy T, Bartek J, Massague J, Hanel F, Eilers M (2001) Repression of p15INK4b expression by Myc through association with Miz-1. Nat Cell Biol 3:392–399

von der Lehr N, Johansson S, Wu S, Bahram F, Castell A, Cetinkaya C, Hydbring P, Weidung I, Nakayama K, Nakayama KI, et al (2003) The F-box protein Skp2 participates in c-Myc proteasomal degradation and acts as a cofactor for c-Myc-regulated transcription. Mol Cell 11:1189–1200

Watson JD, Oster SK, Shago M, Khosravi F, Penn LZ (2002) Identifying genes regulated in a Myc-dependent manner. J Biol Chem 277:36921–36930

Wechsler DS, Shelly CA, Petroff CA, Dang CV (1997) MXI1, a putative tumor suppressor gene, suppresses growth of human glioblastoma cells. Cancer Res 57:4905–4912

Wonsey DR, Zeller KI, Dang CV (2002) The c-Myc target gene PRDX3 is required for mitochondrial homeostasis and neoplastic transformation. Proc Natl Acad Sci U S A 99:6649–6654

Wu KJ, Polack A, Dalla-Favera R (1999) Coordinated regulation of iron-controlling genes, H-ferritin and IRP2, by c-MYC. Science 283:676–679

Wu S, Cetinkaya C, Munoz-Alonso MJ, von der Lehr N, Bahram F, Beuger V, Eilers M, Leon J, Larsson LG (2003) Myc represses differentiation-induced p21CIP1 expression via Miz-1-dependent interaction with the p21 core promoter. Oncogene 22:351–360

Yang BS, Geddes TJ, Pogulis RJ, de Crombrugghe B, Freytag SO (1991) Transcriptional suppression of cellular gene expression by c-Myc. Mol Cell Biol 11:2291–2295

Yin XY, Grove LE, Prochownik EV (2001) Mmip-2/Rnf-17 enhances c-Myc function and regulates some target genes in common with glucocorticoid hormones. Oncogene 20:2908–2917

Zeller KI, Haggerty TJ, Barrett JF, Guo Q, Wonsey DR, Dang CV (2001) Characterization of nucleophosmin (B23) as a Myc target by scanning chromatin immunoprecipitation. J Biol Chem 276:48285–48291

Zervos AS, Gyuris J, Brent R (1993) Mxi1, a protein that specifically interacts with Max to bind Myc-Max recognition sites. Cell 72:223–232

CTMI (2006) 302:169–203

c-Myc, Genome Instability, and Tumorigenesis: The Devil Is in the Details

M. Wade · G. M. Wahl (✉)

Gene Expression Lab, The Salk Institute, 10010N. Torrey Pines Rd., La Jolla, CA 92037, USA
wahl@salk.edu

Abstract The c-myc oncogene acts as a pluripotent modulator of transcription during normal cell growth and proliferation. Deregulated c-myc activity in cancer can lead to excessive activation of its downstream pathways, and may also stimulate changes in gene expression and cellular signaling that are not observed under non-pathological conditions. Under certain conditions, aberrant c-myc activity is associated with the appearance of DNA damage-associated markers and karyotypic abnormalities. In this chapter, we discuss mechanisms by which c-myc may be directly or indirectly associated with the induction of genomic instability. The degree to which c-myc-induced genomic instability influences the initiation or progression of cancer is likely to depend on other factors, which are discussed herein.

1 Introduction

1.1 Overview

Cells must overcome multiple barriers designed to limit growth and proliferation to become tumorigenic [1–3]. The aim of this chapter is to discuss accumulating evidence that expression of c-Myc and other oncoproteins can compromise genomic integrity, how this may contribute to tumorigenesis, and to consider some of the potential mechanisms involved. In addition to other chapters in this volume, we refer the reader to the following excellent reviews detailing the diverse biological effects of the c-Myc protein on cell growth, proliferation, apoptosis, and differentiation [4–10].

1.2 Genetic Instability and Cancer Progression

The genesis of a malignant cell is a multistage process requiring the progressive accumulation of genetic and epigenetic changes [3]. Debates have arisen over whether the large number of changes required for malignancy (typically 6–10) arise spontaneously or whether events occur during tumor progression that increase genomic instability [11–13]. Consistent with the latter idea, many human tumors exhibit structural chromosomal aberrations such as amplifications that harbor increased copies of the c-*myc* oncogene [14, 15], and this type of genetic instability is not detected at measurable frequencies in normal cells [16]. This suggests that the mechanisms that maintain structural chromosome integrity are compromised during tumor progression. Consistent with this, loss of p53 function occurs frequently during cancer progression and creates a permissive environment for gene amplification [17, 18].

Vogelstein and colleagues have suggested subdividing tumors with genomic instability into two broad categories; those displaying chromosomal instability (CIN) and those with microsatellite instability (MIN) [19]. CIN represents a numerical and/or structural change in the karyotype, while MIN describes the expansion or contraction of homopolymers or tandem short repeats throughout the genome [20, 21]. CIN may occur due to mutations in genes required for the partitioning of chromosomes during mitosis, in genes that control cell-cycle checkpoints, or in genes that participate in DNA metabolism and repair [22]. Structural aberrations leading to CIN-like chromosomal abnormalities can also occur following break-induced translocations. These translocations can be balanced, such as the Ig:myc translocation in Burkitt's lymphoma (BL) [23] or unbalanced, such as non-reciprocal translocations generated as a result of bridge-breakage-fusion cycles [24]. MIN is typically caused by mutation or epigenetic inactivation of genes encoding proteins that participate in mismatch repair [25, 26]. As technology has improved the resolution at which karyotypic differences between normal and tumor can be determined, it has become clear that virtually all tumors exhibit abnormalities at the DNA level. In this review, other changes in the genome including point mutations, deletions, and base modifications will be included as manifestations of genomic instability.

Induction of cell-cycle arrest and activation of apoptosis are parts of the normal cellular defenses against oncogene-driven proliferation [27, 28]. It follows that inactivation of either of these two processes could enhance the likelihood of tumorigenesis. For example, variants with defective arrest or apoptotic machinery are more likely to survive oncogene activation than their "normal" counterparts. Chemical carcinogens and ionizing radiation, which accelerate tumorigenesis by increasing the frequency of somatic mutation [29], can increase the probability of generating such variants. Mutation rates are accelerated in mice following topical application of carcinogens [30]. Carcinomas arising in such mice frequently display mutations in the *H-ras* oncogene, a mutation also associated with human carcinomas [31, 32]. This strongly implicates induction of somatic mutations as an important factor in cancer progression. Viruses can also increase tumorigenicity, but for many years physical agents and oncogenic viruses were thought to work by different mechanisms [33]. Four decades ago, Nichols suggested that the mechanisms of radiation, chemical, and virus-driven oncogenesis may be shared, when he stated: "... it is possible that one of the earliest changes in tumor cells involves activation of a gene locus which increases the likelihood of non-disjunction or other mitotic error" [33]. Thus, Nichols proposed that, like chemical carcinogens and ionizing radiation, viruses might increase mutation frequency. This provided a conceptual framework expanded upon by Nowell [34] and Loeb [35] who suggested that genetic lability could accelerate tumor progression through mutation of genes that are essential for maintaining chromosomal integrity. Lesions in such genes would give rise to a "mutator phenotype" able to fuel further instability. The MIN phenotype (see above) is one specific example of the mutator phenotype. While the MIN phenotype was first identified in Lynch syndrome (hereditary non-polyposis colon cancer) [36], microsatellite instability has subsequently been observed in a variety of other cancers [37–39].

1.3 Viruses, Oncogenes, and Connections to Genome Destabilization

The link between tumor-associated viruses and perturbation of the genome is clear in birds and rodents, and accumulating data suggest viruses may have a similar impact on genome stability in human cancer. Early work in this field by Nichols demonstrated that infection of cells with the oncogenic Rous sarcoma virus (RSV) induced strand breaks and chromosomal abnormalities [40]. RSV-induced tumorigenesis is attributed to expression of the oncogene *v-src* [41], and overexpression of cellular *c-src* can promote genomic instability [42]. Together these data indicate that oncogene activation by viruses and consequent genome destabilization may be important in tumorigenesis. Viruses can also induce neoplasia by deregulating the expression of endogenous proto-oncogenes [43]. Integration of retroviruses near the *c-myc* promoter leads to aberrant *c-myc* expression in avian and murine tumors [44, 45]. Similarly, retroviral integration increases transcription of *ras*, an oncogene implicated in the initiation or progression of human cancer [46].

Many human tumors associated with oncogenic viruses also display genomic instability. For example, chromosomal instability is observed in human papillomavirus (HPV)-associated cancers [47]. HPV-induced perturbation of the genome appears to precede the invasive stage of cancer [48]. Instability is almost certainly due to the virally encoded E6 and E7 proteins, which inactivate the tumor suppressors p53, pRB, and pocket proteins related to pRB [49]. Oncogenic HPV has been implicated in inducing strand breaks [51, 50], which are precursors of diverse types of structural chromosomal alterations (e.g., see Windle et al. [52]). Furthermore, activation of oncogenic ras in murine fibroblasts induces structural and numerical chromosomal aberrations within one cell cycle [53], as does Mos, an oncogene that activates the mitogen-activated protein kinase (MAPK) pathway [54].

Considerable data therefore indicate that oncogene activation may be a common mechanism by which genomic instability arises in tumors. In the following sections we will discuss the diverse mechanisms by which aberrant c-myc expression may also lead to genomic instability.

1.4
Activation of c-myc and Initiation of Instability

Many mechanisms can lead to the activation of c-myc during tumorigenesis, including enhanced transcription by other oncogenic signaling pathways [56, 55], chromosomal rearrangements [15, 57], and resistance of Myc protein to ubiquitin-mediated proteolysis [58, 59]. c-myc is deregulated in the majority of breast carcinomas and in the early and late stages of colorectal cancer [60–64]. Overexpression of *c-myc* is also associated with the etiology of hepatocellular carcinoma (HCC) [65].

Elevated *c-myc* expression and genomic instability appear to be correlated in the solid tumor types mentioned above [66–68]. This raises the intriguing possibility that high-level *c-myc* expression in some situations might actually contribute to genome destabilization. In vitro and in vivo studies over the past decade strengthen this possibility. For example, Mai and colleagues showed that elevated c-myc increases the frequency of obtaining variants resistant to the antimetabolites *N*-(phosphonacetyl)-L-aspartate (PALA) and methotrexate via amplification of their respective target genes, *CAD* and *DHFR* [69–71]. This was recently confirmed by Felsher and Bishop [72]. *Cyclin D* and *ribonucleotide reductase R2* are also amplified following activation of c-myc in the absence of drug selection [73, 74], implying that c-myc function, and not the genome destabilizing effects of the selective agents [75], explains the observed increase in amplification frequency. While it has not been determined whether preferred regions are destabilized by *c-myc* overexpression, fluorescent in situ hybridization (FISH) and spectral karyotypic analyses indicate that *c-myc* overexpression may induce alterations at multiple genomic regions [74, 76]. This could have significant physiological impact since amplification of genes such as *mdm2*, *cyclin D*, and *c-erbB2* occur frequently in human cancers as the overproduced gene products provide cells with growth and survival advantages [77–79].

In vivo models of tumorigenesis support the notion that c-myc-induced instability contributes to the neoplastic phenotype. For example, Felsher and Bishop demonstrated that induction of instability in Rat1a fibroblasts by activation of c-myc rendered them tumorigenic in mice [72]. Importantly, c-myc was activated in cells under conditions where apoptosis would not be expected to occur (e.g., complete medium). Furthermore, cell lines derived from such tumors retained the ability to undergo c-myc-induced apoptosis. These data suggest that induction of genomic instability by c-myc does not always require a selection against apoptotic pathways. Transient activation of c-myc was sufficient to induce tumorigenesis and gene amplification. Therefore, initiation of genomic instability by c-myc likely contributes to neoplastic progression in this cell type. The genetic changes that occur following activation of c-myc also appear to be important during liver and breast carcinogenesis in vivo. For example, pre-neoplastic cells from both tissues contain non-random chromosomal rearrangements, including translocations and deletions that persist in late-stage HCC and mammary carcinomas [67, 80]. The early appearance of instability in these models correlates with deregulated c-myc activity. Persistence of chromosomal rearrangements into "mature" tumors suggests that a combination of c-myc-induced instability and subsequent selective pressure are important factors in the HCC and breast carcinoma models.

In other tumor types, it appears that inhibition of p53-induced apoptosis, rather than induction of instability, is the main block to c-myc-driven tumorigenesis. To illustrate, expression of *c-myc* under the control of the IgH [81] or

Igκ or γ [82] enhancers leads to B cell lymphoma with pre-B cell and B cell phenotypes, respectively. Both models show a protracted latency prior to onset of lymphoma, suggesting secondary events are required for c-myc-induced B cell tumors. Various genetic lesions that decrease p53 function, or that prevent induction of apoptosis, accelerate c-myc-induced lymphomagenesis [83–85]. It seems that large-scale genomic instability is not required in the Eμ-myc model of B cell lymphoma, since tumors in which c-myc-induced apoptosis was inhibited by dominant-negative caspase-9 were pseudodiploid [86]. Using an integrated LacZ reporter, Rockwood and colleagues analyzed the mutation and rearrangement rates in c-myc-driven lymphomas [87]. Strikingly, they found that chromosomal rearrangement but not mutation rate was enhanced in lymphomas compared to normal tissue, and that the p16Ink4a/p19arf locus was deleted. These data indicate that deregulated c-myc activity likely selects for cells with defects in the retinoblastoma (Rb) and p53 tumor suppressor pathways. While BL biopsies are usually pseudodiploid, comparative genomic hybridization and spectral karyotypic analysis have found that, similar to mouse models, numerous chromosomal aberrations, including deletions are present [88]. In summary, it appears that selection for somatic mutations in tumor suppressor pathways is the primary determinant in c-myc-induced B cell lymphomagenesis. Once cells resistant to apoptosis emerge, the growth and proliferative functions of c-myc are able to drive tumorigenesis.

2
Possible Mechanisms of c-myc-Induced Instability

The complex karyotype that is observed in biopsies from human tumors is a footprint of multiple genetic changes that have occurred during tumorigenesis. Therefore, it is not possible to conclude when during tumor progression such changes arose, and whether the instability is a continuing process or a reflection of a historic event. Consequently, it is not possible to derive cause and effect relationships between genomic instability and *c-myc* overexpression by analyzing archival human tumor samples. However, an examination of gene amplification mechanisms suggests how excess myc activity and genomic instability might be causally linked. The two mechanisms for amplification in mammalian cells are re-replication of target loci and induction of strand breaks [24, 52, 89–91]. Re-replication involves the initiation of multiple rounds of DNA replication within a single S-phase. Recent data demonstrate that high-level overexpression of cdc6 and cdt1 proteins, which are required for replication origin licensing, can induce re-replication at some frequency in cancer cell lines [92]. Since c-myc can transactivate genes encoding replication origin licensing proteins ([93, 94] and Sect. 2.5 below), it remains possible that it could induce amplification by a re-replication mechanism.

The second mechanism for gene amplification involves chromosome breakage, which can be induced in a number of ways [52, 95, 96]. Importantly, recent data show that elevated *c-myc* expression can lead to metaphase chromosome abnormalities including those that harbor amplified genes and that usually reflect breakage during G1 or S-phase [72]. Breakage has also been observed in G0/G1 arrested cells expressing the c-Myc/estrogen receptor fusion protein (Myc-ER) under conditions where apoptosis was not induced [97]. The same study showed phosphorylation of p53 on Ser15, an indicator of DNA damage. Finally, c-myc activation can lead to a delay in G2, which usually occurs in cells that have experienced DNA damage during

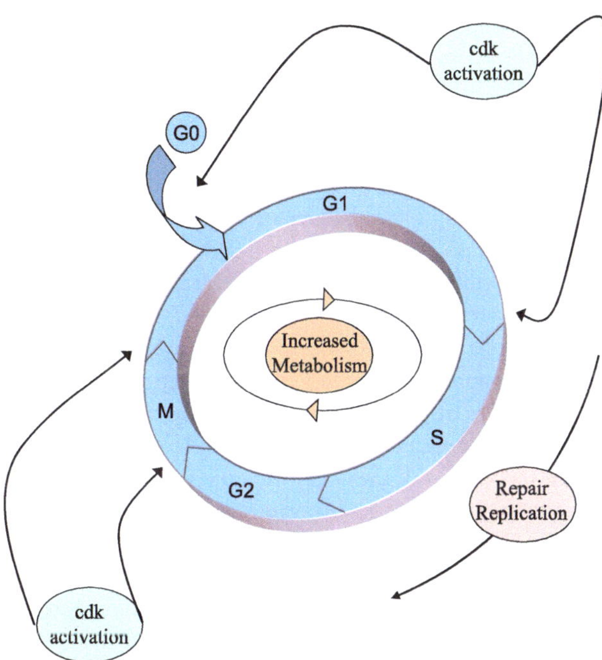

Fig. 1 Summary of potential sites of c-myc-induced DNA damage. Activation of cyclin/cdk complexes by c-myc can lead to premature entry into S-phase or exit from G2/M. Both these events may induce DNA damage as described in the text and in the following figures. Additionally, increased metabolic activity induced by c-myc can generate reactive oxygen species, which can contribute to DNA damage. High level c-myc expression also activates the transcription of DNA replication and repair components, which may impact the fidelity of these processes

S-phase and have arrested for repair [98]. Together, these data support the conclusion that elevated levels of c-myc can induce the types of DNA damage that precede gene amplification and other structural chromosome alterations. The available literature suggests that c-myc may destabilize the genome by multiple mechanisms. This section focuses on five we consider most likely: (1) cell growth and metabolism, (2) unscheduled entry into S-phase, (3 and 4) abrogation of stress-induced cell-cycle checkpoints at G1/S and G2/M, and (5) modulation of DNA damage response and repair pathways (Fig. 1).

2.1 Increased Metabolism and Induction of ROS

The mechanisms by which c-myc couples mitogenic stimulation to growth and proliferation are gradually being elucidated. Physiological activation of c-myc can be achieved in several ways. In quiescent B cells, *c-myc* expression can be activated by nuclear factor (NF)-κB and protein kinase C (PKC) signaling [99], whereas *c-myc* transcription is controlled by src and signal transducer and activator of transcription (STAT) signaling in platelet-derived growth factor (PDGF)-stimulated fibroblasts [100, 101]. Activation of c-myc induces growth of B cells in the absence of proliferation, and c-myc overexpression can increase cell size throughout the cell cycle [102, 103]. Concordant with these results, c-myc gene targets include rate-limiting enzymes in the glycolytic and respiratory pathways and in biosynthetic pathways [104–106].

The metabolic burst associated with emergence from quiescence and entry into S-phase is a potential source of reactive oxygen species (ROS). ROS are essential mediators of proliferative signals, but at high levels can cause oxidative base modifications and single- or double-stranded DNA breaks. If such lesions are not repaired, they may become fixed in the genome during DNA replication. ROS are estimated to induce up to 10,000 lesions per cell per day [107]. However, the mutagenic potential of these lesions is limited by a combination of antioxidants and DNA repair enzymes. It follows that since oncogenes such as *ras* and *c-myc* are key players in mitogenic pathways, aberrant signaling from either might create an oxidative burden. In support of this, activation of oncogenic ras can induce ROS in various cell lines in vitro [108, 109]. Adding to these data, other groups have found that activation of c-myc can increase intracellular ROS [110, 97]. While activation of c-myc is associated with induction of DNA damage in serum-deprived and cycling normal human fibroblasts, preincubation with antioxidant only appears to reduce damage in the former case [97, 111]. These data indicate that although ROS can contribute to c-myc-induced DNA damage under certain circumstances, other mechanisms are also likely to be involved. Data from other studies also highlight the complex role of ROS as mediators of c-myc-induced effects. For example, ROS induced by c-myc in NIH3T3 cells do not appear to be cytotoxic unless the cells are cultured in low serum [110]. Additionally, ROS are mediators of c-myc-induced apoptosis in some human cell lines but are associated with induction of an arrested state resembling senescence in normal human fibroblasts [97, 112]. A similar senescent-like state has been described in normal human fibroblasts exposed to ionizing radiation [113], oxidative stress [114], and following telomere shortening [115]. Taken together, the data support the idea that in some normal cell types, inappropriate c-myc activation can induce sufficient DNA damage to elicit a stress response resulting in some cells undergoing permanent cell-cycle exit.

Elevated ROS are found in some human tumors and tumor-derived cell lines [116, 117]. In addition to their role in mitogenic signaling mentioned above, there is evidence that ROS can also contribute to mutations associated with tumor initiation or progression. For example, many of the point mutations found in tumor suppressor genes in human cancer can be induced by oxidative stress [118–121]. Furthermore, elevated frequency of such lesions can be found in the p53 gene in normal hepatocytes of individuals with Wilson's disease, a disorder associated with elevated ROS and increased risk of hepatocellular carcinoma [122]. There is also an elevated frequency of oxidative stress-related p53 mutations in ulcerative colitis, another disease that is linked to an increased risk of cancer [123].

Induction of MIN occurs predominantly through mutation of mismatch repair genes, but excessive ROS can also lead to MIN in vitro [124, 125]. MIN can generate frameshift mutations in tumor suppressor genes [126], such as those that inactivate the type II transforming growth factor-β receptor (TGF-βRII) [127]. This may allow colon epithelial cells to escape growth restriction mediated by ligation of TGF-β to TGF-βRII. Furthermore, oxidative stress can increase the frequency of frameshift mutations in lung and colorectal carcinoma cell lines [128, 129]. Together these data suggest that ROS may contribute to destabilization of the genome in certain malignancies.

Although many human cancers are associated with environmental agents such as those inhaled by smoking, the age-specific incidence of sporadic cancers of the ovary, pancreas, and colon does not vary significantly between populations [130]. This suggests that endogenous cellular processes may be involved in the initiation of some tumors. The ability of *c-myc* and other oncogenes to activate metabolic pathways leading to oxidative stress suggests they could be considered candidate pro-mutagens. However, whether ROS induced by c-myc in vivo is sufficient to induce somatic mutation remains untested. This is likely to be determined by the contributions of multiple signaling pathways in the cell, which in turn will be influenced by cell type and the surrounding environment. As one example, in a mouse model of HCC, *c-myc* overexpression in hepatocytes results in liver tumors, with a latency of more than 1 year, suggesting that multiple changes are required for c-myc-induced HCC [131]. By contrast, when *TGF-α* is co-expressed with *c-myc*, the latency for tumor onset is decreased dramatically. Concomitantly, ROS levels and chromosomal and mitochondrial genome instability increased [133, 132]. Supplementing the diet of these mice with the antioxidant vitamin E reduced ROS levels and also reduced proliferation. Coincident with the block to proliferation, the amount of genomic instability was also significantly decreased. Additional data showed that mitochondrial DNA deletions were also reduced by vitamin E in this study, providing compelling evidence that ROS produced as a result of a combination of deregulated *c-myc* and *TGF-α* expression can induce DNA damage in vivo. These data suggest that inhibition of proliferation and DNA damage by antioxidants can prevent c-myc-induced instability and tumor progression.

2.2
Unscheduled Entry into S-Phase

In mammalian cells, c-myc activation can increase cell number as well as cell size, which may depend on the cell type [102, 134]. Studies in rodent cells demonstrate that the G1 interval is longer in *c-myc*-null cells when compared to wildtype [135]. These data suggest that c-myc facilitates progression through G1 into S-phase. In part, these observations may be explained by the ability of c-myc to downregulate inhibitors of cyclin/cdk complexes or to stimulate transcription of genes encoding cyclins. The activation of cyclin/cdk complexes removes the block to the transition from G1 to S-phase, which is mediated, at least in part, by the Rb protein [136]. Briefly, hypophosphorylated Rb prevents transcription of genes required for S-phase in two ways. First, Rb can sequester the transcription factor E2F1, which has been implicated in the control of S-phase entry [137]. Second, Rb can form a complex with E2F1 (and other E2F family members) that actively represses S-phase gene transcription [138]. This section will focus only on bypass of the cell-cycle checkpoints associated with the transition from G0/G1 to S-phase in the absence of exogenous stresses. The bypass of DNA damage-induced checkpoints will be addressed in Sects. 2.3 and 2.4).

Numerous mechanisms may promote the transition into S-phase [139–143]. For simplicity, the following illustrates a linear pathway in which c-myc activates cyclin E/cdk2 leading to S-phase entry independently of Rb status. Activation of cyclin E/cdk2 is important for entry into S-phase, although the critical downstream targets are unknown [144–146]. c-myc can activate the cyclin E/cdk2 complex, primarily by altering the levels or distribution of the cyclin E/cdk2 inhibitor, p27. p27 loss is a poor prognostic indicator in tumors of the breast and in gastric and colon carcinoma; a feature of all these cancers is overexpression of *c-myc* [147, 148]. Furthermore, deletion of p27 reduces the latency to tumor onset in *c-myc* transgenic mice [149]. Cdk-2 dependent phosphorylation at threonine 187 is required for degradation of p27 [150]. The phosphorylation allows binding of the Skp1/Cul1/F-box (SCF) ligase complex, which ubiquitinates p27 and targets it for proteasome-mediated degradation [151–153]. Cul1, a component of the SCF ligase complex, is also required for efficient ubiquitination and degradation of p27 [154, 153]. In some systems, c-myc can induce Cul1, leading to p27 degradation and S-phase entry [155]. Together these data provide one explanation for the ability of c-myc to overcome a p27-induced cell-cycle block. Additionally, c-myc can directly target cyclin D2, leading to the sequestration of p27 into heat-labile complexes and permitting cyclin E/cdk2 activation [156]. The activation of cyclin E/cdk2 by c-myc is also sufficient to bypass the G1/S block imposed by hypophosphorylated Rb and p16 [157]. These data indicate one mechanism by which c-myc can bypass Rb-mediated checkpoints without Rb hyperphosphorylation.

Inappropriate cyclin E expression can induce genomic perturbations. For example, the bypass of an Rb-imposed cell-cycle block by c-myc and cyclin E is associated with endoreduplication [141], and cyclin E/cdk2 activity can induce chromosomal instability [158]. Although the mechanism for this is unknown, it is possible that excessive cdk activity might perturb replication origin licensing, which has been linked to instability [159–161]. Interestingly, inappropriate cyclin E/cdk activity appears to accelerate S-phase entry but actually slows replication [158, 162], raising the possibility that DNA damage and activation of the S-phase checkpoint may occur under such conditions. Studies in yeast indicate that precocious cyclin/cdk activity can delay firing of replication origins, leading to strand breakage and chromosomal abnormalities [163]. Whether this can occur in mammalian cells has yet to be shown. However, a reasonable speculation is that inappropriate entry to S-phase induced by c-myc in the absence of correct origin licensing might lead to DNA damage (Fig. 2).

2.3
Abrogation of G1/S Arrest Induced by DNA Damage

DNA damage activates checkpoints throughout the cell cycle that prevent the replication and transmission of mutated DNA [164]. Activation of a p53-dependent checkpoint at or prior to the restriction point can prevent entry

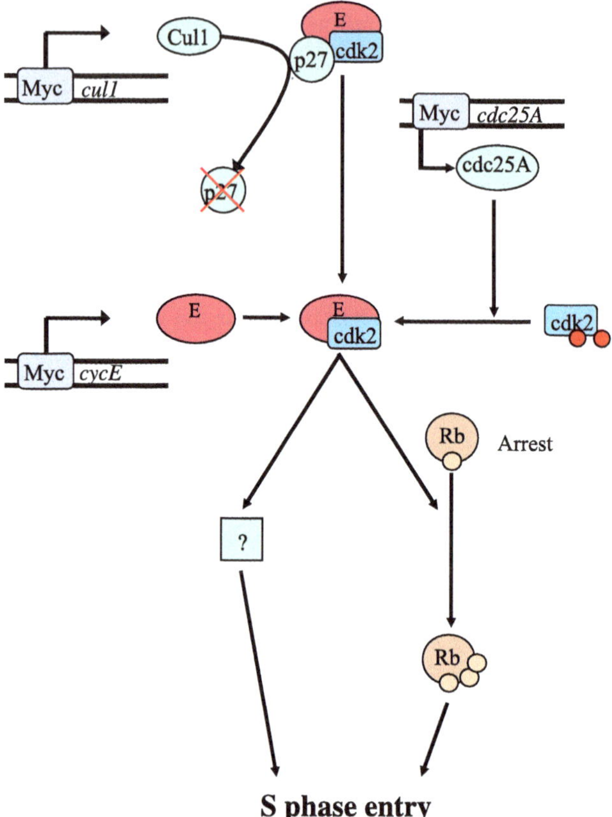

Fig. 2 c-myc can induce restriction point bypass by mulitple mechanisms. c-myc can activate cul1 transcription in some cell types, leading to degradation of the cyclin/cdk2 inhibitor, p27. Additionally, c-myc can transactivate cyclin E and cdc25A, a phosphatase which activates cdk2. Together, these activities activate cyclin E/cdk2 kinase, which in turn should inactivate Rb, release E2Fs and enable S-phase progression. c-myc can also activate a parallel pathway for S-phase progression, which requires cyclin E/cdk2 activation, but does not require inactivation of Rb. The downstream targets of cyclin E/cdk2 in this pathway are unknown

of cells with as few as one unrepaired double-strand break into S-phase [165, 166]. DNA lesions are recognized by specific protein complexes, which transduce the DNA damage signal to downstream effectors to elicit arrest. Below we briefly describe the activation of p53 in response to DNA strand breaks and present experimental data demonstrating that c-myc can attenuate this pathway in some cell strains.

Mre11/Rad50/Nbs1 (MRN) complexes are recruited rapidly to sites of breakage [167]. This termolecular complex is involved in the processing of DNA lesions that arise during replication and following DNA damage [168, 169]. Activation of the ATM kinase also occurs rapidly after strand breakage as a result of an intramolecular phosphorylation event [170]. However, the mechanism by which the break is detected and subsequently activates ATM remains to be determined. Although MRN is phosphorylated by ATM, it can be recruited to sites of damage in the absence of ATM activity, indicating that these two events are not linked [171]. ATM induces direct phosphorylation of p53 at Ser15, and indirectly induces phosphorylation of p53 at Ser20 by activating the damage checkpoint kinase chk2 [172, 173]. These modifications can activate p53 either by decreasing p53 binding to its negative regulator, mdm2, or by increasing association with the transcriptional co-activator p300/CBP [175, 174]. Activated p53 then regulates the transcription of numerous target genes leading to cell-cycle arrest, apoptosis, or increased repair, depending on the cell type and type of damage induced [166]. The inhibition of Rb phosphorylation by p21 is partially responsible for p53-dependent G1 arrest [176].

Constitutive overexpression of *c-myc* in epithelial cells can compromise ionizing radiation-induced arrest, forcing cells into S-phase prematurely [177]. The escape from radiation-induced G1 arrest is a direct result of c-myc action, and not the result of selection for checkpoint-deficient variants, as it occurs in a significant fraction of normal fibroblasts and epithelial cells expressing an inducible *c-myc-ER* construct [97, 177]. The replication of DNA strand breaks during S-phase is a potential source of continuing genomic instability, since break repair could generate dicentric chromosomes, which can then enter into bridge-breakage-fusion cycles (see Sect. 1.2 and [24]). Therefore, c-myc's ability to attenuate damage-induced checkpoints is likely to contribute to genomic instability.

The abrogation of p53-dependent arrest by c-myc can lead to apoptosis in some cell types [178], which could provide a backup mechanism for limiting the emergence of genetically unstable variants. Recent data indicate that regulation of *p21* expression by c-myc is a determinant of the apoptotic response. For example, c-myc can specifically block the DNA damage-induced accumulation of p21 normally observed in colon carcinoma cells [179]. Concomitant with the decrease in p21 levels, the response of the cells to DNA damage was switched from arrest to apoptosis. These data suggest that in the context of a DNA damage signal, *p21* induction should be able to prevent apoptosis. A corollary is that the ability of c-myc to override a damage-induced arrest should require *p21* downregulation, and S-phase entry should induce apoptosis. However, cells overexpressing c-myc can escape damage-induced arrest and enter S-phase with elevated p21 levels [97, 180]. Other studies show that the anti-apoptotic function of p21 does not necessarily require its ability to inhibit the cell cycle [181, 182]. This raises the possibility that cells with damaged DNA that enter the cell cycle due to deregulated *c-myc* expression may evade apoptosis if p21 levels are sustained. In turn, this may increase the possibility that DNA lesions become fixed in the genome during replication or repair.

Felsher and Bishop showed that aneuploidy could be induced by c-myc in exponentially growing Rat1a fibroblasts and normal human fibroblasts, but that damage associated with strand breakage (i.e., double minutes, polycentric chromosomes) was only observed in the Rat1a cells [72]. This is presumably because normal cells respond to strand breaks induced by c-myc by undergoing a p53-dependent arrest resembling senescence [183]. The Rat1a cells are immortal and have no p21 function due to methylation of the promoter [184]. A lack of p53-mediated arrest in rodent cells may create a permissive environment for a wide range of c-myc-induced chromosomal aberrations. Conversely, in human cells, activation of p53 may restrict the emergence of certain types of chromosomal defects, as noted. However, c-myc activity is still able to induce aneuploidy in normal human cells, indicating that it can compromise the fidelity of events associated with mitosis (see Sect. 2.4).

Fig. 3a, b Activation of c-myc can override damage-induced checkpoints. **a** The signaling pathway downstream of DNA damage is simplified for clarity. Following strand breakage, the ATM kinase is activated, although the mechanism by which break detection occurs is unknown. p53 is stabilized and activated by ATM-induced phosphorylation. Activated p53 induces the transcription of numerous target genes, among which are several that induce apoptosis, stimulate DNA repair, or promote cell-cycle arrest. For example, induction of the cyclin/cdk inhibitor, p21 inhibits cyclin-cdks such as cyclin E/cdk2, which prevents Rb hyperphosphorylation and inactivation, thereby blocking S-phase entry. Excess myc activity can attenuate the DNA damage response and induce cell-cycle progression downstream of p53 activation by inhibiting p21 function in some cell types, although in other situations c-myc-induced bypass occurs without apparent alterations of p21 levels (see **b**). For discussion of other components up- and downstream of p53 activation, see Wahl and Carr [166]. **b** Override of the p53-dependent DNA damage response by c-myc. DNA damage can lead to simultaneous, p53-dependent transcription of cell-cycle arrest and pro-apoptotic genes. In some cell types, the induction of p21 can inhibit p53-dependent apoptosis. c-myc can selectively inhibit p21 induction when bound to Miz protein at the p21 promoter, resulting in apoptosis (*1* and [179]). (*2*) In other cell types, c-myc-mediated inhibition of p21 appears to lead to cell-cycle entry, which is dependent on cyclin E/cdk2 activity, but does not involve Miz, and may rather be related to sequestration of p21 into other cyclin-cdk complexes. Under these conditions, replication of damaged DNA may lead to chromosomal abnormalities, which could trigger apoptosis or give rise to genetic variants

Activation of cyclin/cdk complexes by c-myc may also be involved in the abrogation of damage-induced checkpoints. The indirect activation of cyclin E by Myc could potentially participate in this process. Cyclin E and c-myc appear to activate some common elements of the DNA damage response. For example, activation of c-myc or overexpression of cyclin E in the absence of exogenous stress leads to an increase in p53 Ser15 phosphorylation in pri-

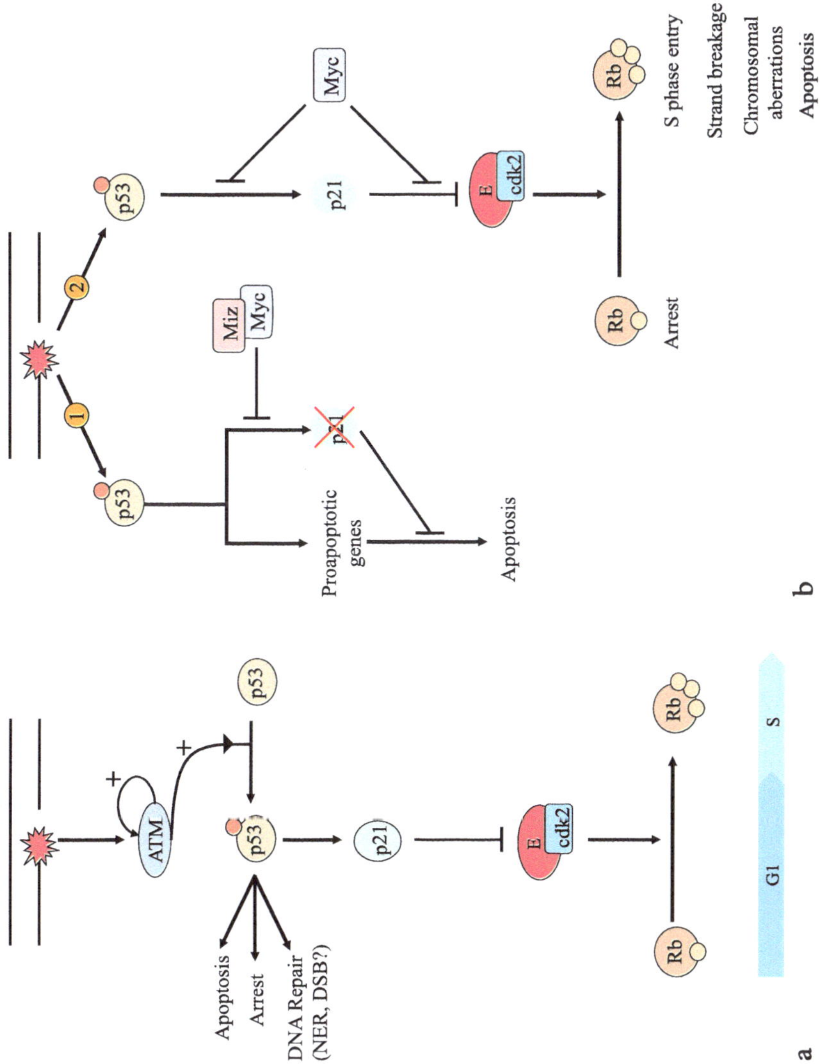

mary cells [97, 185]. This demonstrates that inappropriate proliferative signals induce DNA damage and elicit a classical p53-dependent damage response. However, the mechanisms by which c-myc and cyclin E override DNA damage-induced checkpoints are likely to be distinct. To illustrate, expression of cyclin E induces genomic instability in normal human fibroblasts and immortalized epithelial cells [158, 185]. However, induction of chromosomal instability by cyclin E requires abrogation of p53 or p21 function [185]. In contrast, c-myc can induce chromosomal instability in primary human cells with an intact p53 pathway [72]. Furthermore, c-myc can abrogate ionizing radiation-induced arrest, but cyclin E overexpression is unable to do so [97, 177, 185]. Taken together, these data suggest that activation of cyclin E may contribute to induction of genomic instability by c-myc, but that other activities of c-myc are likely required to bypass damage-induced checkpoints (Fig. 3).

2.4
Abrogation of Arrest at G2/M

The tight coupling of mitosis and DNA replication ensures the replication and faithful segregation to each daughter of only one complete genome per cell cycle [186]. Cell-cycle checkpoints in G2 and M function to maintain the structural integrity of the duplicated chromosomes and ensure their equal partitioning at cell division. Defective processes during mitosis can lead to an abnormal karyotype. For example, aneuploidy occurs following defects in chromosomal segregation. Additionally, abrogation of arrest induced at G2/M can also lead to endoreduplication (re-replication of the genome without cell division) [187–189].

Overexpression of c-myc has been correlated with endoreduplication and aneuploidy in several models. Prolonged arrest at mitosis following exposure to agents that perturb the mitotic spindle results in "mitotic slippage," leaving cells arrested with 4N DNA content in a G1-like biochemical state [190–193]. Overexpression of *c-myc* compromises this arrest, leading to endoreduplication [180]. In addition to drug-induced perturbation of microtubules, sequestration of E2F transcription factors can also lead to mitotic slippage, and c-myc is able to induce endoreduplication under these conditions [141, 177]. In primary cells, endoreduplication is countered by apoptosis [180]. However, in cells that are resistant to apoptosis, such genomic instability can be tolerated [194]. In summary, for cells that have reduced apoptotic responses, c-myc activation could induce cell-cycle progression and lead to endoreduplication, which could perpetuate instability and accelerate tumor progression.

The ability of c-myc activation alone to induce accumulation of cells with 4N DNA content [98] is consistent with its ability to induce sufficient DNA damage to provoke a G2/M checkpoint arrest response. However, G2/M arrest in *c-myc*-expressing cells also seems to lead to increased ploidy. One potential explanation is that under these conditions elevated *c-myc* expression in cells arrested at G2/M may enable DNA synthesis to reinitiate in the absence of cell division to induce polyploidy. Although the mechanism for this is unclear, the data summarized above raise the possibility that it could involve premature activation of cyclin/cdk complexes and other factors involved in replication origin licensing and initiation of S-phase (Fig. 4; see also Sects. 2 and 2.3 above).

2.5
Modulation of DNA Damage Response and Repair Pathways

DNA damage response and repair pathways are present to ensure the faithful replication and segregation of genetic material. Conversely, attenuation of damage response or repair pathways contributes to genomic instability. A link between c-myc activation and DNA metabolism is particularly attractive when the effects of c-myc on replication and genomic instability are considered. This section summarizes recent analyses indicating that c-myc regulates the expression of genes involved in DNA replication and the DNA damage response and repair pathways.

Microarray analyses indicate that c-myc can upregulate genes involved in DNA replication including Topoisomerase I (*TOP1*), *mcm4*, *mcm6*, *mcm7* and *cdt1* [93, 94, 103, 155]. TOP1 is required during DNA replication to relax supercoils that are generated by passing replication forks [195]. Therefore, the induction of this enzyme by c-myc might facilitate S-phase progression. However, overexpression of *TOP1* can induce illegitimate recombination, and trigger instability [196]. Mcm6, mcm7, and cdt1 are required for firing of replication origins and can also induce genomic instability when expressed at high levels ([197] and see Sect. 1).

Although these data show a correlation between myc activation and gene expression, at present their biological significance is unclear. However, two recent reports suggest that components of the DNA repair machinery may be involved in the response to activation of c-myc. The first report focused on the Nbs1 protein, a component of the MRN complex involved in repair of replication and damage-associated breaks ([169] and Sect. 2.3 above). Chiang et al. [198] showed that small interfering (si)RNA-mediated knockdown of *c-myc* decreases Nbs1 levels, and they postulate that induction of *Nbs1* by c-myc is required during DNA replication. However, the length of S-phase is unaffected in *c-myc*-null rat fibroblasts compared to the parental line [135, 199]. Additionally, Nbs1 deficiency in transformed fibroblasts does not affect the rate of DNA synthesis [200]. Further work is therefore required to determine

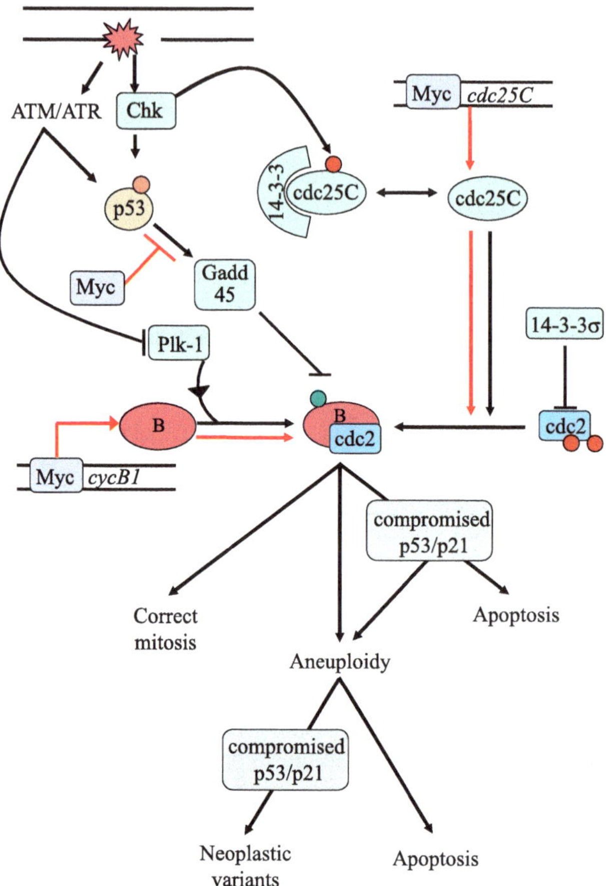

Fig. 4 c-myc Activation of cyclin B/cdc2 may contribute to chromosomal instability. Multiple regulatory pathways converge at cyclin B/cdc2 to control its mitosis-promoting activity. Inhibitory phosphorylations are removed from the cdc2 subunit by cdc25C phosphatase, and Plk-1 kinase phosphorylates cyclin B, leading to activation of the holoenzyme. Following induction of DNA damage by exogenous stresses or oncogene activation, several pathways lead to arrest at G2/M, presumably by inhibiting cyclin B/cdc2. Arrest pathways involve sequestration of cdc25C and cdc2 in the cytosol by 14-3-3 and 14-3-3σ proteins, respectively, and upregulation of the cyclin B/cdc2 inhibitor, Gadd45. c-myc can upregulate cyclin B and cdc25C, leading to activation of cyclin B/cdc2, which should lead to mitotic entry. Additionally, c-myc can attenuate p53 function, which has been implicated by several studies in the G2/M checkpoint. Since c-myc has been reported to induce aneuploidy and can activate cyclin B/cdc2, it is possible that c-myc overexpression perturbs events in G2-M to reduce the fidelity of chromosome segregation. See Sect. 2 for further details

whether c-myc and Nbs1 interact in pathways that affect DNA metabolism. A second study indicated that loss of the WRN protein (a DNA helicase involved in repair) leads to senescence in cells overexpressing *c-myc* [201]. The authors speculate that WRN activity may be required in certain cellular contexts to facilitate c-myc-driven proliferation during tumorigenesis.

It is unclear how c-myc-induced upregulation of DNA repair genes such as *Nbs1* or *WRN* might affect genomic integrity. During normal proliferation, induction of repair enzymes by c-myc might facilitate the resolution of breaks arising during replication and thus contribute to replication fork progression. However, it has also been suggested that inappropriate induction of repair enzymes during S-phase could promote unscheduled repair of replication intermediates and increase the probability of generating chromosomal aberrations [202]. Conversely, inhibition of scheduled DNA repair during the cell cycle can also lead to chromosomal defects. Interestingly, a recent report indicates that c-myc activation may suppress the repair of double-strand breaks in normal human cells [111]. The authors suggest that this may explain the increased frequency of chromosomal rearrangements following activation of c-myc. Whether c-myc inhibits repair directly via transactivation or repression of DNA damage response or repair genes or via a more indirect mechanism remains to be determined. Finally, conditions that accelerate or retard replication fork progression can induce chromosome breakage, suggesting that perturbation of S-phase progression could also increase the probability of chromosomal rearrangement. It is conceivable that *c-myc* overexpression could affect S-phase progression given the number of target genes it regulates with functions in DNA replication [93, 94].

3 Reversible Activation of Oncogenes and Genomic Instability

Loeb postulated that induction of a mutator phenotype initiates a genetically irreversible tumor progression [203]. This is because once genes critical for maintenance of genomic stability are mutated, re-establishment of a normal genome becomes impossible. Therefore, if c-myc is acting as an endogenous activator of the mutator phenotype, turning off *c-myc* expression should not lead to the re-emergence of cells with a normal karyotype. Furthermore, if the gene expression changes resulting from the rearrangements induced by *c-myc* overexpression were sufficient to sustain growth, turning *c-myc* off should not lead to tumor regression. Felsher and Bishop [72] showed that c-myc-induced gene amplification and tumorigenicity persisted in Rat1a cells following c-myc inactivation. These data suggest that, at least in the Rat1a cells, c-myc-driven instability correlated with a durable tumorigenic phenotype that persists in the absence of the initiating event (i.e., c-myc activation).

By contrast with these data, other studies show a requirement for persistent c-myc activity to maintain tumor cells in vivo. T cell lymphomas initiated by *c-myc* activation undergo apoptosis and regression when *c-myc* is turned off [204]. Similarly, inactivation of c-myc in the skin and pancreas leads to regression of papillomatosis and β-cell hyperplasia, respectively, which are accompanied by apoptosis [205, 206]. Osteosarcomas and mammary carcinomas initiated by c-myc also revert after the *myc* transgene is turned off [207]. Mutations in the Wnt pathway leading to excessive Wnt signaling are associated with a number of human cancers [208, 209]. c-myc is positively regulated by the Wnt signaling pathway and may be required for Wnt-induced tumorigenesis [210, 211]. In support of this, activation of Wnt in the breast leads to carcinoma concomitant with elevated c-myc [212]. Similar to the reversible activation of c-myc, inactivation of Wnt is sufficient to induce tumor regression [213].

The regression mechanisms have not been elucidated. Loss of c-myc functions such as proliferation, angiogenesis, and inhibition of differentiation are likely to be important. Another possibility is that genomic instability could be a trigger for apoptosis once c-myc is inactivated. Perhaps c-myc can attenuate signaling from the damaged genome to the apoptotic machinery. Alternatively, c-myc may activate some enzymes involved in DNA metabolism (see Sect. 2), which would prevent apoptosis at the expense of initiating irregular repair. DNA damage could induce apoptosis and regression, but the downstream effectors of apoptosis remain unknown. To illustrate, inactivation of Wnt in the breast leads to regression regardless of p53 status, implying the involvement of p53-independent apoptotic mechanisms [213].

The studies outlined above suggest that *c-myc* expression is required for sustained tumorigenesis. Furthermore these data seem to indicate that genomic destabilization may not be sufficient to maintain tumorigenic potential in these models. Therefore, one might conclude that c-myc is not able to engender the classical mutator phenotype as described by Loeb (see above), since the tumorigenicity is reversible. However, following a period of remission, some tumors resumed growth in the absence of oncogene activity [204, 207]. Murine mammary carcinomas that relapsed in the absence of c-myc activity frequently exhibited ras mutations [207]. Complex chromosomal rearrangements were also observed in relapsed lymphoid tumors that had escaped dependence on c-myc [76]. Interestingly, all relapsed tumors displayed novel karyotypic aberrations compared to primary tumors. It is possible that in the breast model, pre-existing *ras* mutations are present in some of the c-myc-induced tumors and that these cells provide a selective advantage for regrowth in the absence of c-myc activity. In contrast, there was no genetic lesion common to all relapsed lymphoid tumors. This raises the intriguing possibility that acquisition of specific genetic lesions induced by c-myc enhance the propensity for relapse in some tumors. Studies of Wnt-driven tumorigenesis indicate that p53 status is an important determinant of relapse. For example, loss of one p53 allele leads to a sevenfold increase in relapse frequency of breast tumors [213]. This suggests that attenuation of p53 function may be one mechanism by which genomically unstable tumors initiated by oncogenes could relapse. Does this mean that relapsed tumors are those that have sustained somatic mutations in p53 and now provide a selective advantage in the context of c-myc-induced chromosomal changes? Preliminary data from Karlsson et al. [76] suggests that *p53* and *arf* loci are intact in relapsed tumors, indicating that genetic inactivation of these tumor suppressors is not required for escape from oncogene dependence. However, it is possible that epigenetic inactivation of the p53 pathway may contribute to tumor progression in this model.

It is important to note that some tumors do not relapse once c-myc is turned off. For example, full regression of c-myc-driven hyperplasia is observed in the pancreas and skin [205, 206]. Furthermore, osteosarcomas driven by c-myc regress when the transgene is inhibited [214]. Therefore, in some cell types, genomic instability may be insufficient to phenocopy the required functions of c-myc. The basis of these differences is not understood. However, it is possible that hyperplasia in some tissues remains dependent on other functions of c-myc such as its role in stimulating angiogenesis. In addition, c-myc activation in the skin can inhibit or promote differentiation, depending on the cell type, further underscoring the complex response to c-myc in vivo [205, 215].

4
Summary

Oncogenic activation of c-myc affects multiple intracellular pathways, culminating in neoplastic transformation in many cell types. Frequently associated with deregulated c-myc activity are numerical and structural alterations of the karyotype. In certain tumors, comparison of normal and pre-neoplastic tissues reveals chromosomal aberrations specifically associated with c-myc activation. The persistence of these lesions during tumor progression indicates that they are selected for during tumorigenesis. Due to its ability to impact numerous biological functions, c-myc is carefully controlled in the non-pathological state. By extension, deregulated c-myc activity is potentially catastrophic for the cell. Activation of apoptosis in response to c-myc plays a critical role in limiting its deleterious effects. However, should this pathway become disabled or desensitized, c-myc has the potential to wreak havoc on the genome. Mechanistic links between c-myc activation and genome destabilization are beginning to emerge from in vitro and in vivo studies. For example, disruption of cell-cycle checkpoints by c-myc can lead to aberrant DNA replication, a source of genomic instability. Other data indicate that metabolic effects of c-myc, which may be independent of its cell-cycle promoting ability, might also lead to DNA damage. Specifically, oxygen radicals produced following c-myc activation could precipitate genomic changes including break-induced rearrangements and oxidative base modifications. The ability of c-myc to compromise p53-dependent cell-cycle checkpoints indicates that, under certain conditions, genomic perturbations may occur even in the presence of tumor suppressor genes.

In vivo models have provided great insight into the complexities involved in c-myc-induced tumorigenesis. The reversible activation models have demonstrated that many tumors remain dependent on c-myc expression and undergo apoptosis once c-myc is turned off. These data indicate that there is a functional inactivation of the apoptotic pathway in the presence of c-myc activity, rather than a selection for cells that have lost the ability to induce cell death. The mechanism of apoptosis induction following c-myc inactivation is incompletely understood. Many explanations have been put forward, based on some of the known biological effects of c-myc. These include regression of vasculature, which would reduce tumor nutrient supply and re-establishment of differentiation, which may sensitize cells to programmed cell death. However, the link between genome destabilization and apoptosis might offer an alternative explanation. Perhaps DNA damage signaling pathways, which normally initiate apoptosis in response to karyotypic abnormalities, are attenuated while c-myc is expressed. Re-activation of these pathways once c-myc is switched off might lead to the rapid elimination of cells with abnormal genomes. Further studies that address the interaction of c-myc with components of the DNA damage response pathway are likely to provide valuable data in this emerging area of c-myc research. Determining the effect of c-myc expression in the context of DNA damage response/repair pathway deficiencies in vivo may provide further insight into the role of c-myc-induced instability in tumorigenesis.

References

1. Bishop JM (1991) Molecular themes in oncogenesis. Cell 64:235–248
2. Bodmer WF, Tomlinson I (1996) Population genetics of tumours. Ciba Found Symp 197:181–189; discussion 189–193

3. Hanahan D, Weinberg RA (2000) The hallmarks of cancer. Cell 100:57–70
4. Potter M, Melchers F (eds) (1997) Proceedings of the 14th Mechanisms in B-cell Neoplasia meeting. Bethesda, Maryland, 21–23 October 21–23 1996. Current Topics in Microbiology and Immunology, vol 224. Springer-Verlag, Heidelberg Berlin New York, pp 1–291
5. Watson PH, Singh R, Hole AK (1996) Influence of c-myc on the progression of human breast cancer. In: Gunthert U, Birchmeier B (eds) Current Topics in Microbiology and Immunology, vol 213. Springer-Verlag, Heidelberg Berlin New York, pp 267–283
6. Mushinski JF, Hanley-Hyde J, Rainey GJ, Kuschak TI, Taylor C, Fluri M, Stevens LM, Henderson DW, Mai S (1999) Myc-induced cyclin D2 genomic instability in murine B cell neoplasms. In: Melchers F, Potter M (eds) Current Topics in Microbiology and Immunology, vol 246. Springer-Verlag, Heidelberg Berlin New York, pp 183–189; discussion 190–192
7. Eisenman RN (2001) Deconstructing myc. Genes Dev 15:2023–2030
8. Oster SK, Ho CS, Soucie EL, Penn LZ (2002) The myc oncogene: MarvelouslY complex. Adv Cancer Res 84:81–154
9. Lutz W, Leon J, Eilers M (2002) Contributions of Myc to tumorigenesis. Biochim Biophys Acta 1602:61–71
10. Pelengaris S, Khan M, Evan G (2002) c-myc: more than just a matter of life and death. Nat Rev Cancer 2:764–776
11. Armitage P, Doll R (1954) The age distribution of cancer and multistage theory of carcinogenesis. Br J Cancer 8:1–12
12. Tomlinson I, Bodmer W (1999) Selection, the mutation rate and cancer: ensuring that the tail does not wag the dog. Nat Med 5:11–12
13. Loeb KR, Loeb LA (2000) Significance of multiple mutations in cancer. Carcinogenesis 21:379–385
14. Savelyeva L, Schwab M (2001) Amplification of oncogenes revisited: from expression profiling to clinical application. Cancer Lett 167:115–123
15. Popescu NC, Zimonjic DB (2002) Chromosome-mediated alterations of the MYC gene in human cancer. J Cell Mol Med 6:151–159
16. Wright JA, Smith HS, Watt FM, Hancock MC, Hudson DL, Stark GR (1990) DNA amplification is rare in normal human cells. Proc Natl Acad Sci U S A 87:1791–1795
17. Livingstone LR, White A, Sprouse J, Livanos E, Jacks T, Tlsty TD (1992) Altered cell cycle arrest and gene amplification potential accompany loss of wild-type p53. Cell 70:923–935
18. Yin Y, Tainsky MA, Bischoff FZ, Strong LC, Wahl GM (1992) Wild-type p53 restores cell cycle control and inhibits gene amplification in cells with mutant p53 alleles. Cell 70:937–948
19. Lengauer C, Kinzler KW, Vogelstein B (1997) Genetic instability in colorectal cancers. Nature 386:623–627
20. Ionov Y, Peinado MA, Malkhosyan S, Shibata D, Perucho M (1993) Ubiquitous somatic mutations in simple repeated sequences reveal a new mechanism for colonic carcinogenesis. Nature 363:558–561
21. Thibodeau SN, Bren G, Schaid D (1993) Microsatellite instability in cancer of the proximal colon. Science 260:816–819

22. Jallepalli PV, Lengauer C (2001) Chromosome segregation and cancer: cutting through the mystery. Nat Rev Cancer 1:109–117
23. Klein G (1999) Immunoglobulin gene associated chromosomal translocations in B-cell derived tumors. In: Melchers F, Potter M (eds) Current Topics in Microbiology and Immunology, vol 246. Springer-Verlag, Heidelberg Berlin New York, pp 161–167
24. Stark GR (1993) Regulation and mechanisms of mammalian gene amplification. Adv Cancer Res 61:87–113
25. Parsons R, Li GM, Longley M, Modrich P, Liu B, Berk T, Hamilton SR, Kinzler KW, Vogelstein B (1995) Mismatch repair deficiency in phenotypically normal human cells. Science 268:738–740
26. Fishel R, Lescoe MK, Rao MR, Copeland NG, Jenkins NA, Garber J, Kane M, Kolodner R (1993) The human mutator gene homolog MSH2 and its association with hereditary nonpolyposis colon cancer. Cell 75:1027–1038
27. Wright WE, Shay JW (2001) Cellular senescence as a tumor-protection mechanism: the essential role of counting. Curr Opin Genet Dev 11:98–103
28. Hickman JA (2002) Apoptosis and tumourigenesis. Curr Opin Genet Dev 12:67–72
29. Balmain A, Harris CC (2000) Carcinogenesis in mouse and human cells: parallels and paradoxes. Carcinogenesis 21:371–377
30. Ross JA, Nesnow S (1999) Polycyclic aromatic hydrocarbons: correlations between DNA adducts and ras oncogene mutations. Mutat Res 424:155–166
31. Bos JL (1988) The ras gene family and human carcinogenesis. Mutat Res 195:255–271
32. Balmain A, Brown K (1988) Oncogene activation in chemical carcinogenesis. Adv Cancer Res 51:147–182
33. Nichols WW (1963) Relationships of viruses, chromosomes and carcinogenesis. Hereditas 50:53–80
34. Nowell PC (1965) Chromosome changes in primary tumors. Prog Exp Tumor Res 7:83–103
35. Loeb LA, Springgate CF, Battula N (1974) Errors in DNA replication as a basis of malignant changes. Cancer Res 34:2311–2321
36. Bronner CE, Baker SM, Morrison PT, Warren G, Smith LG, Lescoe MK, Kane M, Earabino C, Lipford J, Lindblom A, et al (1994) Mutation in the DNA mismatch repair gene homologue hMLH1 is associated with hereditary non-polyposis colon cancer. Nature 368:258–261
37. Duval A, Hamelin R (2002) Mutations at coding repeat sequences in mismatch repair-deficient human cancers: toward a new concept of target genes for instability. Cancer Res 62:2447–2454
38. Hussein MR, Sun M, Roggero E, Sudilovsky EC, Tuthill RJ, Wood GS, Sudilovsky O (2002) Loss of heterozygosity, microsatellite instability, and mismatch repair protein alterations in the radial growth phase of cutaneous malignant melanomas. Mol Carcinog 34:35–44
39. Simpson AJ, Caballero OL, Pena SD (2001) Microsatellite instability as a tool for the classification of gastric cancer. Trends Mol Med 7:76–80
40. Nichols WW, Levan A, Heneen WK, Peluse M (1965) Synergism of the Schmidt-Ruppin strain of the Rous sarcoma virus and cytidine triphosphate in the induction of chromosome breaks in human cultured leukocytes. Hereditas 54:213–236

41. Stehelin D, Guntaka RV, Varmus HE, Bishop JM (1976) Purification of DNA complementary to nucleotide sequences required for neoplastic transformation of fibroblasts by avian sarcoma viruses. J Mol Biol 101:349–365
42. Nanus DM, Lynch SA, Rao PH, Anderson SM, Jhanwar SC, Albino AP (1991) Transformation of human kidney proximal tubule cells by a src-containing retrovirus. Oncogene 6:2105–2111
43. Neil JC, Cameron ER (2002) Retroviral insertion sites and cancer: fountain of all knowledge? Cancer Cell 2:253–255
44. Tsichlis PN (1987) Oncogenesis by Moloney murine leukemia virus. Anticancer Res 7:171–180
45. Hsu T, Moroy T, Etiemble J, Louise A, Trepo C, Tiollais P, Buendia MA (1988) Activation of c-myc by woodchuck hepatitis virus insertion in hepatocellular carcinoma. Cell 55:627–635
46. Kim R, Trubetskoy A, Suzuki T, Jenkins NA, Copeland NG, Lenz J (2003) Genome-based identification of cancer genes by proviral tagging in mouse retrovirus-induced T-cell lymphomas. J Virol 77:2056–2062
47. zur Hausen H (1991) Human papillomaviruses in the pathogenesis of anogenital cancer. Virology 184:9–13
48. Bibbo M, Montag AG, Lerma-Puertas E, Dytch HE, Leelakusolvong S, Bartels PH (1989) Karyometric marker features in tissue adjacent to invasive cervical carcinomas. Anal Quant Cytol Histol 11:281–285
49. Munger K, Howley PM (2002) Human papillomavirus immortalization and transformation functions. Virus Res 89:213–228
50. White AE, Livanos EM, Tlsty TD (1994) Differential disruption of genomic integrity and cell cycle regulation in normal human fibroblasts by the HPV oncoproteins. Genes Dev 8:666–677
51. Duensing S, Munger K (2002) The human papillomavirus type 16 E6 and E7 oncoproteins independently induce numerical and structural chromosome instability. Cancer Res 62:7075–7082
52. Windle B, Draper BW, Yin YX, O'Gorman S, Wahl GM (1991) A central role for chromosome breakage in gene amplification deletion formation, and amplicon integration. Genes Dev 5:160–174
53. Denko NC, Giaccia AJ, Stringer JR, Stambrook PJ (1994) The human Ha-ras oncogene induces genomic instability in murine fibroblasts within one cell cycle. Proc Natl Acad Sci U S A 91:5124–5128
54. Fukasawa K, Vande Woude GF (1997) Synergy between the Mos/mitogen-activated protein kinase pathway and loss of p53 function in transformation and chromosome instability. Mol Cell Biol 17:506–518
55. Kolligs FT, Kolligs B, Hajra KM, Hu G, Tani M, Cho KR, Fearon ER (2000) γ-Catenin is regulated by the APC tumor suppressor and its oncogenic activity is distinct from that of β-catenin. Genes Dev 14:1319–1331
56. van Es JH, Barker N, Clevers H (2003) You Wnt some, you lose some: oncogenes in the Wnt signaling pathway. Curr Opin Genet Dev 13:28–33
57. Boxer LM, Dang CV (2001) Translocations involving c-myc and c-myc function. Oncogene 20:5595–5610

58. Bahram F, von der Lehr N, Cetinkaya C, Larsson LG (2000) c-Myc hot spot mutations in lymphomas result in inefficient ubiquitination and decreased proteasome-mediated turnover. Blood 95:2104–2110
59. Gregory MA, Hann SR (2000) c-Myc proteolysis by the ubiquitin-proteasome pathway: stabilization of c-Myc in Burkitt's lymphoma cells. Mol Cell Biol 20:2423–2435
60. Bonilla M, Ramirez M, Lopez-Cueto J, Gariglio P (1988) In vivo amplification and rearrangement of c-myc oncogene in human breast tumors. J Natl Cancer Inst 80:665–671
61. Escot C, Theillet C, Lidereau R, Spyratos F, Champeme MH, Gest J, Callahan R (1986) Genetic alteration of the c-myc protooncogene (MYC) in human primary breast carcinomas. Proc Natl Acad Sci U S A 83:4834–4838
62. Sikora K, Chan S, Evan G, Gabra H, Markham N, Stewart J, Watson J (1987) c-myc oncogene expression in colorectal cancer. Cancer 59:1289–1295
63. Smith DR, Myint T, Goh HS (1993) Over-expression of the c-myc proto-oncogene in colorectal carcinoma. Br J Cancer 68:407–413
64. Stewart J, Evan G, Watson J, Sikora K (1986) Detection of the c-myc oncogene product in colonic polyps and carcinomas. Br J Cancer 53:1–6
65. Nagy P, Evarts RP, Marsden E, Roach J, Thorgeirsson SS (1988) Cellular distribution of c-myc transcripts during chemical hepatocarcinogenesis in rats. Cancer Res 48:5522–5527
66. Sargent LM, Sanderson ND, Thorgeirsson SS (1996) Ploidy and karyotypic alterations associated with early events in the development of hepatocarcinogenesis in transgenic mice harboring c-myc and transforming growth factor alpha transgenes. Cancer Res 56:2137–2142
67. McCormack SJ, Weaver Z, Deming S, Natarajan G, Torri J, Johnson MD, Liyanage M, Ried T, Dickson RB (1998) Myc/p53 interactions in transgenic mouse mammary development, tumorigenesis and chromosomal instability. Oncogene 16:2755–2766
68. Fearon ER, Gruber SB (2001) Molecular abnormalities in colon and rectal cancer. In: Mendelson J, Howley PM, Israel MA, Liotta LA (eds) The molecular basis of cancer. W.B. Saunders, Philadelphia, pp 289–312
69. Denis N, Kitzis A, Kruh J, Dautry F, Corcos D (1991) Stimulation of methotrexate resistance and dihydrofolate reductase gene amplification by c-myc. Oncogene 6:1453–1457
70. Mai S (1994) Overexpression of c-myc precedes amplification of the gene encoding dihydrofolate reductase. Gene 148:253–260
71. Mai S, Hanley-Hyde J, Fluri M (1996) c-Myc overexpression associated DHFR gene amplification in hamster, rat, mouse and human cell lines. Oncogene 12:277–288
72. Felsher DW, Bishop JM (1999) Transient excess of MYC activity can elicit genomic instability and tumorigenesis. Proc Natl Acad Sci U S A 96:3940–3944
73. Mai S, Hanley-Hyde J, Rainey GJ, Kuschak TI, Paul JT, Littlewood TD, Mischak H, Stevens LM, Henderson DW, Mushinski JF (1999) Chromosomal and extrachromosomal instability of the cyclin D2 gene is induced by Myc overexpression. Neoplasia 1:241–252

74. Kuschak TI, Taylor C, McMillan-Ward E, Israels S, Henderson DW, Mushinski JF, Wright JA, Mai S (1999) The ribonucleotide reductase R2 gene is a non-transcribed target of c-Myc-induced genomic instability. Gene 238:351–365
75. Paulson TG, Almasan A, Brody LL, Wahl GM (1998) Gene amplification in a p53-deficient cell line requires cell cycle progression under conditions that generate DNA breakage. Mol Cell Biol 18:3089–3100
76. Karlsson A, Giuriato S, Tang F, Fung-Weier J, Levan G, Felsher DW (2003) Genomically complex lymphomas undergo sustained tumor regression upon MYC inactivation unless they acquire novel chromosomal translocations. Blood 101:2797–2803
77. Hynes NE (1993) Amplification and overexpression of the erbB-2 gene in human tumors: its involvement in tumor development, significance as a prognostic factor, and potential as a target for cancer therapy. Semin Cancer Biol 4:19–26
78. Jiang W, Kahn SM, Tomita N, Zhang YJ, Lu SH, Weinstein IB (1992) Amplification and expression of the human cyclin D gene in esophageal cancer. Cancer Res 52:2980–2983
79. Oliner JD, Kinzler KW, Meltzer PS, George DL, Vogelstein B (1992) Amplification of a gene encoding a p53-associated protein in human sarcomas. Nature 358:80–83
80. Sargent LM, Zhou X, Keck CL, Sanderson ND, Zimonjic DB, Popescu NC, Thorgeirsson SS (1999) Nonrandom cytogenetic alterations in hepatocellular carcinoma from transgenic mice overexpressing c-Myc and transforming growth factor-alpha in the liver. Am J Pathol 154:1047–1055
81. Adams JM, Harris AW, Pinkert CA, Corcoran LM, Alexander WS, Cory S, Palmiter RD, Brinster RL (1985) The c-myc oncogene driven by immunoglobulin enhancers induces lymphoid malignancy in transgenic mice. Nature 318:533–538
82. Kovalchuk AL, Qi CF, Torrey TA, Taddesse-Heath L, Feigenbaum L, Park SS, Gerbitz A, Klobeck G, Hoertnagel K, Polack A, Bornkamm GW, Janz S, Morse HC 3rd (2000) Burkitt lymphoma in the mouse. J Exp Med 192:1183–1190
83. Schmitt CA, McCurrach ME, de Stanchina E, Wallace-Brodeur RR, Lowe SW (1999) INK4a/ARF mutations accelerate lymphomagenesis and promote chemoresistance by disabling p53. Genes Dev 13:2670–2677
84. Eischen CM, Roussel MF, Korsmeyer SJ, Cleveland JL (2001) Bax loss impairs Myc-induced apoptosis and circumvents the selection of p53 mutations during Myc-mediated lymphomagenesis. Mol Cell Biol 21:7653–7662
85. Strasser A, Harris AW, Bath ML, Cory S (1990) Novel primitive lymphoid tumours induced in transgenic mice by cooperation between myc and bcl-2. Nature 348:331–333
86. Schmitt CA, Fridman JS, Yang M, Baranov E, Hoffman RM, Lowe SW (2002) Dissecting p53 tumor suppressor functions in vivo. Cancer Cell 1:289–298
87. Rockwood LD, Torrey TA, Kim JS, Coleman AE, Kovalchuk AL, Xiang S, Ried T, Morse HC 3rd, Janz S (2002) Genomic instability in mouse Burkitt lymphoma is dominated by illegitimate genetic recombinations, not point mutations. Oncogene 21:7235–7240
88. Lindstrom MS, Wiman KG (2002) Role of genetic and epigenetic changes in Burkitt lymphoma. Semin Cancer Biol 12:381–387
89. Schimke RT (1986) Methotrexate resistance and gene amplification. Mechanisms and implications. Cancer 57:1912–1917

90. Schimke RT, Sherwood SW, Hill AB, Johnston RN (1986) Overreplication and recombination of DNA in higher eukaryotes: potential consequences and biological implications. Proc Natl Acad Sci U S A 83:2157–2161
91. Windle BE, Wahl GM (1992) Molecular dissection of mammalian gene amplification: new mechanistic insights revealed by analyses of very early events. Mutat Res 276:199–224
92. Vaziri C, Saxena S, Jeon Y, Lee C, Murata K, Machida Y, Wagle N, Hwang DS, Dutta A (2003) A p53-dependent checkpoint pathway prevents rereplication. Mol Cell 11:997–1008
93. Watson JD, Oster SK, Shago M, Khosravi F, Penn LZ (2002) Identifying genes regulated in a Myc-dependent manner. J Biol Chem 277:36921–36930
94. Fernandez PC, Frank SR, Wang L, Schroeder M, Liu S, Greene J, Cocito A, Amati B (2003) Genomic targets of the human c-Myc protein. Genes Dev 17:1115–1129
95. Smith KA, Agarwal ML, Chernov MV, Chernova OB, Deguchi Y, Ishizaka Y, Patterson TE, Poupon MF, Stark GR (1995) Regulation and mechanisms of gene amplification. Philos Trans R Soc Lond B Biol Sci 347:49–56
96. Galloway SM (1994) Chromosome aberrations induced in vitro: mechanisms, delayed expression, and intriguing questions. Environ Mol Mutagen 23 Suppl 24:44–53
97. Vafa O, Wade M, Kern S, Beeche M, Pandita TK, Hampton GM, Wahl GM (2002) c-Myc can induce DNA damage increase reactive oxygen species, and mitigate p53 function: a mechanism for oncogene-induced genetic instability. Mol Cell 9:1031–1044
98. Felsher DW, Zetterberg A, Zhu J, Tlsty T, Bishop JM (2000) Overexpression of MYC causes p53-dependent G2 arrest of normal fibroblasts. Proc Natl Acad Sci U S A 97:10544–10548
99. Snow EC (1997) The role of c-myc during normal B cell proliferation, and as B cells undergo malignant transformation. In: Potter M, Melchers F (eds) Current Topics in Microbiology and Immunology, vol 224. Springer-Verlag, Heidelberg Berlin New York, pp 211–220
100. Bowman T, Broome MA, Sinibaldi D, Wharton W, Pledger WJ, Sedivy JM, Irby R, Yeatman T, Courtneidge SA, Jove R (2001) Stat3-mediated Myc expression is required for Src transformation and PDGF-induced mitogenesis. Proc Natl Acad Sci U S A 98:7319–7324
101. Courtneidge SA (2002) Role of Src in signal transduction pathways. The Jubilee Lecture. Biochem Soc Trans 30:11–17
102. Iritani BM, Eisenman RN (1999) c-Myc enhances protein synthesis and cell size during B lymphocyte development. Proc Natl Acad Sci U S A 96:13180–13185
103. Schuhmacher M, Staege MS, Pajic A, Polack A, Weidle UH, Bornkamm GW, Eick D, Kohlhuber F (1999) Control of cell growth by c-Myc in the absence of cell division. Curr Biol 9:1255–1258
104. Gomez-Roman N, Grandori C, Eisenman RN, White RJ (2003) Direct activation of RNA polymerase III transcription by c-Myc. Nature 421:290–294
105. O'Connell BC, Cheung AF, Simkevich CP, Tam W, Ren X, Mateyak MK, Sedivy JM (2003) A large scale genetic analysis of c-Myc-regulated gene expression patterns. J Biol Chem 278:12563–12573

106. Shim H, Dolde C, Lewis BC, Wu CS, Dang G, Jungmann RA, Dalla-Favera R, Dang CV (1997) c-Myc transactivation of LDH-A: implications for tumor metabolism and growth. Proc Natl Acad Sci U S A 94:6658–6663
107. Helbock HJ, Beckman KB, Shigenaga MK, Walter PB, Woodall AA, Yeo HC, Ames BN (1998) DNA oxidation matters: the HPLC-electrochemical detection assay of 8-oxo-deoxyguanosine and 8-oxo-guanine. Proc Natl Acad Sci U S A 95:288–293
108. Irani K, Xia Y, Zweier JL, Sollott SJ, Der CJ, Fearon ER, Sundaresan M, Finkel T, Goldschmidt-Clermont PJ (1997) Mitogenic signaling mediated by oxidants in Ras-transformed fibroblasts. Science 275:1649–1652
109. Yang JQ, Li S, Huang Y, Zhang HJ, Domann FE, Buettner GR, Oberley LW (2001) V-Ha-Ras overexpression induces superoxide production and alters levels of primary antioxidant enzymes. Antioxid Redox Signal 3:697–709
110. Tanaka H, Matsumura I, Ezoe S, Satoh Y, Sakamaki T, Albanese C, Machii T, Pestell RG, Kanakura Y (2002) E2F1 and c-Myc potentiate apoptosis through inhibition of NF-kappaB activity that facilitates MnSOD-mediated ROS elimination. Mol Cell 9:1017–1029
111. Karlsson A, Deb-Basu D, Cherry A, Turner S, Ford J, Felsher DW (2003) Defective double-strand DNA break repair and chromosomal translocations by MYC overexpression. Proc Natl Acad Sci U S A 100(17):9974–9979
112. Xu Y, Nguyen Q, Lo DC, Czaja MJ (1997) c-myc-Dependent hepatoma cell apoptosis results from oxidative stress and not a deficiency of growth factors. J Cell Physiol 170:192–199
113. Linke SP, Clarkin KC, Wahl GM (1997) p53 mediates permanent arrest over multiple cell cycles in response to gamma-irradiation. Cancer Res 57:1171–1179
114. Chen Q, Ames BN (1994) Senescence-like growth arrest induced by hydrogen peroxide in human diploid fibroblast F65 cells. Proc Natl Acad Sci U S A 91:4130–4134
115. Harley CB, Futcher AB, Greider CW (1990) Telomeres shorten during ageing of human fibroblasts. Nature 345:458–460
116. Matsui A, Ikeda T, Enomoto K, Hosoda K, Nakashima H, Omae K, Watanabe M, Hibi T, Kitajima M (2000) Increased formation of oxidative DNA damage 8-hydroxy-2′-deoxyguanosine, in human breast cancer tissue and its relationship to GSTP1 and COMT genotypes. Cancer Lett 151:87–95
117. Gupta A, Rosenberger SF, Bowden GT (1999) Increased ROS levels contribute to elevated transcription factor and MAP kinase activities in malignantly progressed mouse keratinocyte cell lines. Carcinogenesis 20:2063–2073
118. Feig DI, Reid TM, Loeb LA (1994) Reactive oxygen species in tumorigenesis. Cancer Res 54:1890s–1894s
119. Shibutani S, Takeshita M, Grollman AP (1991) Insertion of specific bases during DNA synthesis past the oxidation-damaged base 8-oxodG. Nature 349:431–434
120. Rodin SN, Rodin AS (2000) Human lung cancer and p53: the interplay between mutagenesis and selection. Proc Natl Acad Sci U S A 97:12244–12249
121. Cheng KC, Cahill DS, Kasai H, Nishimura S, Loeb LA (1992) 8-Hydroxyguanine, an abundant form of oxidative DNA damage, causes G-T and A-C substitutions. J Biol Chem 267:166–172

122. Hussain SP, Raja K, Amstad PA, Sawyer M, Trudel LJ, Wogan GN, Hofseth LJ, Shields PG, Billiar TR, Trautwein C, Hohler T, Galle PR, Phillips DH, Markin R, Marrogi AJ, Harris CC (2000) Increased p53 mutation load in nontumorous human liver of Wilson disease and hemochromatosis: oxyradical overload diseases. Proc Natl Acad Sci U S A 97:12770–12775
123. Hussain SP, Amstad P, Raja K, Ambs S, Nagashima M, Bennett WP, Shields PG, Ham AJ, Swenberg JA, Marrogi AJ, Harris CC (2000) Increased p53 mutation load in noncancerous colon tissue from ulcerative colitis: a cancer-prone chronic inflammatory disease. Cancer Res 60:3333–3337
124. Jackson AL, Chen R, Loeb LA (1998) Induction of microsatellite instability by oxidative DNA damage. Proc Natl Acad Sci U S A 95:12468–12473
125. Jackson AL, Loeb LA (2000) Microsatellite instability induced by hydrogen peroxide in Escherichia coli. Mutat Res 447:187–198
126. Huang J, Papadopoulos N, McKinley AJ, Farrington SM, Curtis LJ, Wyllie AH, Zheng S, Willson JK, Markowitz SD, Morin P, Kinzler KW, Vogelstein B, Dunlop MG (1996) APC mutations in colorectal tumors with mismatch repair deficiency. Proc Natl Acad Sci U S A 93:9049–9054
127. Markowitz S, Wang J, Myeroff L, Parsons R, Sun L, Lutterbaugh J, Fan RS, Zborowska E, Kinzler KW, Vogelstein B, et al (1995) Inactivation of the type II TGF-beta receptor in colon cancer cells with microsatellite instability. Science 268:1336–1338
128. Gasche C, Chang CL, Rhees J, Goel A, Boland CR (2001) Oxidative stress increases frameshift mutations in human colorectal cancer cells. Cancer Res 61:7444–7448
129. Zienolddiny S, Ryberg D, Haugen A (2000) Induction of microsatellite mutations by oxidative agents in human lung cancer cell lines. Carcinogenesis 21:1521–1526
130. Muir CS, Wagner G, Demaret E, Nagy-Tiborcz A, Schlaefer K, Villhauer-Lehr M, Whelan S (1982) Directory of on-going research in cancer epidemiology 1982. IARC Sci Publ, Lyon, pp 1–715
131. Murakami H, Sanderson ND, Nagy P, Marino PA, Merlino G, Thorgeirsson SS (1993) Transgenic mouse model for synergistic effects of nuclear oncogenes and growth factors in tumorigenesis: interaction of c-myc and transforming growth factor alpha in hepatic oncogenesis. Cancer Res 53:1719–1723
132. Factor VM, Kiss A, Woitach JT, Wirth PJ, Thorgeirsson SS (1998) Disruption of redox homeostasis in the transforming growth factor-alpha/c-myc transgenic mouse model of accelerated hepatocarcinogenesis. J Biol Chem 273:15846–15853
133. Factor VM, Laskowska D, Jensen MR, Woitach JT, Popescu NC, Thorgeirsson SS (2000) Vitamin E reduces chromosomal damage and inhibits hepatic tumor formation in a transgenic mouse model. Proc Natl Acad Sci U S A 97:2196–2201
134. Trumpp A, Refaeli Y, Oskarsson T, Gasser S, Murphy M, Martin GR, Bishop JM (2001) c-Myc regulates mammalian body size by controlling cell number but not cell size. Nature 414:768–773
135. Mateyak MK, Obaya AJ, Adachi S, Sedivy JM (1997) Phenotypes of c-Myc-deficient rat fibroblasts isolated by targeted homologous recombination. Cell Growth Differ 8:1039–1048
136. Blagosklonny MV, Pardee AB (2002) The restriction point of the cell cycle. Cell Cycle 1:103–110

137. Chellappan SP, Hiebert S, Mudryj M, Horowitz JM, Nevins JR (1991) The E2F transcription factor is a cellular target for the RB protein. Cell 65:1053–1061
138. Weintraub SJ, Chow KN, Luo RX, Zhang SH, He S, Dean DC (1995) Mechanism of active transcriptional repression by the retinoblastoma protein. Nature 375:812–815
139. Beier R, Burgin A, Kiermaier A, Fero M, Karsunky H, Saffrich R, Moroy T, Ansorge W, Roberts J, Eilers M (2000) Induction of cyclin E-cdk2 kinase activity, E2F-dependent transcription and cell growth by Myc are genetically separable events. EMBO J 19:5813–5823
140. Mateyak MK, Obaya AJ, Sedivy JM (1999) c-Myc regulates cyclin D-Cdk4 and -Cdk6 activity but affects cell cycle progression at multiple independent points. Mol Cell Biol 19:4672–4683
141. Santoni-Rugiu E, Falck J, Mailand N, Bartek J, Lukas J (2000) Involvement of Myc activity in a G(1)/S-promoting mechanism parallel to the pRb/E2F pathway. Mol Cell Biol 20:3497–3509
142. Lasorella A, Noseda M, Beyna M, Yokota Y, Iavarone A (2000) Id2 is a retinoblastoma protein target and mediates signalling by Myc oncoproteins. Nature 407:592–598
143. Leone G, Sears R, Huang E, Rempel R, Nuckolls F, Park CH, Giangrande P, Wu L, Saavedra HI, Field SJ, Thompson MA, Yang H, Fujiwara Y, Greenberg ME, Orkin S, Smith C, Nevins JR (2001) Myc requires distinct E2F activities to induce S phase and apoptosis. Mol Cell 8:105–113
144. Dulic V, Lees E, Reed SI (1992) Association of human cyclin E with a periodic G1-S phase protein kinase. Science 257:1958–1961
145. Hatakeyama M, Brill JA, Fink GR, Weinberg RA (1994) Collaboration of G1 cyclins in the functional inactivation of the retinoblastoma protein. Genes Dev 8:1759–1771
146. Hinds PW, Mittnacht S, Dulic V, Arnold A, Reed SI, Weinberg RA (1992) Regulation of retinoblastoma protein functions by ectopic expression of human cyclins. Cell 70:993–1006
147. Loda M, Cukor B, Tam SW, Lavin P, Fiorentino M, Draetta GF, Jessup JM, Pagano M (1997) Increased proteasome-dependent degradation of the cyclin-dependent kinase inhibitor p27 in aggressive colorectal carcinomas. Nat Med 3:231–234
148. Catzavelos C, Bhattacharya N, Ung YC, Wilson JA, Roncari L, Sandhu C, Shaw P, Yeger H, Morava-Protzner I, Kapusta L, Franssen E, Pritchard KI, Slingerland JM (1997) Decreased levels of the cell-cycle inhibitor p27Kip1 protein: prognostic implications in primary breast cancer. Nat Med 3:227–230
149. Martins CP, Berns A (2002) Loss of p27(Kip1) but not p21(Cip1) decreases survival and synergizes with MYC in murine lymphomagenesis. EMBO J 21:3739–3748
150. Montagnoli A, Fiore F, Eytan E, Carrano AC, Draetta GF, Hershko A, Pagano M (1999) Ubiquitination of p27 is regulated by Cdk-dependent phosphorylation and trimeric complex formation. Genes Dev 13:1181–1189
151. Vlach J, Hennecke S, Amati B (1997) Phosphorylation-dependent degradation of the cyclin-dependent kinase inhibitor p27. EMBO J 16:5334–5344
152. Nguyen H, Gitig DM (1999) Koff Cell-free degradation of p27(kip1), a G1 cyclin-dependent kinase inhibitor, is dependent on CDK2 activity and the proteasome. Mol Cell Biol 19:1190–1201

153. Carrano AC, Eytan E, Hershko A, Pagano M (1999) SKP2 is required for ubiquitin-mediated degradation of the CDK inhibitor p27. Nat Cell Biol 1:193–199
154. Tsvetkov LM, Yeh KH, Lee SJ, Sun H, Zhang H (1999) p27(Kip1) ubiquitination and degradation is regulated by the SCF(Skp2) complex through phosphorylated Thr187 in p27. Curr Biol 9:661–664
155. O'Hagan RC, Ohh M, David G, de Alboran IM, Alt FW, Kaelin WG Jr, DePinho RA (2000) Myc-enhanced expression of Cul1 promotes ubiquitin-dependent proteolysis and cell cycle progression. Genes Dev 14:2185–2191
156. Vlach J, Hennecke S, Alevizopoulos K, Conti D, Amati B (1996) Growth arrest by the cyclin-dependent kinase inhibitor p27Kip1 is abrogated by c-Myc. EMBO J 15:6595–6604
157. Alevizopoulos K, Vlach J, Hennecke S, Amati B (1997) Cyclin E and c-Myc promote cell proliferation in the presence of p16INK4a and of hypophosphorylated retinoblastoma family proteins. EMBO J 16:5322–5333
158. Spruck CH, Won KA, Reed SI (1999) Deregulated cyclin E induces chromosome instability. Nature 401:297–300
159. Walter J, Sun L, Newport J (1998) Regulated chromosomal DNA replication in the absence of a nucleus. Mol Cell 1:519–529
160. Pihan GA, Purohit A, Wallace J, Knecht H, Woda B, Quesenberry P, Doxsey SJ (1998) Centrosome defects and genetic instability in malignant tumors. Cancer Res 58:3974–3985
161. Hua XH, Yan H, Newport J (1997) A role for Cdk2 kinase in negatively regulating DNA replication during S phase of the cell cycle. J Cell Biol 137:183–192
162. Angus SP, Wheeler LJ, Ranmal SA, Zhang X, Markey MP, Mathews CK, Knudsen ES (2002) Retinoblastoma tumor suppressor targets dNTP metabolism to regulate DNA replication. J Biol Chem 277:44376–44384
163. Lengronne A, Schwob E (2002) The yeast CDK inhibitor Sic1 prevents genomic instability by promoting replication origin licensing in late G(1). Mol Cell 9:1067–1078
164. Nyberg KA, Michelson RJ, Putnam CW, Weinert TA (2002) Toward maintaining the genome: DNA damage and replication checkpoints. Annu Rev Genet 36:617–656
165. Huang LC, Clarkin KC, Wahl GM (1996) Sensitivity and selectivity of the DNA damage sensor responsible for activating p53-dependent G1 arrest. Proc Natl Acad Sci U S A 93:4827–4832
166. Wahl GM, Carr AM (2001) The evolution of diverse biological responses to DNA damage: insights from yeast and p53. Nat Cell Biol 3:E277–E286
167. D'Amours D, Jackson SP (2002) The Mre11 complex: at the crossroads of DNA repair and checkpoint signalling. Nat Rev Mol Cell Biol 3:317–327
168. Mirzoeva OK, Petrini JH (2001) DNA damage-dependent nuclear dynamics of the Mre11 complex. Mol Cell Biol 21:281–288
169. Petrini JH (1999) The mammalian Mre11-Rad50-nbs1 protein complex: integration of functions in the cellular DNA-damage response. Am J Hum Genet 64:1264–1269
170. Bakkenist CJ, Kastan MB (2003) DNA damage activates ATM through intermolecular autophosphorylation and dimer dissociation. Nature 421:499–506

171. Mirzoeva OK, Petrini JH (2003) DNA replication-dependent nuclear dynamics of the mre11 complex. Mol Cancer Res 1:207–218
172. Banin S, Moyal L, Shieh S, Taya Y, Anderson CW, Chessa L, Smorodinsky NI, Prives C, Reiss Y, Shiloh Y, Ziv Y (1998) Enhanced phosphorylation of p53 by ATM in response to DNA damage. Science 281:1674–1677
173. Matsuoka S, Rotman G, Ogawa A, Shiloh Y, Tamai K, Elledge SJ (2000) Ataxia telangiectasia-mutated phosphorylates Chk2 in vivo and in vitro. Proc Natl Acad Sci U S A 97:10389–10394
174. Lambert PF, Kashanchi F, Radonovich MF, Shiekhattar R, Brady JN (1998) Phosphorylation of p53 serine 15 increases interaction with CBP. J Biol Chem 273:33048–33053
175. Unger T, Juven-Gershon T, Moallem E, Berger M, Vogt Sionov R, Lozano G, Oren M, Haupt Y (1999) Critical role for Ser20 of human p53 in the negative regulation of p53 by Mdm2. EMBO J 18:1805–1814
176. Harper JW, Adami GR, Wei N, Keyomarsi K, Elledge SJ (1993) The p21 Cdk-interacting protein Cip1 is a potent inhibitor of G1 cyclin-dependent kinases. Cell 75:805–816
177. Sheen JH, Dickson RB (2002) Overexpression of c-Myc alters G(1)/S arrest following ionizing radiation. Mol Cell Biol 22:1819–1833
178. Hermeking H, Eick D (1994) Mediation of c-Myc-induced apoptosis by p53. Science 265:2091–2093
179. Seoane J, Le HV, Massague J (2002) Myc suppression of the p21(Cip1) Cdk inhibitor influences the outcome of the p53 response to DNA damage. Nature 419:729–734
180. Li Q, Dang CV (1999) c-Myc overexpression uncouples DNA replication from mitosis. Mol Cell Biol 19:5339–5351
181. Shim J, Lee H, Park J, Kim H, Choi EJ (1996) A non-enzymatic p21 protein inhibitor of stress-activated protein kinases. Nature 381:804–806
182. Asada M, Yamada T, Ichijo H, Delia D, Miyazono K, Fukumuro K, Mizutani S (1999) Apoptosis inhibitory activity of cytoplasmic p21(Cip1/WAF1) in monocytic differentiation. EMBO J 18:1223–1234
183. Di Leonardo A, Linke SP, Clarkin K, Wahl GM (1994) DNA damage triggers a prolonged p53-dependent G1 arrest and long-term induction of Cip1 in normal human fibroblasts. Genes Dev 8:2540–2551
184. Allan LA, Duhig T, Read M, Fried M (2000) The p21(WAF1/CIP1) promoter is methylated in Rat-1 cells: stable restoration of p53-dependent p21(WAF1/CIP1) expression after transfection of a genomic clone containing the p21(WAF1/CIP1) gene. Mol Cell Biol 20:1291–1298
185. Minella AC, Swanger J, Bryant E, Welcker M, Hwang H, Clurman BE (2002) p53 and p21 form an inducible barrier that protects cells against cyclin E-cdk2 deregulation. Curr Biol 12:1817–1827
186. Nasmyth K (2002) Segregating sister genomes: the molecular biology of chromosome separation. Science 297:559–565
187. Niculescu AB 3rd, Chen X, Smeets M, Hengst L, Prives C, Reed SI (1998) Effects of p21(Cip1/Waf1) at both the G1/S and the G2/M cell cycle transitions: pRb is a critical determinant in blocking DNA replication and in preventing endoreduplication. Mol Cell Biol 18:629–643

188. Bates S, Ryan KM, Phillips AC, Vousden KH (1998) Cell cycle arrest and DNA endoreduplication following p21Waf1/Cip1 expression. Oncogene 17:1691–1703
189. Stewart ZA, Leach SD, Pietenpol JA (1999) p21(Waf1/Cip1) inhibition of cyclin E/Cdk2 activity prevents endoreduplication after mitotic spindle disruption. Mol Cell Biol 19:205–215
190. Andreassen PR, Margolis RL (1994) Microtubule dependency of p34cdc2 inactivation and mitotic exit in mammalian cells. J Cell Biol 127:789–802
191. Khan SH, Wahl GM (1998) p53 and pRb prevent rereplication in response to microtubule inhibitors by mediating a reversible G1 arrest. Cancer Res 58:396–401
192. Kung AL, Sherwood SW, Schimke RT (1990) Cell line-specific differences in the control of cell cycle progression in the absence of mitosis. Proc Natl Acad Sci U S A 87:9553–9557
193. Lanni JS, Jacks T (1998) Characterization of the p53-dependent postmitotic checkpoint following spindle disruption. Mol Cell Biol 18:1055–1064
194. Minn AJ, Boise LH, Thompson CB (1996) Expression of Bcl-xL and loss of p53 can cooperate to overcome a cell cycle checkpoint induced by mitotic spindle damage. Genes Dev 10:2621–2631
195. Pourquier P, Pommier Y (2001) Topoisomerase I-mediated DNA damage. Adv Cancer Res 80:189–216
196. Zhu J, Schiestl RH (1996) Topoisomerase I involvement in illegitimate recombination in Saccharomyces cerevisiae. Mol Cell Biol 16:1805–1812
197. Labib K, Diffley JF (2001) Is the MCM2-7 complex the eukaryotic DNA replication fork helicase? Curr Opin Genet Dev 11:64–70
198. Chiang YC, Teng SC, Su YN, Hsieh FJ, Wu KJ (2003) c-MYC directly regulates the transcription of NBS1 gene involved in DNA double-strand break repair. J Biol Chem 13:13
199. Schorl C, Sedivy JM (2003) Loss of protooncogene c-Myc function impedes G(1) phase progression both before and after the restriction point. Mol Biol Cell 14:823–835
200. Williams BR, Mirzoeva OK, Morgan WF, Lin J, Dunnick W, Petrini JH (2002) A murine model of Nijmegen breakage syndrome. Curr Biol 12:648–653
201. Grandori C, Wu KJ, Fernandez P, Ngouenet C, Grim J, Clurman BE, Moser MJ, Oshima J, Russell DW, Swisshelm K, Frank S, Amati B, Dalla-Favera R, Monnat RJ Jr (2003) Werner syndrome protein limits MYC-induced cellular senescence. Genes Dev 17:1569–1574
202. Schar P (2001) DNA damage genome instability Spontaneous, and cancer—when DNA replication escapes control. Cell 104:329–332
203. Loeb LA (2001) A mutator phenotype in cancer. Cancer Res 61:3230–3239
204. Felsher DW, Bishop JM (1999) Reversible tumorigenesis by MYC in hematopoietic lineages. Mol Cell 4:199–207
205. Pelengaris S, Littlewood T, Khan M, Elia G, Evan G (1999) Reversible activation of c-Myc in skin: induction of a complex neoplastic phenotype by a single oncogenic lesion. Mol Cell 3:565–577
206. Pelengaris S, Khan M, Evan GI (2002) Suppression of Myc-induced apoptosis in beta cells exposes multiple oncogenic properties of Myc and triggers carcinogenic progression. Cell 109:321–334

207. D'Cruz CM, Gunther EJ, Boxer RB, Hartman JL, Sintasath L, Moody SE, Cox JD, Ha SI, Belka GK, Golant A, Cardiff RD, Chodosh LA (2001) c-MYC induces mammary tumorigenesis by means of a preferred pathway involving spontaneous Kras2 mutations. Nat Med 7:235–239
208. Ilyas M, Straub J, Tomlinson IP, Bodmer WF (1999) Genetic pathways in colorectal and other cancers. Eur J Cancer 35:1986–2002
209. Ugolini F, Charafe-Jauffret E, Bardou VJ, Geneix J, Adelaide J, Labat-Moleur F, Penault-Llorca F, Longy M, Jacquemier J, Birnbaum D, Pebusque MJ (2001) WNT pathway and mammary carcinogenesis: loss of expression of candidate tumor suppressor gene SFRP1 in most invasive carcinomas except of the medullary type. Oncogene 20:5810–5817
210. Smalley MJ, Dale TC (1999) Wnt signalling in mammalian development and cancer. Cancer Metastasis Rev 18:215–230
211. You Z, Saims D, Chen S, Zhang Z, Guttridge DC, Guan KL, MacDougald OA, Brown AM, Evan G, Kitajewski J, Wang CY (2002) Wnt signaling promotes oncogenic transformation by inhibiting c-Myc-induced apoptosis. J Cell Biol 157:429–440
212. Smalley MJ, Dale TC (2001) Wnt signaling and mammary tumorigenesis. J Mammary Gland Biol Neoplasia 6:37–52
213. Gunther EJ, Moody SE, Belka GK, Hahn KT, Innocent N, Dugan KD, Cardiff RD, Chodosh LA (2003) Impact of p53 loss on reversal and recurrence of conditional Wnt-induced tumorigenesis. Genes Dev 17:488–501
214. Jain M, Arvanitis C, Chu K, Dewey W, Leonhardt E, Trinh M, Sundberg CD, Bishop JM, Felsher DW (2002) Sustained loss of a neoplastic phenotype by brief inactivation of MYC. Science 297:102–104
215. Arnold I, Watt FM (2001) c-Myc activation in transgenic mouse epidermis results in mobilization of stem cells and differentiation of their progeny. Curr Biol 11:558–568

CTMI (2006) 302:205–234

Lessons Learned from Myc/Max/Mad Knockout Mice

M. Pirity[1,2] · J. K. Blanck[1] · N. Schreiber-Agus[1] (✉)

[1]Department of Molecular Genetics, Albert Einstein College of Medicine, 1300 Morris Park Avenue, Ullmann 809, Bronx, NY 10461, USA
Agus@aecom.yu.edu

[2]Institute of Genetics, Biological Research Center, Hungarian Academy of Sciences, Szeged,Hungary

Abstract The past two decades of gene targeting experiments have allowed us to make significant strides towards understanding how the Myc/Max/Mad network influences multiple aspects of cellular behavior during development. Here we summarize the findings obtained from the *myc/max/mad* knockout mice generated to date, namely those in which the N-*myc*, c-*myc*, L-*myc*, *mad1*, *mxi1*, *mad3*, *mnt*, or *max* genes have been targeted. A compilation of lessons we have learned from these *myc/max/mad* knockout mouse models, and suggestions as to where future efforts could be focused, are also presented.

1 Overview

Members of the Myc/Max/Mad network of related basic region helix-loop-helix/leucine zipper (bHLH/LZ) transcription factors have been implicated as important regulators of many aspects of cellular behavior. Insight into the precise roles of these proteins in normal physiological processes and in tumorigenesis has emerged from studies employing cell culture-based systems and genetically modified animals. Here we summarize the findings obtained from the Myc/Max/Mad knockout mice generated to date, namely those in which the N-*myc*, c-*myc*, L-*myc*, *mad1*, *mad3*, *mxi1*, *mnt*, or *max* genes have been targeted. We also include a compilation of lessons learned from these loss-of-function models about the in vivo functions of this complex protein network. The reader is directed to other reviews (Baudino and Cleveland 2001; Dang et al. 1999; Eisenman 2000; Grandori et al. 2000; Luscher 2001; Morgenbesser and DePinho 1994; Nesbit et al. 1999; Oster et al. 2002; Pelengaris et al. 2000; Schreiber-Agus and DePinho 1998; Zhou and Hurlin 2001) for complementary information garnered from gain-of-function systems.

2 N-*myc* Knockout Mice

Since the early 1990s, investigators have been generating loss-of-function models for N-Myc (and c-Myc; see next section) with the goals of elucidating its function during mouse embryogenesis and gaining insight into how its dysregulation contributes to the pathogenesis of certain types of human cancers including those of the nervous system (for recent reviews on Myc and human cancers see Lutz et al. 2002; Nesbit et al. 1999). In the earliest studies, N-*myc* gene inactivation was accomplished by disrupting or replacing the coding exons of the N-*myc* gene with a neomycin resistance cassette (a so-called "conventional knockout"), resulting in a non-functional (null) N-*myc* allele after gene targeting and homologous recombination. As gene-targeting technologies became more sophisticated, other types of loss-of-function models were generated including hypomorphic mutations and tissue-specific ("conditional") deletions (for a review of gene targeting strategies and techniques see Joyner 2000). Together, these models, which are discussed sequentially here, have advanced our understanding of the function of N-Myc in the context of a whole organism.

Conventional N-*myc* knockout mice were generated concomitantly by three independent groups in the earliest years of gene targeting experiments (Charron et al. 1990, 1992; Sawai et al. 1991, 1993; Stanton et al. 1990, 1992). The consensus phenotype for all three versions of N-*myc* homozygous null mice was embryonic lethality between embryonic days (E) 10.5 and 12.5 of gestation, with abnormalities including developmental delay and a decrease in size/cellularity of certain organs [including central and peripheral nervous system (CNS and PNS), mesonephros, lung, liver, heart, and gut]. Notably, the most severely affected organs appeared to be those that normally exhibit high levels of N-*myc* expression (with liver being an exception to this generalization). As an aside, it is likely that the cardiovascular phenotype most severely compromises the embryos and leads to their death. Differences in the reported phenotypes of the various N-*myc*-deficient mouse lines could have resulted from variations in targeting strategy or genetic background of the mice or from the subjectivity of morphological and histological assessments (reviewed in Davis and Bradley 1993).

The embryonic lethality of these mice in a narrow timeframe indicated that there was a non-redundant, critical role for N-Myc during this developmental stage. In contrast, the survival of N-Myc-deficient mice until midgestation suggested that N-Myc function is dispensable until that time. This dispensability was also observed when embryonic stem cells (ES) homozygous null for N-*myc* were generated by one of the groups by sequential disruption; these ES cells did not exhibit altered growth or morphology in the undifferentiated or differentiated states (Sawai et al. 1991). This was surprising given that N-*myc* expression is detectable in the inner cell mass of the developing blastocyst at E3.5 (Sawai et al. 1993); it is this compartment from which ES cells are derived. In trying to understand this dichotomy, it was realized that an essential role for N-Myc becomes apparent around the time that c-*myc* and N-*myc* assume specified expression patterns (onset of organogenesis; Charron et al. 1992; Sawai et al. 1993; Stanton et al. 1992). More specifically, in pre-implantation embryos all three *myc* family members (c-, N-, and L-*myc*) are expressed together. During midgestation, c-*myc* is found to be broadly expressed, while N and L-*myc* exhibit more tissue-restricted patterns (e.g., brain and kidney) of expression. Moreover, c-*myc* expression generally is found to correlate with actively proliferating cells (both embryonic and mature), while N- and L-*myc* expression is associated more with cells that are undergoing differentiation (primarily embryonic). (For reviews on *myc* family gene expression see DePinho et al. 1991; Zimmerman and Alt 1990.)

Given these expression patterns, the question then arose as to whether, in the knockout mice, compensation among highly related Myc family members was occurring during the extensive proliferation events of early embryogenesis (when the *myc* family genes are co-expressed), but not later on when the members have assumed tissue- and cell type-specific expression (and, possibly functions). The concept that the unique expression patterns assumed by N- and c-*myc* during later embryogenesis underlies their apparently nonredundant roles is supported by more recent studies (Malynn et al. 2000) that are described in the next section of this review. Moreover, the theme of functional redundancy/compensation will emerge time and again in the explanation of phenotypes of *myc*/*max*/*mad* knockout mice.

On a related note, it is likely that the regulatory mechanisms controlling c- and N-*myc* expression patterns are linked, since in the absence of N-Myc, c-*myc* is upregulated in a compensatory manner in specific tissues/cell types (e.g., neuroepithelium, where c-*myc* expression is normally absent; Stanton et al. 1992). Despite this upregulation, CNS abnormalities are still present in the N-*myc*-null mice, perhaps suggesting that (1) the cellular subtypes in which c-*myc* is upregulated are not those in which N-*myc* normally is expressed, (2) the timing of the c-*myc* compensatory upregulation may be too late to rescue the neuronal tissue defects (c-*myc* ectopic expression in the neuroepithelium was documented only at E10.5), or (3) the degree of c-*myc* upregulation is not sufficient to rescue the neural tissue defects. With respect to the latter point, the concept of reaching threshold levels of *myc* expression for the proper execution of growth/development has been supported by other studies (see below and Sect. 3; Moens et al. 1992; Moens et al. 1993; Nagy et al. 1998; Trumpp et al. 2001).

The original N-*myc* knockout studies suggested a normal role for N-Myc on the cellular level in the regulation of proliferation/prevention of terminal differentiation in maturing tissues. In addition, a role for N-Myc in tissue–tissue interactions was also proposed based on studies such as one where N-*myc*$^{+/+}$ versus −/− organ cultures of lung buds were compared beyond the point of the embryonic lethality (Sawai et al. 1993). In that study, the development of bronchial branches from the N-*myc*$^{-/-}$ (but not N-*myc*$^{+/+}$) lung buds was shown to be dependent upon the addition of serum to the media. This suggested that N-Myc deficiency impairs signaling pathways necessary for normal lung morphogenesis. As will be discussed below, this theory has gained support from the nature of the lung defects observed in N-*myc* hypomorphic mutants (Moens et al. 1992).

A follow-up study upon the conventional N-*myc*-null mutant animals helped to further our understanding of how N-Myc participates in this function of signaling during tissue morphogenesis. In that study, the effects of N-Myc deficiency in the developing liver in particular were examined more closely (Giroux and Charron 1998). N-Myc-deficient hepatocytes at E11.5 (near the time of lethality) exhibited massive apoptosis and reduced cell number, phenotypes that were surprising given that N-*myc* is not expressed at significant levels in hepatocytes (although c-*myc* is). However, N-*myc* expression levels have been shown to be significant in structures that are involved in the induction of the liver primordium (namely cardiac mesoderm and mesenchyme of the septum transversum). Most likely, in the absence of N-*myc* expression, these structures are unable to send the necessary signals (extracellular growth factors) for hepatocyte development and survival, thus leading to the hepatic phenotype observed in N-*myc*-null embryos. Another possibility is that the N-Myc-deficient hematopoietic cells that populate the liver are not sending out growth/survival signals; these cells are reduced in number in the N-Myc-deficient mice (Giroux and Charron 1998). Of note is the finding that, in contrast to the liver, other compromised organs in the N-Myc-deficient mouse did not exhibit considerable levels of apoptosis. This suggests that, in those compromised organs, other cellular processes may be affected by the absence of N-Myc in a tissue autonomous or tissue interaction-dependent manner (Giroux and Charron 1998; Sawai et al. 1993).

In addition to the conventional N-*myc* knockout mice, a hypomorphic (leaky) mutant was generated after an unexpected recombination event led to the insertion of a targeting vector into the first intron of the N-*myc* locus as opposed to the expected replacement of the N-*myc* second exon coding region sequences (Moens et al. 1992). As a result of this integration event, the targeted N-*myc* allele was capable of producing, via alternative splicing, either a normal N-*myc* transcript or a truncated transcript; the predominance of one transcript form over the other varied among various tissue types tested. Mice homozygous for this leaky mutation (termed N-$myc^{9a/9a}$ mice) survived beyond the time of death of the conventional knockout mice (E10.5–E12.5), but died immediately after birth because they were unable to oxygenate their blood. This phenotype suggested a lung defect, a theory that was confirmed histologically by the impaired proliferation and branching of the pulmonary epithelium. Although several other phenotypes were observed in the N-$myc^{9a/9a}$ mice (including their smaller overall body size and spleen size), a majority of the organs tested appeared grossly and histologically normal—and these organs tended to have higher levels of the normal N-*myc* transcript and lower levels of the mutant transcript relative to what was seen in the lung. Taken together, these findings suggested that N-Myc plays an important role in the epithelial–mesenchymal interactions that underlie lung morphogenesis. Moreover, since reducing the levels of N-Myc in the lung led to marked defects therein, this was the first in vivo indication that a threshold level of N-Myc might be crucial for the proper development of certain lineages/tissues. This threshold may be achieved in the other tissues in which N-Myc normally functions that show no apparent phenotype.

The generation of the N-*myc* hypomorphic mutant provided the opportunity to produce an allelic series of mutations wherein the levels of N-Myc could be incrementally reduced from wildtype to zero. As a first step towards producing this series, a compound heterozygote between the leaky (Moens et al. 1992) and null N-*myc* mutations (Stanton et al. 1992) was generated and analyzed (Moens et al. 1993); these compound heterozygotes were shown to have roughly 15% of normal N-Myc levels. The phenotype of these mice was embryonic lethality at midgestation (by E14.5) due to cardiac failure (note that the timing of the lethality is between that of the N-*myc*-null homozygotes and that of the N-$myc^{9a/9a}$ homozygotes). Histological analyses of the mutant hearts revealed hypoplasia in the compact layer of the ventricular myocardium, suggesting a role for N-Myc in maintaining the proliferation and/or inhibiting the differentiation of myocytes in this compartment. Consistent with what was shown in the first study of the N-$myc^{9a/9a}$ hypomorphic homozygotes (Moens et al. 1992), the compound N-$myc^{9a/null}$ heterozygotes displayed defects in lung branching morphogenesis, albeit to a greater degree (Moens et al. 1993). Other tissues that were shown to be affected in the null mice appear to be normal in these compound heterozygotes (e.g., brain), suggesting that the residual N-Myc therein is sufficient for their growth and development (although the possibility that some form of compensation may be occurring in these tissues cannot be excluded). Finally, in a follow-up study, kidney development was studied further in the context of wildtype, null, and hypomorphic levels of N-Myc (and various combinations thereof; Bates et al. 2000). Using whole embryonic kidneys when obtainable, as well as organ explants, the authors were able to demonstrate a dose-dependent role for N-Myc in kidney development. With decreasing levels of N-Myc, the phenotypes of kidney hypoplasia, reduced ureteric bud tips, and reduced developing glomeruli and nephrons increased in severity. Notably, the kidney hypoplasia was shown to be due to reduced proliferation as opposed to increased apoptosis, suggesting that N-Myc normally is regulating the expression/activity of cell-cycle regulators or growth factors, possibly in both autocrine and paracrine manners (Bates et al. 2000). Note that these findings in the kidney differ from those made in hepatocytes, wherein N-Myc deficiency led to massive apoptosis (Giroux and Charron 1998). This suggests that N-Myc may be participating in different cellular processes in different cell types.

A second allelic series for N-*myc* was described later; this series was generated with a single targeting vector during an attempt to introduce a point mutation into the N-Myc leucine zipper in vivo (Nagy et al. 1998). Mice homozygous for the point mutation died at E11.5 of phenotypes similar to those described in the conventional knockout reports. This indicated that crippling the N-Myc leucine zipper appears to be equivalent to the complete loss of N-Myc. Within the allelic series that was generated as a "byproduct" of this study, novel hypomorphs were characterized and shown to have phenotypes similar to those described in earlier studies (e.g., lung and heart defects), with severity correlating with the levels of N-Myc relative to those in wild-type tissues. In addition, new phenotypes emerged among the hypomorph or compound mutant mice (e.g., vertebral fusion possibly due to failure of chondrocytes to condense properly), suggesting additional roles for N-Myc in other cell types (Nagy et al. 1998).

The hypomorphic mutations and allelic series have been powerful tools for elucidating the multiple roles of N-Myc over the course of development. A complementary approach that can provide significant insight along these lines involves "conditional knockouts," which allows gene function to be investigated in a specific cell type and/or at a specific time (Gertsenstein et al. 2002; Joyner 2000; Kwan 2002). The majority of conditional knockout studies take advantage of the Cre/loxP system in which animals are generated carrying a silent mutation in the gene of interest (e.g., they carry loxP sites in introns around an exon or exons to be deleted). These animals are then mated (in a binary scheme) to transgenic animals harboring the Cre recombinase driven by a cell type-/tissue-specific promoter of interest. The offspring of this mating will be deleted for the region between the loxP sites when Cre has catalyzed recombination between these sites (Gertsenstein et al. 2002; Joyner 2000; Kwan 2002).

Recently, a conditional knockout of N-*myc* has been generated (Knoepfler et al. 2002) for the purpose of assessing complete N-Myc loss of function in the mouse without the complication of embryonic lethality of the conventional knockouts. In that study, nestin-Cre transgenic mice were employed for the generation of mice specifically deleted for N-*myc* in neuronal progenitor cells. The choice of these Cre transgenic mice was dictated by the desire to analyze N-Myc function during neurogenesis, especially in light of the facts that the CNS/PNS of the conventional N-Myc knockout mice were severely affected, and that N-myc is frequently deregulated in nervous system tumors such as neuroblastoma. Mice deleted for N-*myc* in this manner exhibited a marked reduction in brain size (in particular cerebellum and neocortex) likely resulting from reduced proliferation as well as reduced growth. Neuronal differentiation was also shown to be inappropriately triggered, while changes in apoptotic indices were not observed. On the molecular level, the effects on proliferation/differentiation could be correlated with changes in cell-cycle regulators (including cyclin D2, p18^{ink4c}, and p27^{Kip1}). On the organismal level, these defects resulted in multiple abnormalities including ataxia, behavioral problems, and tremors. Together, these findings point to a critical (and nonredundant) role for N-Myc in normal neurogenesis, possibly in regulating the switch between proliferation and differentiation of neuronal progenitor cells (Knoepfler et al. 2002). This level of insight could not have been gleaned from the original knockout studies, and the findings call for additional conditional experiments for the dissection of the cellular functions of N-Myc (and other Myc/Max/Mad members) in specific cell types.

3 c-*myc* Knockout Mice

Mice made deficient for c-Myc through conventional knockout approaches were reported (Davis et al. 1993) soon after the original N-*myc* knockout papers emerged. These c-*myc*$^{-/-}$ mice died between E9.5 and 10.5 (slightly earlier than those deficient for N-Myc), and exhibited pleiotropic phenotypes including small size, developmental delay, enlarged heart and pericardium, decreased yolk sac circulation, and aberrant neural tube closure. Of note, many of these phenotypes are reminiscent of those reported for the N-*myc* conventional knockout mice (see above). To differentiate between the c- and N-*myc* knockout phenotypes, side-by-side comparisons of mice rendered deficient for each of these genes in equivalent manners may need to be performed.

Similar to what was described for N-Myc above, the lethality of these mice in this narrow developmental window indicated that there was a nonredundant, critical role for c-Myc around this stage of embryogenesis. However, an experiment suggested in this original c-*myc* knockout paper (Davis et al. 1993), and ultimately carried out in 2000 (Malynn et al. 2000), showed that this was not entirely the case. In that later study, the highly related N-*myc* gene was knocked into the c-*myc* locus, generating a situation where N-*myc*, instead of c-*myc*, was expressed under the control of c-*myc* regulatory sequences. Mice homozygous for this replacement mutation were healthy, fertile, and for the most part normal (despite subtle phenotypes including periodic dystrophy of skeletal muscles). Moreover, two different cell types isolated from these mice, namely lymphocytes and fibroblasts, also did not show significant alterations in phenotype. These findings suggest that N-Myc could indeed substitute for c-Myc throughout embryogenesis and beyond, provided that it was expressed in a "c-Myc manner." As such, this study provides strong experimental support for the hypothesis outlined above which states that there is a considerable degree of functional overlap/redundancy between the Myc family members in their roles as developmental regulators. Moreover, their preservation as distinct genes throughout evolution likely results from their specific temporal and spatial expression patterns, which impart upon them the appearance of nonredundancy (Krakauer and Nowak 1999). This line of reasoning may also explain why L-Myc-deficient mice have been shown to be viable and fertile with no discernable phenotypes, since L-*myc* is typically co-expressed with either c-*myc* or N-*myc* (Hatton et al. 1996). Because of this co-expression, L-Myc may be evolving into a functionally inactive form, a theory supported by the weak transformation and transactivation activities of L-Myc (relative to the activities of c- or N-Myc) in cell culture-based assays (Cole and McMahon 1999; Nesbit et al. 1999). Alternatively, is possible that subtle phenotypic changes that may be emerging with L-Myc deficiency were not detected by the histological/immunohistochemical techniques employed in the L-*myc* knockout study (Hatton et al. 1996). Further studies are needed to clarify the role of L-Myc in the context of the whole organism (see Sect. 5 below).

Returning to the c-*myc* conventional knockout mice, their phenotype was revisited recently with re-derived mice. These studies showed that c-Myc deficiency was associated with marked defects in vasculogenesis, angiogenesis, and primitive erythropoiesis in the yolk sac and in the embryo proper (Baudino et al. 2002). These defects could be the primary cause of the embryonic lethality of c-*myc*$^{-/-}$ mice, and could be extrapolated to explain the prevalence of Myc deregulation in human tumors [estimated to contribute to one-seventh of cancer-related deaths (Baudino et al. 2002; Dang 1999)]. More specifically, as a transcription factor, c-Myc may be regulating growth factors/cytokines that are necessary both for vascular and hematopoietic development and for the angiogenic switch that permits neovascularization during tumor progression (Baudino et al. 2002; and reviewed in Oster et al. 2002).

As discussed for N-Myc above, limitations associated with the analysis of embryonic lethal phenotypes prompted several groups to resort to conditional targeting approaches for the further elucidation of the consequences of c-Myc deficiency in vivo. In one such study, an allelic series was generated for the comparison of mice with levels of c-Myc ranging from wildtype to reduced (hypomorphic and heterozygous and various combinations thereof) to zero (Trumpp et al. 2001). On the organismal level, the mice in this series displayed an increasing degree of multi-organ hypoplasia (i.e., reduction in cell number) leading to reduced body size, with decreasing levels of Myc. As has been discussed for the N-*myc* knockout mice above, the augmented phenotype resulting from reductions in c-Myc levels argues for a threshold level of Myc expression necessary for proper growth and development. On the cellular level, mouse embryonic fibroblasts (MEFs) derived from conditional c-*myc* knockout mice generated in that study (Trumpp et al. 2001; and in a related one: de Alboran et al. 2001) showed reduced proliferation with decreasing Myc levels and ultimately growth arrest in G0/G1 with the complete absence of Myc. Furthermore, both T cells (Trumpp et al. 2001) and B cells (de Alboran et al. 2001) specifically deleted for c-*myc* after breeding of conditionally targeted mice to the appropriate Cre-transgenic mice displayed impaired proliferative responses after stimulation in culture.

An essential and non-redundant role for c-Myc in T and B lineages was supported further in a study in which c-*myc*-null ES cells were injected into wildtype or Rag1-deficient recipient blastocysts for the generation of chimeric mice (Douglas et al. 2001). This blastocyst complementation approach allows one to evaluate the capability of deficient ES cells to contribute to lymphocytic lineages (reviewed in Chen 1996; Spanopoulou 1996). In the wildtype chimeras, c-*myc*$^{-/-}$ cells did not contribute to the white blood cell compartment, since these cells were outcompeted by their wildtype counterparts. In the absence of functional wildtype counterparts in the *rag1*$^{-/-}$ chimeric mice, the requirement for c-Myc was defined more precisely. Specifically, c-*myc*$^{-/-}$ thymocytes were shown to be impaired in maturation beyond the double negative stage when significant growth and proliferation is required and when N-*myc* expression is downregulated dramatically (Douglas et al. 2001). That it is c-Myc, and not N-Myc, per se that may be playing an essential role in lymphocyte development is supported by findings from a separate study utilizing similar blastocyst complementation approaches. There, N-Myc deficiency did not lead to any marked changes in B or T cell differentiation/function (Malynn et al. 1995). Together, the results from all of the above studies suggest that mammalian c-Myc plays an integral role in the decisions of cell-cycle entry and exit, although an additional role in cell survival cannot be ruled out.

Interestingly, a role for Myc in the regulation of cellular growth (i.e., accumulation of mass) has been gaining increasing support from *Drosophila* (see below), from cell culture-based studies, and from gene target identification efforts. However, defects in cell growth were reported in only a minority of the N-*myc*/c-*myc* knockout mouse papers discussed above. For instance, changes in cell growth (in addition to proliferation) were documented in neurons of the N-*myc* conditional knockout mice (Knoepfler et al. 2002), and in c-*myc*$^{-/-}$ thymocytes generated via the Rag1-deficient blastocyst approach (Douglas et al. 2001). In contrast, in another study (Trumpp et al. 2001), the effects of c-Myc loss were assessed in the process of primary T cell activation, which allows cell-cycle versus cell growth changes to be dissected one from another;

there, c-Myc did not appear to be required for cell growth. The notion that the Myc/Max/Mad network is indeed regulating growth/metabolism clearly needs to be revisited in the gain- and loss-of-function mouse models that exist and in ones that will be developed in the future.

With respect to *Drosophila myc*, when it was cloned initially (Gallant et al. 1996; Schreiber-Agus et al. 1997; P. Gallant, this volume), it was realized that it may correspond to the *diminutive* (*dm*) gene, a mutation which results in small body size, among other abnormalities. Indeed, *dm* was shown to be a c-*myc* hypomorph resulting from the insertion of a gypsy transposon upstream of the Myc translation initiation site (Gallant et al. 1996). Later, mosaic analyses in the *Drosophila* wing with more severe *dmyc* hypomorphic alleles showed that reductions in dMyc levels led to decreased cell size (but not cell number), body size, and viability (Johnston et al. 1999). In the reciprocal gain-of-function mosaic studies, overexpressed dMyc was shown to increase cellular growth rate by promoting G1/S progression, thus increasing cell size (mass; Johnston et al. 1999). The findings of these studies on *Drosophila* Myc support a model wherein, in response to environmental cues, dMyc is regulating targets involved in cell growth, biosynthesis, and metabolism. Indeed, several gene targets that fall into these categories have been identified in the fly (Orian et al. 2003; Zaffran et al. 1998). Notably, similar types of gene targets have been identified for mammalian c-Myc with the aid of a rat1 immortalized fibroblast cell line disrupted for both copies of c-*myc* [(cell line developed by Sedivy and colleagues (Mateyak et al. 1997)] among other cell-culture systems. (Relevant gene target papers include Guo et al. 2000; O'Connell et al. 2003; Shiio et al. 2002.)

The aforementioned c-*myc*-deficient rat1 system has been an important tool for furthering our understanding of the function of Myc on the cellular and molecular levels. As such, a discussion of this system is included in this review of loss-of-function models. c-*myc*-null somatic cells were generated to circumvent the problems associated with (1) the embryonic lethality of c-Myc-deficient mice (and the fact that c-Myc-deficient cells derived from the conventional knockout mice were unobtainable) and (2) the redundancy of Myc family members that may be masking critical functions. These c-*myc*$^{-/-}$ rat1 cells displayed a reduced cell division rate and biosynthetic rate, as well as morphological changes consistent with cytoskeletal alterations (Mateyak et al. 1997). This phenotype differed considerably from that of c-*myc*$^{-/-}$ ES cells generated in the original c-*myc* knockout study (Davis et al. 1993); these ES cells did not display any marked changes in cell cycle or growth even after extensive passaging, presumably due to functional compensation by N-*myc* which is expressed therein. Notably, in the c-*myc*$^{-/-}$ rat1 cells, expression of N- and L-*myc* is believed to be absent. This being said, the fact that the c-*myc*$^{-/-}$ rat1 cells were still capable of cell division indicates that Myc function is not absolutely critical for cell viability (albeit with the caveat that during the establishment of these knockout cells they may have learned how to bypass the Myc requirement). However, the marked phenotypic alterations that did occur in the absence of Myc reflect its essential contribution to multiple cellular processes. Notably, these alterations could be rescued effectively only by other Myc family members, suggesting that there are neither functional equivalents of Myc nor predominating downstream gene targets mediating Myc function (Berns et al. 2000; Landay et al. 2000; Nikiforov et al. 2000, 2002b; Xiao et al. 1998).

The c-*myc*$^{-/-}$ rat1 rat1 cell system has allowed for the reinforcement of the role of c-Myc in controlling the accumulation of cellular mass as well as the decision to enter/exit the cell cycle (Holzel et al. 2001; Mateyak et al. 1997, 1999; Schorl and Sedivy 2003). For example, in one study (Schorl and Sedivy 2003), the finding of delayed passage through the restriction point in G1 and delayed progression into S suggested roles for Myc in regulating the cell-cycle machinery and metabolic processes, respectively (albeit that it is still somewhat ambiguous as to whether one of these functions is secondary to the other). In addition to these roles, the c-*myc*$^{-/-}$ rat1 system has supported a function for Myc in sensitizing cells to apoptotic signals, perhaps at the level of the mitochondria (Adachi et al. 2001; Soucie et al. 2001). This function has been described in overexpression studies in cell culture (Nasi et al. 2001; Oster et al. 2002; Pelengaris et al. 2002) but does not appear to have been sufficiently addressed/documented in the *myc* knockout mouse models.

Finally, as mentioned above, the rat1 c-*myc*$^{-/-}$ cells have been employed in various efforts to identify gene targets downstream of Myc. In some of these studies, these c-Myc-deficient cells were utilized to confirm candidate targets identified through other approaches such as inducible overexpression systems coupled with differential expression analysis, functional screens, educated guessing, etc. (Bush et al. 1998; Felton-Edkins et al. 2003; Nikiforov et al. 2002a; among others). In other studies, parental rat1 fibroblasts were compared to the c-*myc*$^{-/-}$ fibroblasts (as well as to c-*myc*$^{-/-}$ fibroblasts reconstituted with Myc) by complementary (c)DNA microarray (Guo et al. 2000; O'Connell et al. 2003; Watson et al. 2002) or quantitative proteomic analysis (Shiio et al. 2002). Although the lists of gene targets identified in these larger-scale efforts exhibit significant variability, there is some consensus with respect to Myc activating genes involved in macromolecular synthesis and metabolism and repressing genes involved in a cell's interaction with the external environment (O'Connell et al. 2003; Shiio et al. 2002; for more on Myc gene targets see chapter by L.A. Lee and C.V. Dang, this volume). For the future, it would be important to compare loss-of-function models for selected Myc gene targets (or combinations thereof) to loss-of-function models for Myc family members to assess the contributions of these gene targets to mediating various functions assigned to Myc.

4
Other *myc/max/mad* Family Knockout Mice

4.1
mad Family Knockout Mice

Like the Myc family proteins, members of the Mad family (Mad1, Mxi1, Mad3, and Mad4) bear bHLH/LZ motifs that mediate heterodimerization to Max and sequence-specific DNA binding ability (reviewed in Baudino and Cleveland 2001; Eisenman 2000; Grandori et al. 2000; Luscher 2001; Schreiber-Agus and DePinho 1998; Zhou and Hurlin 2001; S. Rottmann and B. Lüscher, this volume). Through their interaction with co-repressors such as Sin3/histone deacetylase (HDAC), Mad family proteins can repress transcription at E-boxes present in a set of gene targets that overlaps partially or completely with the set of targets recognized by the transactivating Myc/Max complex (James and Eisenman 2002; Nikiforov et al. 2003; O'Hagan et al. 2000). Based on their ability to antagonize Myc function in transactivation and transformation, their induced expression during the approach to the terminally differentiated state, and their chromosomal mapping to hotspots for deletion in a variety of human cancers, Mad family proteins have been promoted as growth/tumor suppressors. Support for their roles as growth suppressors comes from cell culture-based studies, for example those that have shown that these proteins can cause dramatic reductions in foci number in the Myc/Ras cotransformation assay. However, support for the roles of Mad family proteins as potent tumor suppressors has been less forthcoming. More specifically, it has been difficult to document functional inactivation of these genes in human tumors or tumor-prone phenotypes of the respective knockout mice, barring several exceptions. Examples of these exceptions are increased susceptibility of Mxi1-deficient mice to induced tumorigenesis (Schreiber-Agus et al. 1998; and see below), mutations/allele specific loss of *MXI1* in human prostate cancers (Eagle et al. 1995; Prochownik et al. 1998), and downregulation of MAD1 protein levels in invasive breast cancers (Han et al. 2000). In addition, as discussed next, the developmental phenotypes of the individual *mad* family member knockout mice are relatively mild—all of these mice have been shown to be viable, fertile, and grossly normal. This was somewhat surprising given the induced expression of these genes during organogenesis, with *mad3* being expressed in proliferating cells before their differentiation, *mxi1* and *mad4* being expressed in early differentiation, and *mad1* in later differentiation (Queva et al. 1998). The mildness of the knockout phenotypes suggests several possibilities including (1) the activities of these proteins and their roles as Myc antagonists are not as physiologically important as originally thought, (2) there is functional compensation/redundancy between the individual family members in most tissues, or (3) there are other proteins within and beyond the Myc/Max/Mad network that have overlapping roles with Mad family proteins.

Because of the subtle gross phenotypes of *mad* family member knockout mice, various investigators have capitalized on the accessibility of the hematopoietic lineages to elucidate the consequences of Mad/Mxi1 loss of function in vivo. In the original *mad1* knockout study, this type of analysis revealed a function for Mad1 in mediating cell-cycle withdrawal during terminal differentiation of granulocytes (Foley et al. 1998). Specifically, granulocyte precursors in Mad1-deficient mice displayed increased proliferation and delayed differentiation (although their concomitant enhanced sensitivity to apoptosis ultimately resulted in normal numbers of bone marrow and circulating granulocytes). Additional experimental support for Mad1's function in regulating cell-cycle exit derives from transgenic studies in which *mad1* was targeted to various organs/compartments via generalized or tissue-specific promoters (Iritani et al. 2002; Queva et al. 1999; Rudolph et al. 2001). In one transgenic model, Mad1 was expressed ubiquitously using the β-*actin* (*BAP*) promoter; these *BAP–mad1* mice presented on the gross level with early postnatal lethality and dwarfism and on the cellular level with proliferative defects in multiple cell types (Queva et al. 1999). Similarly, transgenic mice specifically expressing Mad1 in T cells (*lck-mad1)* (Rudolph et al. 2001) or B and T cells (*Eμ-lck-mad1)* (Iritani et al. 2002) showed profound proliferative defects in and reduced cellularity of these hematopoietic compartments. This suggests that Mad1 regulates cell-cycle withdrawal by promoting G1 arrest, possibly through the regulation of cyclin/cyclin-dependent kinase (cdk) complexes and cdk inhibitory proteins (ckis) (similar to what was shown with the N-*myc* conditional knockout mice and the c-$myc^{-/-}$ rat1 fibroblasts). The level/activity of these regulators has been assessed in *mad1* transgenic mice, wherein the data suggest that Mad1 may be influencing different cyclin/cdk (including cyclin D1, cyclin A, cdk4) or cki (including $p21^{Cip1}$ and $p27^{Kip1}$) complexes in different cell types (Queva et al. 1999; Rudolph et al. 2001). In addition, the status of various regulators was assessed in cells derived from $mad1^{-/-};p27^{-/-}$ compound mice that were generated to determine whether Mad1 and p27 may cooperate to regulate the cell cycle during differentiation (McArthur et al. 2002). On the gross phenotypic level, these double-knockout mice exhibited a partially penetrant embryonic or perinatal lethality, with surviving animals presenting with kidney and bone marrow hyperplasia. These observed synthetic effects in certain tissues suggest both a genetic interaction between Mad1 and p27 and the possibility that these two proteins have partly redundant functions. This redundancy may allow for p27 to compensate for Mad1 loss in the specific cell types that are not affected in the *mad1* single-knockout mice but are affected in the *mad1/p27* double-knockout mice. Aside from the phenotypes already mentioned, the $mad1^{-/-};p27^{-/-}$ compound mice displayed an impaired differentiation of granulocyte precursors that was more severe than that observed in the *mad1* single-knockout mice (Foley et al. 1998; McArthur et al. 2002). Finally, an analysis of doubly deficient granulocyte lines showed enhanced cyclin E:cdk2 kinase activity when these lines were induced to differentiate. These observations raise the possibility that p27 and Mad1 cooperate to downregulate cyclin E:cdk2 activity in the promotion of cell-cycle exit during differentiation of specific cell types (McArthur et al. 2002).

In addition to these changes, another finding in the Mad1/p27 doubly deficient granulocytic cells was dramatically increased c-*myc* messenger (m)RNA and protein levels. In this case, the upregulation of *myc* accompanying Mad1 and p27 loss could indicate that these two proteins normally cooperate to downregulate *myc* during cell-cycle exit (McArthur et al. 2002). In several of the other knockout mouse models, however, there is a different type of crosstalk among the Myc/Max/Mad family members wherein there is induced expression of one family member in an attempt to compensate for the loss of another [e.g., c-*myc* is upregulated in the neuroepithelium of N-*myc* knockout mice (Stanton et al. 1992) or *mad3* and *mxi1* are upregulated in Mad1-deficient spleens (Foley et al. 1998)]. Additionally, in fibroblasts derived from the Mnt-deficient mice (discussed in the following section), c-*myc* expression is downregulated, perhaps in response to the loss of this potent Myc antagonist (Hurlin et al. 2003).

Similar to the *mad1* knockout mice described above, Mxi1-deficient mice displayed enhanced proliferative potentials in specific cell types (Schreiber-Agus et al. 1998). Hyperplastic changes were observed histologically in the splenic white pulp as well as the prostate, this supporting a normal role for Mxi1 in growth control. Moreover, several cell types isolated from the Mxi1-deficient mouse exhibited enhanced proliferative potential under certain conditions as compared to their wildtype counterparts. For instance, $mxi1^{-/-}$ fibroblasts were more readily able than their wildtype counterparts to generate actively growing colonies of clonal origin when subjected to low-density seeding assays. In addition, in response to mitogenic stimulation, Mxi1-deficient splenic T cells showed increased proliferation due to an increased exit from G1. A further assessment of whether Mxi1 is regulating the levels/activity of cyclin/cdk or cki complexes may shed light on the common and distinct molecular mechanisms by which Mxi1 and Mad1 can induce G1 arrest. Finally, despite these findings of increased proliferation in various cell types of the Mxi1-deficient mice (Schreiber-Agus et al. 1998), the rate of spontaneous tumor formation therein was not increased to a significant extent [nor was it increased in the Mad1-deficient mice (Foley et al. 1998)]. However, *mxi1*$^{-/-}$ animals did display an increased rate of tumorigenesis when it was induced by carcinogens or by the concomitant loss of the *ink4a/arf* tumor suppressor locus. This suggests that other cooperating mutations may be necessary to potentiate tumor formation in the Mxi1-deficient mice, and the same could hold true for other Mad-deficient mice as well.

In contrast to the *mad1* and *mxi1* knockout mice, *mad3* homozygous null mice did not display any evident phenotype associated with cell-cycle exit or differentiation in the cell types examined including hematopoietic cells, fibroblasts, and neural precursors cells (Queva et al. 2001). This lack of alterations in these cellular processes was somewhat surprising given that *mad3* is unique among the Mad members in that it is expressed in proliferating cells committed to differentiation (Queva et al. 1998, 2001). A phenotype that was apparent in the Mad3-deficient mouse was that *mad3*$^{-/-}$ thymocytes and neural progenitor cells were more sensitive than their wildtype counterparts to apoptosis induced by gamma irradiation (Queva et al. 2001). Since Myc has been proposed to sensitize cells to apoptosis (reviewed in Oster et al. 2002; Pelengaris et al. 2002), it is tempting to speculate that Mad3 (and possibly other Mad family proteins) may participate in the inhibition of apoptosis. Indeed, as hinted to above, Mad1-deficient granulocyte precursor cells grown under limiting amounts of cytokines showed a decreased rate of survival, and Mad1-deficient granulocytic cluster-forming cells exhibited an increased sensitivity to other apoptosis-inducing stimuli such as gamma irradiation (Foley et al. 1998). Conversely, decreased sensitivity to apoptotic stimuli was observed with the Mad1 transgenic mice (Iritani et al. 2002; Queva et al. 1999). Further studies are needed to assess whether or not the loss of Mxi1 can sensitize cells to apoptosis as has been shown for Mad1 and Mad3. It is possible that the different Mad family members perform this function in specified cell types under specific apoptosis-inducing conditions.

Taken together, data from the *mad* family member knockout mice have supported the roles for Mad family proteins as antagonists of Myc and have suggested that there may be functional diversity within the family in different tissue types/cellular contexts. However, the full physiological significance of their roles remains to be elucidated. A better understanding of this could benefit from the generation of additional *mad* family knockouts (e.g., *mad4*), compound knockouts (double, triple, or quadruple), and possibly knockouts of other, related Myc/Max/Mad network members (in this regard, see next section).

4.2
mnt Knockout Mice

Mnt is a more recently described bHLH/LZ protein of the Myc/Max/Mad network that shares many properties with the Mad family proteins. These properties include the abilities to heterodimerize with Max to bind to the E box, to repress transcription at the E-box through interaction with Sin3, and to suppress Myc/Ras cotransformation [(Hurlin et al. 1997); Mnt is also known as Rox (Meroni et al. 1997)]. Despite these shared functions, a number of differences exist between Mnt and the Mad family members. Apart from its SID (Sin3-interacting domain)/repression domain, Mnt contains several proline-rich regions that are reminiscent of transcriptional activation domains. Indeed, deletion of Mnt's SID converts Mnt from a transcriptional repressor into a transcriptional activator. Furthermore, in addition to being expressed in differentiating cells, *mnt* can be found co-expressed with *myc* and *max* in proliferating cells (Hurlin et al. 1997). In contrast to the Mad1 ubiquitous gain-of-function mice, which survive until the early post-natal stage (Queva et al. 1999), transgenic mice ubiquitously expressing *mnt* exhibit a marked developmental delay and die during midgestation (between E8.5 and E10.5; Hurlin et al. 1997). Finally, it is possible that an ancestral version of Mnt gave rise to Mnt as well as the entire Mad/Mxi1 family in vertebrates, since in the fly only a single ortholog of *mnt* can be found (Peyrefitte et al. 2001; Orian et al. 2003).

Mnt-deficient mice were generated recently and shown to exhibit runting and early postnatal lethality (Hurlin et al. 2003). Although the basis for this lethality remains to be described further, this developmental phenotype is more severe than that seen in any of the Mad family knockout mice. The difference in the knockout phenotypes may be attributable to the aforementioned unique properties of Mnt, and may also assign it a particularly important role as a Myc antagonist. Support for this theory also stems from conditional knockout mice in which *mnt* was specifically deleted from the mammary epithelium by crossing a mouse with a floxed *mnt* gene to an MMTV-Cre mouse (Hurlin et al. 2003). These conditional knockout mice developed mammary adenocarcinomas with variable latency, similar to what has been reported for mice overexpressing Myc in the same compartment. The conditional approach was employed in this case to circumvent the postnatal lethality associated with conventional Mnt knockout mice, and the choice of the MMTV-Cre mouse was dictated by the fact that *MNT* maps to a hotspot for deletion in human breast and other cancers.

MEFs derived from the conventional Mnt knockout mice were studied further to elucidate Mnt function on the cellular level. These *mnt*$^{-/-}$ MEFs were shown to enter S-phase prematurely and to have an accelerated proliferative rate. Similar to what was shown with the Mad1 mouse models (see previous section), these cell-cycle effects could be correlated with the ability of Mnt to regulate the levels and activities of specific cyclins and cdks (e.g., cyclin E1, cdk4, and cyclin E1:cdk2 complexes). In addition, Mnt-deficient MEFs displayed increased sensitivity to apoptosis under apoptosis-stimulating conditions, reminiscent of what has been described for Mad3-deficient thymocytes and neural progenitor cells and Mad1-deficient granulocyte precursor cells (see previous section). Finally, Mnt-deficient MEFs were more readily able than their wildtype counterparts to escape senescence, and they were also capable of being transformed by oncogenic Ras alone (i.e., they have bypassed the requirement for Myc to cooperate in this process; Hurlin et al. 2003).

Taken together, the phenotype of the Mnt-deficient cells and mammary epithelium strongly resembles that resulting from Myc overexpression, points to Mnt's significant role as a Myc antagonist, and raises questions as to whether Mnt is a bona fide tumor suppressor involved in human cancer pathogenesis.

4.3
max Knockout Mice

The ability of Myc proteins to activate transcription at the E-box, to promote proliferation and transformation, and to induce apoptosis is highly dependent upon their dimerization with Max. Likewise, interaction of the Mad members and Mnt with Max is necessary for their ability to repress transcription at the E-box, to stimulate differentiation, and to inhibit transformation (Baudino and Cleveland 2001; Eisenman 2000; Grandori et al. 2000; Luscher 2001). Because Max is pivotal to the network and potentiates the activities of opposing transcriptional regulators, one would predict that Max should have an essential role in development. Indeed, Max deficiency in the mouse results in embryonic lethality at E6.5–E7.5 accompanied by developmental arrest in embryonic and extraembryonic tissues (Shen-Li et al. 2000). This developmental arrest most likely results from a reduction in cellular proliferation as opposed to an increase in apoptosis. The timing of the embryonic lethality, in that it is earlier than the lethality observed for c- or N-Myc-deficient mice (see Sects. 2 and 3), is consistent with the centrality of Max in the Myc/Max/Mad network. In addition, the severity of the Max-deficient phenotype is suggestive of an essential role for the Myc proteins even in the post-implantation/pre-organogenesis stages of development; this role could be masked in the Myc knockout mice due to functional complementation by Myc paralogs. Finally, a role for Myc even during the early cleavage and pre-implantation stages of development cannot be ruled out, since maternal stores of Max in the Max-deficient embryos may be sufficient to rescue them through these stages (Shen-Li et al. 2000 and references therein). All this being said, the extrapolation of roles for Myc family members from the Max-null phenotype must be done with some caution, since Max can associate with other proteins including itself, the Mad family members, Mnt, and Mga (reviewed in Baudino and Cleveland 2001; Eisenman 2000; Grandori et al. 2000; Luscher 2001; Oster et al. 2002). Elucidating the individual and collective contributions of these interactions to developmental processes would be a significant challenge.

5
Lessons Learned and Questions for the Future

A compilation of lessons we have learned to date from the *myc/max/mad* knockout mouse models, and suggestions as to where future efforts could be focused, follows.

5.1
Myc Function Is Essential for Normal Embryonic Development

Functions of the Myc family proteins (at least N-Myc and c-Myc) are required, and unable to be compensated for, from midembryogenesis and beyond, when these proteins are expressed in distinct cellular compartments. Support for this derives from the individual c- and N-*myc* knockout mice which display embryonic lethality between E9.5 and E12.5 (Charron et al. 1990, 1992; Davis et al. 1993; Sawai et al. 1991, 1993; Stanton et al. 1990, 1992). Since the Max knockout succumbs to lethality even earlier (~E6.5–7.5; Shen-Li et al. 2000), it is likely that there is also a role for Myc family proteins during the postimplantation through midembryogenesis stage. This role could be masked in the single Myc knockouts by functional compensation between c- and N-*myc*, which are expressed in overlapping patterns at this developmental time. However, given the diversity of Max interactions (and the possibility that even Max–Max homodimers may be physiologically important), the possible functions of these other types of complexes in early embryogenesis cannot be excluded. The generation of double or triple knockouts for the *myc* family proteins and their comparison to the *max* knockout may be telling in this regard. Finally, it is conceivable that the Myc/Max/Mad network is also functioning in the earliest stages of embryogenesis (cleavage and pre-implantation). If this is the case, the survival of the *max* knockout through these stages can perhaps be explained by maternal stores of Max (Shen-Li et al. 2000), and the survival of the single *myc* knockouts through these stages can be explained once again by functional compensation between N- and c-Myc. All this being said, one could counter-argue that Myc function may not be absolutely required for cell viability in certain cell types. For instance, in the c-*myc*$^{-/-}$ rat1 system, cells thought to be completely devoid of *myc* expression (since N- and L-*myc* expression is not detectable in these cells) are still capable of dividing (although in an aberrant manner) (Mateyak et al. 1997). However, it is important to note that the c-*myc*$^{-/-}$ rat1 cells were selected for their ability to proliferate, and may have in the selection process figured out a way to bypass the Myc requirement that all cells may possess at the outset.

5.2 Threshold Levels of Myc Expression Must Be Reached for Proper Growth and Development of Certain Tissues

The concept of "threshold Myc levels" was best exemplified in the allelic series generated for N-*myc* and c-*myc* through the use of hypomorphic and null alleles and combinations thereof (Moens et al. 1992, 1993; Nagy et al. 1998; Trumpp et al. 2001). In these allelic series, the severity of the phenotype in certain tissues increased with incremental reductions in the level of Myc. In other words, the emerging phenotypes behaved in a dose-dependent manner. In tissues that were affected in the null mutants but not in mutants with reduced Myc expression, it is possible that the level of Myc was sufficient in the latter animals to ensure proper growth and development. However, crosstalk between the Myc/Max/Mad members and changes in their expression to compensate for the loss of a member cannot be ruled out. This type of compensation has been described in one of the N-*myc* conventional knockout mouse lines wherein c-*myc* was ectopically expressed in the neuroepithelium (Stanton et al. 1992). Despite this upregulation, CNS abnormalities were still present in the N-*myc*-null mice, perhaps suggesting that threshold Myc levels had not been achieved, among other possibilities (discussed in Sect. 2).

While reductions in Myc levels/activity have been shown to impair normal growth and development in the loss-of-function models, excessive Myc is also detrimental as exemplified by numerous gain-of-function mouse models (not discussed here). Achieving the "ideal" spatio-temporal expression levels and activity levels of Myc requires multiple modes of regulation, including that accomplished through the antagonistic actions of other members of the Myc/Max/Mad network.

5.3
N- and c-Myc May Have Evolved to Serve Highly Similar Functions in Different Vertebrate Tissues/Cell Types

The various c- and N-*myc* knockout mouse studies have supported developmental roles for these proteins in multiple processes including cell division, cell growth, cell survival, inhibition of differentiation, tissue–tissue interaction, among others. However, it still remains to be determined if c- and N-Myc are playing *identical* roles in these processes. Complete functional redundancy between these two highly related proteins has gained experimental support from the knockin study (Malynn et al. 2000) in which replacement of the c-*myc* coding sequences with those of N-*myc* resulted in an apparently normal mouse. Perhaps the converse experiment, in which c-*myc* coding sequences replace those of N-*myc* in the N-*myc* locus, would be worthwhile in this regard, given that N-*myc* expression patterns are more restricted than those of c-*myc*. If this type of knockin mouse emerged as phenotypically normal, then the case for complete redundancy would be even more solid (albeit with the caveat that these knockin studies are contrived situations that may force proteins to assume supra-physiological roles). Then, the preservation of c- and N-*myc* as distinct sequences throughout evolution could be attributed (solely?) to the "regulatory asymmetry" (Krakauer and Nowak 1999) imparted upon them by their unique expression patterns. However, it still remains possible that there are functional differences between c- and N-Myc that have escaped our detection, and these differences may relate to their interactions with different cofactors or their regulation of distinct gene targets (although once again these differences may be secondary to their expression patterns). Additional insight into possible differential roles in various tissue types could be gained from the generation of new conditional knockout mice for c- and N-*myc*, and the subsequent combinations thereof for comparison to the individual knockouts.

5.4
L-Myc Appears to Be Dispensable for Normal Growth and Development, but Subtle Roles for L-Myc Cannot Yet Be Excluded

In contrast to the roles of c- and N-Myc, the role of L-Myc and its preservation throughout evolution remains somewhat mysterious, since the L-*myc* knockout mouse did not exhibit any discernable phenotypic abnormalities (Hatton et al. 1996). It has been argued that, since L-*myc* is typically co-expressed with another *myc* family member and has weaker transformation/transactivation activities, L-Myc may temper Myc activity in certain cell types/developmental stages (Hatton et al. 1996) or even may be evolving to be functionally inactive. For additional insight into L-Myc's role, one could consider crossing the L-Myc-deficient mouse onto the c- or N-Myc-deficient backgrounds, challenging the L-myc-deficient mice with chemical or biological agents to elicit latent phenotypes, or specifically inactivating L-*myc* in a cell type that expresses L-*myc* but not c- or N-*myc* (if that exists). With respect to the latter point, clues about this may come from the distinct human cancer types in which L-*myc* deregulation has been implicated to be causal (Lutz et al. 2002; Nesbit et al. 1999). Finally, a knockin study similar to the one discussed above (Malynn et al. 2000), but replacing L-*myc* sequences for c-*myc* sequences in the whole animal or select cell types, could provide insight into L-Myc's functional place within the Myc/Max/Mad network.

5.5 Lack of Dramatic Phenotypes with Single Knockouts of the Mad Family Members

Despite the fact that Mad family proteins have been promoted over the last decade as important Myc antagonists and putative tumor suppressors, Mad1-, Mxi1-, or Mad3-deficient mice did not display dramatic developmental phenotypes or strong predisposition to tumorigenesis (Foley et al. 1998; Queva et al. 2001; Schreiber-Agus et al. 1998). Compensation/redundancy between related Mad family proteins has been offered as a likely explanation for this, and, as with the Myc subfamily of the Myc/Max/Mad network, the preservation of the Mad family members as unique sequences may thus result from their distinct expression patterns during embryogenesis and in adult tissues (Queva et al. 1998). The issue of redundancy could be addressed further via the generation of double, triple, and quadruple knockouts of the Mad family members, studies which are currently ongoing (B. Iritani and R.N. Eisenman, personal communication). Surprisingly, the triple *mad1*, *mxi1*, *mad3* knockout mice are fertile and viable, albeit that they are about 15%–20% larger than age-matched controls. In addition, the splenic B and T cells and thymocytes of these triple-knockout mice exhibit a hyperproliferative phenotype following stimulation (B. Iritani and R.N. Eisenman, personal communication). While the remaining Mad family member (Mad4) may be compensating for the loss of the others and preventing more severe sequelae, it is also possible that even the quadruple-knockout mice will not be grossly compromised. This outcome could suggest compensation by other more distantly related members (e.g., Mnt; see next paragraph) with overlapping, or perhaps even dominant, functions. Moreover, there is also the suggestion that there may be functional compensation/complementation by proteins outside of the Myc/Max/Mad network altogether. Evidence for this has emerged already from the documented synergism between Mad1 and the cki p27 in promoting cell-cycle exit during differentiation of specific cell types (McArthur et al. 2002). These complementary players could also be the ones that may cooperate with the Mad family proteins in the function of tumor suppression. Evidence for this has emerged from the finding that the rate of tumorigenesis in the Mxi1-deficient mouse was increased with the concomitant loss of the *ink4a/arf* tumor suppressor locus (Schreiber-Agus et al. 1998).

The recently generated Mnt-deficient mouse has revealed an important role for this Myc/Max/Mad network member in proliferation, development, and tumor suppression. Specifically, mice rendered deficient for Mnt by conventional means died postnatally, and mice rendered conditionally deficient for Mnt in the mammary epithelium developed mammary adenocarcinomas (Hurlin et al. 2003). These developmental and tumor-prone phenotypes are much more severe than those observed in the Mad family knockout mice, a finding that may indicate a broader role for Mnt in the negative regulation of proliferation and in Myc antagonism. This broader role may result in part from the expression profile of Mnt vis-à-vis that of the Mad family proteins. Notably, *mnt* is ubiquitously expressed and readily detected in quiescent and proliferating cells, while *mad* family gene expression is usually restricted in developing and adult tissues and is found at low levels in proliferating cells. Moreover, Mnt may encode additional functions relative to those encoded by Mad family proteins (perhaps through Mnt's putative transactivation domain) that when lost may contribute to the severity of the knockout phenotype.

Through the analysis of cells derived from the deficient mice, roles for Mnt, like for Mad family proteins, have been established in the negative regulation of cell-cycle entry and in the negative regulation of apoptosis. Additionally, Mnt has been shown to be involved in regulating senescence, a property that has not been fully evaluated in the Mad family knockout mice. Whether or not the Mad family proteins and Mnt have similar roles in different cell types remains to be determined. In this regard, our ability to discern differential cellular functions for related family members of the Myc/Max/Mad network could benefit greatly from applying a certain degree of standardization to loss-of-function studies. This could involve employing equivalent knockout strategies, using mice of the same genetic background, and/or carrying out the phenotypic analysis of the single and compound knockouts in a uniform manner. If single members are expressed in specific cell types, their functions could be elucidated and compared to those of other members expressed alone in other cell types. In cell types where multiple members are expressed, the effects of individual versus combinatorial deficiencies could be compared side-by-side and assessed for augmented phenotypes in the latter.

5.6 The Emerging List of Bona Fide Myc/Max/Mad Gene Targets Should Be Considered in Relation to the Existing Myc/Max/Mad Mouse Models

The emergence of bona fide Myc or Mad downstream gene targets (see chapter by L.A. Lee and C.V. Dang, this volume) raises questions as to whether specific targets are the primary mediators of Myc or Mad functions. Accordingly, loss-of-function (or gain-of-function) models for these gene targets could be compared to those of the Myc/Max/Mad network. Notably, this type of comparison was done for one of the putative *Drosophila* Myc targets, *pitchoune* (*pit*), and the results suggest that Pit may be an important mediator of Myc's ability to regulate cellular growth and metabolism (Zaffran et al. 1998).

Aside from being able to place valid targets downstream of Myc/Max/Mad function in specific cellular processes, the identification of these targets also could prompt revisiting of Myc/Max/Mad loss- or gain-of-function models to assess further the role of this network in additional/alternative cellular processes. For instance, since a role for Myc (and Mad) in the control of cellular growth (accumulation of mass) continues to gain support from a variety of systems, this aspect could be investigated further in the context of the mouse. Preliminary support for this in vivo role comes from the findings of reduced cell growth in neurons of the N-*myc* conditional knockout mice (Knoepfler et al. 2002) and in c-*myc* $^{-/-}$ thymocytes generated via the Rag1-deficient blastocyst approach (Douglas et al. 2001). In addition, certain subsets of thymocytes from *Eμ-lck-Mad1* transgenic mice were shown to be smaller in size than thymocytes from wildtype mice (Iritani et al. 2002). Notably, this reduction in size was accompanied by an enhanced repression of genes involved in the regulation of cell growth. Finally, apart from assessing alterations in cellular growth further in the Myc/Max/Mad mouse models, it may also be worthwhile to evaluate changes in adhesion/cytoskeletal remodeling in specific cell types. The rationale for this stems from the fact that numerous proteins associated with these processes were found to be differentially expressed in a quantitative proteomic analysis of Myc-deficient cells in comparison to controls (Shiio et al. 2002). In addition, among the various phenotypes of the c-*myc*$^{-/-}$ rat1 cells were morphological changes consistent with cytoskeletal alterations (Mateyak et al. 1997).

5.7 New Insights Regarding the Contributions of the Myc/Max/Mad Network to the Pathogenesis of Human Cancers May Lie in the Various Mouse Models

It would be of great value to use the information gleaned from Myc/Max/Mad loss- and gain-of-function mouse models to further our understanding of how dysregulation of this network contributes to cancer pathogenesis. Caution must be exercised though, since the physiological processes in which this network participates, and the gene targets which it regulates, may be different from those that are affected when Myc family proteins are dysregulated (and possibly Mad family proteins are inactivated) in human tumors (this concept is also discussed in Cole and McMahon 1999). Nonetheless, these models could provide important clues. One example of clues of this type emerges from the N-*myc* conditional knockout study in neuronal progenitors (Knoepfler et al. 2002). More specifically, the frequent dysregulation of N-*myc* in nervous system tumors may relate directly to the critical role for N-Myc in controlling growth and proliferation and preventing differentiation of neuronal precursor cells. Another clue may come from mice with a conditional knockout of *mnt* in the mammary epithelium (Hurlin et al. 2003). The finding of mammary adenocarcinomas in a significant proportion of these mice, coupled with the mapping of human *MNT* to a hotspot for loss-of-heterozygosity in breast cancers (17p13.3), warrants further investigation of the possible pathogenetic role of Mnt loss in human tumorigenesis.

6 Conclusion

The past two decades of gene targeting experiments have allowed us to make significant strides towards understanding how the Myc/Max/Mad network influences multiple aspects of cellular behavior during development. With multiple knockout lines, sophisticated approaches, and insights from complementary gain-of-function systems in hand, we are now in a unique position to significantly advance our understanding of Myc/Max/Mad in vivo functions. Among the pressing issues that remain to be addressed in the context of the whole organism are the differential functions of related family members, the nature of the downstream mediators of these functions, the factors controlling Myc/Max/Mad expression/activity, and the relationship of this network to other key regulators of cellular behavior.

Note Added in Proof The reader is directed to the recent literature for additional reports on this topic that were published since the preparation of this review. These reports include those that describe new tissue-specific knockouts for N-*myc* and c-*myc*, conditional inactivation of *myc* in Myc-induced mouse tumor models, and knockouts of various bona fide Myc downstream gene targets.

References

Adachi S, Obaya AJ, Han Z, Ramos-Desimone N, Wyche JH, Sedivy JM (2001) c-Myc is necessary for DNA damage-induced apoptosis in the G(2) phase of the cell cycle. Mol Cell Biol 21:4929–4937

Bates CM, Kharzai S, Erwin T, Rossant J, Parada LF (2000) Role of N-myc in the developing mouse kidney. Dev Biol 222:317–325

Baudino TA, Cleveland JL (2001) The Max network gone mad. Mol Cell Biol 21:691–702

Baudino TA, McKay C, Pendeville-Samain H, Nilsson JA, Maclean KH, White EL, Davis AC, Ihle JN, Cleveland JL (2002) c-Myc is essential for vasculogenesis and angiogenesis during development and tumor progression. Genes Dev 16:2530–2543

Berns K, Hijmans EM, Koh E, Daley GQ, Bernards R (2000) A genetic screen to identify genes that rescue the slow growth phenotype of c-myc null fibroblasts. Oncogene 19:3330–3334

Bush A, Mateyak M, Dugan K, Obaya A, Adachi S, Sedivy J, Cole M (1998) c-myc null cells misregulate cad and gadd45 but not other proposed c-Myc targets. Genes Dev 12:3797–3802

Charron J, Malynn BA, Robertson EJ, Goff SP, Alt FW (1990) High-frequency disruption of the N-myc gene in embryonic stem and pre-B cell lines by homologous recombination. Mol Cell Biol 10:1799–1804

Charron J, Malynn BA, Fisher P, Stewart V, Jeannotte L, Goff SP, Robertson EJ, Alt FW (1992) Embryonic lethality in mice homozygous for a targeted disruption of the N-myc gene. Genes Dev 6:2248–2257

Chen J (1996) Analysis of gene function in lymphocytes by RAG-2-deficient blastocyst complementation. Adv Immunol 62:31–59

Cole MD, McMahon SB (1999) The Myc oncoprotein: a critical evaluation of transactivation and target gene regulation. Oncogene 18:2916–2924

Dang CV (1999) c-Myc target genes involved in cell growth, apoptosis, and metabolism. Mol Cell Biol 19:1–11

Dang CV, Resar LM, Emison E, Kim S, Li Q, Prescott JE, Wonsey D, Zeller K (1999) Function of the c-Myc oncogenic transcription factor. Exp Cell Res 253:63–77

Davis A, Bradley A (1993) Mutation of N-myc in mice: what does the phenotype tell us? Bioessays 15:273–275

Davis AC, Wims M, Spotts GD, Hann SR, Bradley A (1993) A null c-myc mutation causes lethality before 10.5 days of gestation in homozygotes and reduced fertility in heterozygous female mice. Genes Dev 7:671–682

de Alboran IM, O'Hagan RC, Gartner F, Malynn B, Davidson L, Rickert R, Rajewsky K, DePinho RA, Alt FW (2001) Analysis of C-MYC function in normal cells via conditional gene-targeted mutation. Immunity 14:45–55

DePinho RA, Schreiber-Agus N, Alt FW (1991) myc family oncogenes in the development of normal and neoplastic cells. Adv Cancer Res 57:1–46

Douglas NC, Jacobs H, Bothwell AL, Hayday AC (2001) Defining the specific physiological requirements for c-Myc in T cell development. Nat Immunol 2:307–315

Eagle LR, Yin X, Brothman AR, Williams BJ, Atkin NB, Prochownik EV (1995) Mutation of the MXI1 gene in prostate cancer. Nat Genet 9:249–255

Eisenman RN (2000) The Max network: coordinated transcriptional regulation of cell growth and proliferation. Harvey Lect 96:1–32

Felton-Edkins ZA, Kenneth NS, Brown TR, Daly NL, Gomez-Roman N, Grandori C, Eisenman RN, White RJ (2003) Direct regulation of RNA polymerase III transcription by RB, p53 and c-Myc. Cell Cycle 2:181–184

Foley KP, McArthur GA, Queva C, Hurlin PJ, Soriano P, Eisenman RN (1998) Targeted disruption of the MYC antagonist MAD1 inhibits cell cycle exit during granulocyte differentiation. EMBO J 17:774–785

Gallant P, Shiio Y, Cheng PF, Parkhurst SM, Eisenman RN (1996) Myc and Max homologs in Drosophila. Science 274:1523–1527

Gertsenstein M, Lobe C, Nagy A (2002) ES cell-mediated conditional transgenesis. Methods Mol Biol 185:285–307

Giroux S, Charron J (1998) Defective development of the embryonic liver in N-myc-deficient mice. Dev Biol 195:16–28

Grandori C, Cowley SM, James LP, Eisenman RN (2000) The Myc/Max/Mad network and the transcriptional control of cell behavior. Annu Rev Cell Dev Biol 16:653–699

Guo QM, Malek RL, Kim S, Chiao C, He M, Ruffy M, Sanka K, Lee NH, Dang CV, Liu ET (2000) Identification of c-myc responsive genes using rat cDNA microarray. Cancer Res 60:5922–5928

Han S, Park K, Kim HY, Lee MS, Kim HJ, Kim YD, Yuh YJ, Kim SR, Suh HS (2000) Clinical implication of altered expression of Mad1 protein in human breast carcinoma. Cancer 88:1623–1632

Hatton KS, Mahon K, Chin L, Chiu FC, Lee HW, Peng D, Morgenbesser SD, Horner J, DePinho RA (1996) Expression and activity of L-Myc in normal mouse development. Mol Cell Biol 16:1794–1804

Holzel M, Kohlhuber F, Schlosser I, Holzel D, Luscher B, Eick D (2001) Myc/Max/Mad regulate the frequency but not the duration of productive cell cycles. EMBO Rep 2:1125–1132

Hurlin PJ, Queva C, Eisenman RN (1997) Mnt, a novel Max-interacting protein is co-expressed with Myc in proliferating cells and mediates repression at Myc binding sites. Genes Dev 11:44–58

Hurlin PJ, Zhou ZQ, Toyo-oka K, Ota S, Walker WL, Hirotsune S, Wynshaw-Boris A (2003) Deletion of Mnt leads to disrupted cell cycle control and tumorigenesis. EMBO J 22:4584–4596

Iritani BM, Delrow J, Grandori C, Gomez I, Klacking M, Carlos LS, Eisenman RN (2002) Modulation of T-lymphocyte development, growth and cell size by the Myc antagonist and transcriptional repressor Mad1. EMBO J 21:4820–4830

James L, Eisenman RN (2002) Myc and Mad bHLHZ domains possess identical DNA-binding specificities but only partially overlapping functions in vivo. Proc Natl Acad Sci U S A 99:10429–10434

Johnston LA, Prober DA, Edgar BA, Eisenman RN, Gallant P (1999) Drosophila myc regulates cellular growth during development. Cell 98:779–790

Joyner AL (2000) Gene targeting: a practical approach, 2nd edn. (Practical Approach Series). Oxford University Press, Oxford, pp 1–293

Knoepfler PS, Cheng PF, Eisenman RN (2002) N-myc is essential during neurogenesis for the rapid expansion of progenitor cell populations and the inhibition of neuronal differentiation. Genes Dev 16:2699–2712

Krakauer DC, Nowak MA (1999) Evolutionary preservation of redundant duplicated genes. Semin Cell Dev Biol 10:555–559

Kwan KM (2002) Conditional alleles in mice: practical considerations for tissue-specific knockouts. Genesis 32:49–62

Landay M, Oster SK, Khosravi F, Grove LE, Yin X, Sedivy J, Penn LZ, Prochownik EV (2000) Promotion of growth and apoptosis in c-myc nullizygous fibroblasts by other members of the myc oncoprotein family. Cell Death Differ 7:697–705

Luscher B (2001) Function and regulation of the transcription factors of the Myc/Max/Mad network. Gene 277:1–14

Lutz W, Leon J, Eilers M (2002) Contributions of Myc to tumorigenesis. Biochim Biophys Acta 1602:61–71

Malynn BA, Demengeot J, Stewart V, Charron J, Alt FW (1995) Generation of normal lymphocytes derived from N-myc-deficient embryonic stem cells. Int Immunol 7:1637–1647

Malynn BA, de Alboran IM, O'Hagan RC, Bronson R, Davidson L, DePinho RA, Alt FW (2000) N-myc can functionally replace c-myc in murine development, cellular growth, and differentiation. Genes Dev 14:1390–1399

Mateyak MK, Obaya AJ, Adachi S, Sedivy JM (1997) Phenotypes of c-Myc-deficient rat fibroblasts isolated by targeted homologous recombination. Cell Growth Differ 8:1039–1048

Mateyak MK, Obaya AJ, Sedivy JM (1999) c-Myc regulates cyclin D-Cdk4 and -Cdk6 activity but affects cell cycle progression at multiple independent points. Mol Cell Biol 19:4672–4683

McArthur GA, Foley KP, Fero ML, Walkley CR, Deans AJ, Roberts JM, Eisenman RN (2002) MAD1 and p27(KIP1) cooperate to promote terminal differentiation of granulocytes and to inhibit Myc expression and cyclin E-CDK2 activity. Mol Cell Biol 22:3014–3023

Meroni G, Reymond A, Alcalay M, Borsani G, Tanigami A, Tonlorenzi R, Nigro CL, Messali S, Zollo M, Ledbetter DH, Brent R, Ballabio A, Carrozzo R (1997) Rox, a novel bHLHZip protein expressed in quiescent cells that heterodimerizes with Max, binds a non-canonical E box and acts as a transcriptional repressor. EMBO J 16:2892–2906

Moens CB, Auerbach AB, Conlon RA, Joyner AL, Rossant J (1992) A targeted mutation reveals a role for N-myc in branching morphogenesis in the embryonic mouse lung. Genes Dev 6:691–704

Moens CB, Stanton BR, Parada LF, Rossant J (1993) Defects in heart and lung development in compound heterozygotes for two different targeted mutations at the N-myc locus. Development 119:485–499

Morgenbesser SD, DePinho RA (1994) Use of transgenic mice to study myc family gene function in normal mammalian development and in cancer. Semin Cancer Biol 5:21–36

Nagy A, Moens C, Ivanyi E, Pawling J, Gertsenstein M, Hadjantonakis AK, Pirity M, Rossant J (1998) Dissecting the role of N-myc in development using a single targeting vector to generate a series of alleles. Curr Biol 8:661–664

Nasi S, Ciarapica R, Jucker R, Rosati J, Soucek L (2001) Making decisions through Myc. FEBS Lett 490:153–162

Nesbit CE, Tersak JM, Prochownik EV (1999) MYC oncogenes and human neoplastic disease. Oncogene 18:3004–3016

Nikiforov MA, Kotenko I, Petrenko O, Beavis A, Valenick L, Lemischka I, Cole MD (2000) Complementation of Myc-dependent cell proliferation by cDNA expression library screening. Oncogene 19:4828–4831

Nikiforov MA, Chandriani S, O'Connell B, Petrenko O, Kotenko I, Beavis A, Sedivy JM, Cole MD (2002a) A functional screen for Myc-responsive genes reveals serine hydroxymethyltransferase, a major source of the one-carbon unit for cell metabolism. Mol Cell Biol 22:5793–5800

Nikiforov MA, Chandriani S, Park J, Kotenko I, Matheos D, Johnsson A, McMahon SB, Cole MD (2002b) TRRAP-dependent and TRRAP-independent transcriptional activation by Myc family oncoproteins. Mol Cell Biol 22:5054–5063

Nikiforov MA, Popov N, Kotenko I, Henriksson M, Cole MD (2003) The Mad and Myc basic domains are functionally equivalent. J Biol Chem 278:11094–11099

O'Connell BC, Cheung AF, Simkevich CP, Tam W, Ren X, Mateyak MK, Sedivy JM (2003) A large scale genetic analysis of c-Myc-regulated gene expression patterns. J Biol Chem 278:12563–12573

O'Hagan RC, Schreiber-Agus N, Chen K, David G, Engelman JA, Schwab R, Alland L, Thomson C, Ronning DR, Sacchettini JC, Meltzer P, DePinho RA (2000) Gene-target recognition among members of the myc superfamily and implications for oncogenesis. Nat Genet 24:113–119

Orian A, van Steensel B, Delrow J, Bussemaker HJ, Li L, Sawado T, Williams E, Loo LW, Cowley SM, Yost C, Pierce S, Edgar BA, Parkhurst SM, Eisenman RN (2003) Genomic binding by the Drosophila Myc, Max, Mad/Mnt transcription factor network. Genes Dev 17:1101–1114

Oster SK, Ho CS, Soucie EL, Penn LZ (2002) The myc oncogene: MarvelouslY complex. Adv Cancer Res 84:81–154

Pelengaris S, Rudolph B, Littlewood T (2000) Action of Myc in vivo—proliferation and apoptosis. Curr Opin Genet Dev 10:100–105

Pelengaris S, Khan M, Evan G (2002) c-MYC: more than just a matter of life and death. Nat Rev Cancer 2:764–776

Peyrefitte S, Kahn D, Haenlin M (2001) New members of the Drosophila Myc transcription factor subfamily revealed by a genome-wide examination for basic helix-loop-helix genes. Mech Dev 104:99–104

Prochownik EV, Eagle Grove L, Deubler D, Zhu XL, Stephenson RA, Rohr LR, Yin X, Brothman AR (1998) Commonly occurring loss and mutation of the MXI1 gene in prostate cancer. Genes Chromosomes Cancer 22:295–304

Queva C, Hurlin PJ, Foley KP, Eisenman RN (1998) Sequential expression of the MAD family of transcriptional repressors during differentiation and development. Oncogene 16:967–977

Queva C, McArthur GA, Ramos LS, Eisenman RN (1999) Dwarfism and dysregulated proliferation in mice overexpressing the MYC antagonist MAD1. Cell Growth Differ 10:785–796

Queva C, McArthur GA, Iritani BM, Eisenman RN (2001) Targeted deletion of the S-phase-specific Myc antagonist Mad3 sensitizes neuronal and lymphoid cells to radiation-induced apoptosis. Mol Cell Biol 21:703–712

Rudolph B, Hueber AO, Evan GI (2001) Expression of Mad1 in T cells leads to reduced thymic cellularity and impaired mitogen-induced proliferation. Oncogene 20:1164–1175

Sawai S, Shimono A, Hanaoka K, Kondoh H (1991) Embryonic lethality resulting from disruption of both N-myc alleles in mouse zygotes. New Biol 3:861–869

Sawai S, Shimono A, Wakamatsu Y, Palmes C, Hanaoka K, Kondoh H (1993) Defects of embryonic organogenesis resulting from targeted disruption of the N-myc gene in the mouse. Development 117:1445–1455

Schorl C, Sedivy JM (2003) Loss of protooncogene c-Myc function impedes G(1) phase progression both before and after the restriction point. Mol Biol Cell 14:823–835

Schreiber-Agus N, DePinho RA (1998) Repression by the Mad(Mxi1)-Sin3 complex. Bioessays 20:808–818

Schreiber-Agus N, Stein D, Chen K, Goltz JS, Stevens L, DePinho RA (1997) Drosophila Myc is oncogenic in mammalian cells and plays a role in the diminutive phenotype. Proc Natl Acad Sci U S A 94:1235–1240

Schreiber-Agus N, Meng Y, Hoang T, Hou H Jr, Chen K, Greenberg R, Cordon-Cardo C, Lee HW, DePinho RA (1998) Role of Mxi1 in ageing organ systems and the regulation of normal and neoplastic growth. Nature 393:483–487

Shen-Li H, O'Hagan RC, Hou H Jr, Horner JW 2nd, Lee HW, DePinho RA (2000) Essential role for Max in early embryonic growth and development. Genes Dev 14:17–22

Shiio Y, Donohoe S, Yi EC, Goodlett DR, Aebersold R, Eisenman RN (2002) Quantitative proteomic analysis of Myc oncoprotein function. EMBO J 21:5088–5096

Soucie EL, Annis MG, Sedivy J, Filmus J, Leber B, Andrews DW, Penn LZ (2001) Myc potentiates apoptosis by stimulating Bax activity at the mitochondria. Mol Cell Biol 21:4725–4736

Spanopoulou E (1996) Cellular and molecular analysis of lymphoid development using Rag-deficient mice. Int Rev Immunol 13:257–288

Stanton BR, Reid SW, Parada LF (1990) Germ line transmission of an inactive N-myc allele generated by homologous recombination in mouse embryonic stem cells. Mol Cell Biol 10:6755–6758

Stanton BR, Perkins AS, Tessarollo L, Sassoon DA, Parada LF (1992) Loss of N-myc function results in embryonic lethality and failure of the epithelial component of the embryo to develop. Genes Dev 6:2235–2247

Trumpp A, Refaeli Y, Oskarsson T, Gasser S, Murphy M, Martin GR, Bishop JM (2001) c-Myc regulates mammalian body size by controlling cell number but not cell size. Nature 414:768–773

Watson JD, Oster SK, Shago M, Khosravi F, Penn LZ (2002) Identifying genes regulated in a Myc-dependent manner. J Biol Chem 277:36921–36930

Xiao Q, Claassen G, Shi J, Adachi S, Sedivy J, Hann SR (1998) Transactivation-defective c-MycS retains the ability to regulate proliferation and apoptosis. Genes Dev 12:3803–3808

Zaffran S, Chartier A, Gallant P, Astier M, Arquier N, Doherty D, Gratecos D, Semeriva M (1998) A Drosophila RNA helicase gene, pitchoune, is required for cell growth and proliferation and is a potential target of d-Myc. Development 125:3571–3584

Zhou ZQ, Hurlin PJ (2001) The interplay between Mad and Myc in proliferation and differentiation. Trends Cell Biol 11:S10–S14

Zimmerman K, Alt FW (1990) Expression and function of myc family genes. Crit Rev Oncog 2:75–95

CTMI (2006) 302:235–253

Myc/Max/Mad in Invertebrates: The Evolution of the Max Network

P. Gallant (✉)

Universität Zürich, Zoologisches Institut, Winterthurerstrasse 190, 8057 Zürich, Switzerland
gallant@zool.unizh.ch

Abstract The Myc proto-oncogenes, their binding partner Max and their antagonists from the Mad family of transcriptional repressors have been extensively analysed in vertebrates. However, members of this network are found in all animals examined so far. Several recent studies have addressed the physiological function of these proteins in invertebrate model organisms, in particular *Drosophila melanogaster*. This review describes the structure of invertebrate Myc/Max/Mad genes and it discusses their regulation and physiological functions, with special emphasis on their essential role in the control of cellular growth and proliferation.

Abbreviations

BDGP	Berkeley Drosophila Genome Project; http://www.fruitfly.org/
BHLHLZ	Basic-helix 1-loop-helix 2 leucine zipper
EST	Expressed sequence tag
SID	mSin3 interaction domain
Wg	Wingless
Dpp	Decapentaplegic
PI3K	Phosphatidylinositol-3-OH kinase
PIP3	Phosphatidylinositol 3,4,5-triphosphate
TOR	Target of rapamycin
TSC1/2	“Tuberous sclerosis” tumour suppressor gene 1/2

1 Identification of *myc/max*-Related Genes in Invertebrates

The importance of *myc* genes in normal development and disease has been amply documented (Oster et al. 2002). Myc activity has been shown to be required for normal proliferation and growth (Oster et al. 2002); conversely, deregulated activation of Myc contributes to cellular transformation, immortalization and genome instability, and appears to promote growth, cell cycle progression, apoptosis and angiogenesis (Oster et al. 2002). All of these effects are associated with Myc's ability to regulate the expression of a number of target genes, whereby Myc can act as an activator on some targets and as a repressor on others. The mechanism of transcriptional repression by Myc has been recently reviewed and it will not be further addressed here (Wanzel et al. 2003; D. Kleine-Kohlbrecher et al., this volume). Transcriptional activation by Myc is mediated by heterodimers between Myc and Max which bind to specific DNA sequences called E-boxes. These E-boxes can also be bound by heterodimers of Max with Mad proteins, which results in repression of the corresponding genes. Thus, a model has emerged where Max is located at the centre of a network of transcriptional activators and repressors. Since Max levels appear to be fairly constant, it is the relative levels of Myc and Mad proteins which determines the transcriptional status of E-box-containing target genes. The analysis of this network is complicated by a high degree of functional redundancy; mice and humans, where the Max network has been most extensively studied, contain only one *max* gene, but at least 3 partially redundant *myc* genes (c-*myc*, N-*myc*, L-*myc*, plus additional genes derived from processed transcripts) and 5 *mad*-like genes (*mad1*, *mxi1*, *mad3*, *mad4*, *mnt*). To complicate matters further, targeted disruption of either c-*myc* or N-*myc* results in lethality during mid-embryogenesis (Charron et al. 1992; Davis et al. 1993; Sawai et al. 1993).

To circumvent these problems, different approaches were undertaken to identify the Max network in simpler and genetically tractable organisms. Low stringency hybridization approaches led to the cloning of Myc in the sea star *Asterias vulgaris* (Walker et al. 1992), but failed to molecularly identify any *myc* genes in protostomes (see e.g. Shilo and Weinberg 1981; Bishop 1983; Madhavan et al. 1985; Sarid et al. 1987; Blackwood and Eisenman 1991). Instead, the single *Drosophila* Myc orthologue, termed dMyc, was found in yeast two-hybrid screens of a *Drosophila* library where vertebrate Max was used as the bait (Gallant et al. 1996; Schreiber-Agus et al. 1997). *Drosophila* Max (dMax) was cloned in a subsequent yeast two-hybrid screen with dMyc as the bait (Gallant et al. 1996), and the single *Drosophila* Mad/Mnt orthologue (dMnt) was found in yet another yeast two-hybrid screen with dMax as the bait (Loo et al. 2005), and independently by in silico screens of the published *Drosophila* genome sequence (Peyrefitte et al. 2001). The availability of full-genome sequences also allowed the identification of Max network components in *Caenorhabditis elegans* (Yuan et al. 1998), *Anopheles gambiae* (Holt et al. 2002; P. Gallant, unpublished observation) and *Ciona intestinalis* (Dehal et al. 2002; P. Gallant, unpublished observation). In contrast to the situation in metazoans, no *myc*, *max* or *mad* genes are found in fungi or in plants. Two proteins in *Arabidopsis thaliana* called ATmyc1 (Urao et al. 1996) and ATmyc2 (Abe et al. 2003) share sequence similarity with the Myc C-terminus, the BHLHLZ domain (basic-helix 1-loop-helix 2-leucine zipper), but lack the N-terminal hallmarks of animal Myc proteins (Myc Box 1, Myc Box 2; see Sect. 2.2) and therefore probably do not correspond to true Myc orthologs.

2 Analysis of *myc, max, mad* Sequences in Invertebrates

The last common ancestor of insects, nematodes and chordates lived almost 1 billion years ago (Hedges 2002). Any motif that is conserved between orthologous proteins from these different groups is likely to be of functional significance. In the following sections, such evolutionary sequence conservation is discussed for Max network components from different invertebrates and one representative vertebrate, human (for an extensive comparison of vertebrate Myc proteins, see Miyamoto and Freire 2000; Johansson et al. 2001).

2.1 Max

All analysed species encode one Max orthologue, with the exception of *C. elegans*, which contains two *max*-like genes (*mxl-1* and *mxl-3*). As Max needs to interact with Myc and Mad proteins and possibly additional transcription factors such as Mga, TEF-1 and α-Pal (Hurlin et al. 1999; Gupta et al. 1997; Shors et al. 1998), it is not surprising that it is evolutionarily the most conserved component of the network (Atchley and Fitch 1995). The conservation is particularly high in the BHLHLZ domain, which is involved in protein:protein interactions and DNA binding (Fig. 1a, b). Fig. 1a also indicates the positions of exon–exon junctions with respect to the coding sequence; these junctions have been predicted based on comparisons between complementary (c)DNAs and published genomic sequences (Adams et al. 2000; Lander et al. 2001; Venter et al. 2001; Holt et al. 2002). The predicted human gene structure is identical to the published structure of the chicken gene (Sollenberger et al. 1994), sug-

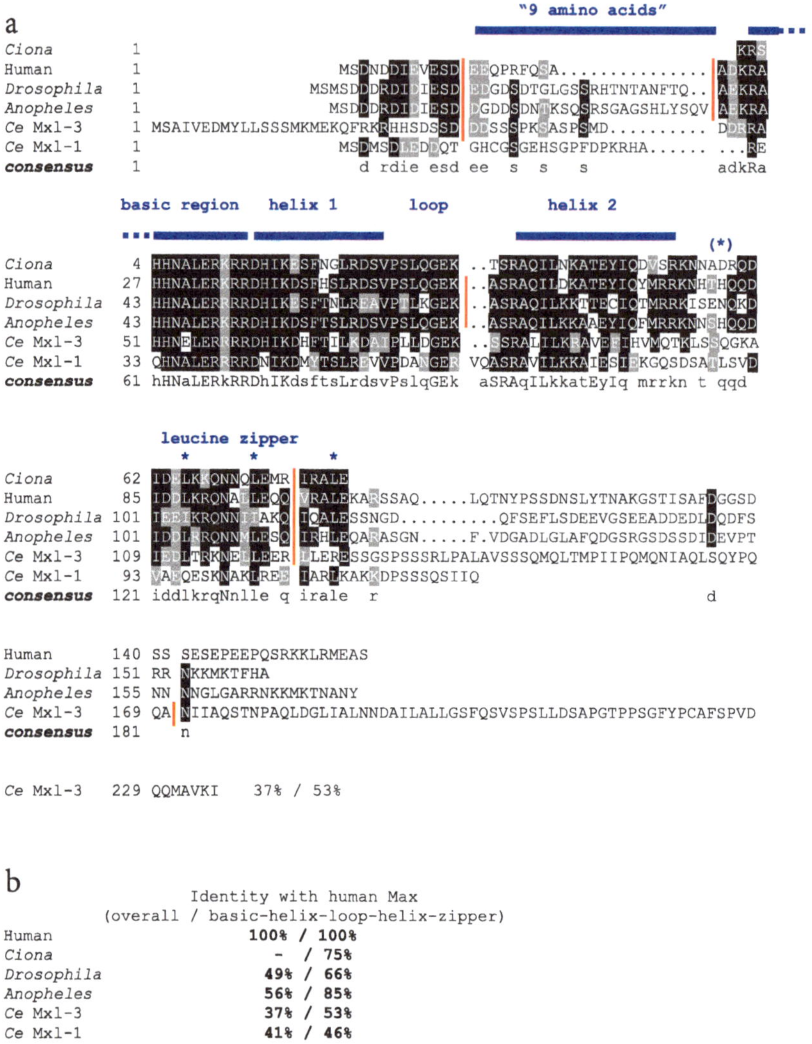

gesting that it reflects a generic vertebrate *max* gene structure. Interestingly, the *max* gene structure is identical in insects (but not in worms; Fig. 1). This evolutionary conservation is particularly intriguing in light of the existence of alternatively spliced *max* messenger (m)RNAs in vertebrates: coding exons 2 (labelled "9 amino acids" in Fig. 1a) and 3 (coding for the "basic-helix 1-loop" domain) are facultatively included in mature *max* mRNAs, as is an exon between the last two indicated coding exons (this facultative exon is not shown

◄

Fig. 1a, b Comparison of Max proteins from different species. **a** amino acid alignment. Shown *above* the sequence are the functional elements of Max; *asterisks* denote hydrophobic amino acids constituting the "leucine zipper". *Red vertical bars* show the positions of exon–exon junctions (except for *Ciona* Max). Full-length proteins are shown, except for *Ciona* Max where only the predicted BHLHLZ region is depicted. **b** Percentage identity of Max proteins with human Max, indicated for the full-length protein and for the BHLHLZ region only. Species shown are: *Ciona intestinalis* (sea squirt); *Drosophila melanogaster* (fruit fly); *Anopheles gambiae* (mosquito); *Caenorhabditis elegans* (nematode worm). Sources of unpublished sequences: *Anopheles gambiae*—accession number BX049732 (EST); *Caenorhabditis elegans* Mxl-3—accession number NP_510223 (protein); *Ciona intestinalis*—genomic scaffold 50, co-ordinates 2920 to 3318 (best match in a TBLASTN search with dMax)

in Fig. 1a) (Blackwood and Eisenman 1991; Makela et al. 1992; King et al. 1993; Vastrik et al. 1993; Koskinen et al. 1994; Tonissen and Krieg 1994; Arsura et al. 1995; FitzGerald et al. 1999). In insects only one mature *max* mRNA has been characterized (Gallant et al. 1996) and one more alternatively spliced EST has been reported (BDGP), but this conservation in gene structure indicates the possible existence of different additional splice isoforms. Furthermore, it suggests that such alternative forms of Max protein might play an essential role in vivo, even though their importance has not been demonstrated so far.

2.2 Myc

Vertebrates contain multiple *myc* genes (see above). They share a three-exon structure, whereby the major translation initiation codon is located at the beginning of the second exon and the open reading frame extends into exon 3 (Spencer and Groudine 1991); a few *myc* genes that deviate from this pattern and are encoded on a single exon probably derive from processed transcripts (e.g. human L2-*myc*). The same three-exon structure has also been found for *Drosophila myc* (P. Gallant, unpublished; however, the existence of additional non-coding exons 3′ of exon 3 has not been rigorously excluded), and the junction between exons 2 and 3 is located at the same codon as in vertebrate *myc* genes (Fig. 2c). This junction is also conserved in the single *myc* gene of *Caenorhabditis intestinalis* (as indicated by a comparison of EST and genomic sequences—P. Gallant, unpublished; Fig. 2c), and presumably also in the *Anopheles gambiae myc* (exon prediction based on the sequence similarity of conceptual translations of genomic DNA with Myc proteins from other species—see Fig. 2c; P. Gallant, unpublished). No *myc* gene has been found in the *C. elegans* genome (Yuan et al. 1998).

a. Myc-Box I

```
Human L-      17 EDFYRSTAPSEDIWKKFELVP
Human N-      29 FGGPDSTPPGEDIWKKFELLP
Human c-      37 QSELQPPAPSEDIWKKFELLP
Ciona         34 SSPTYGACLSEEIWKKFELLP
Asterias      29 SSTLTPPTPSEDIWKKFELYP
Urchin        27 AASPNSTTPSEDIWKKFDDVE
Drosophila    42 QSDLEKIEDMESVFQDYDLEE
Anopheles      7 HWDLIKMEPMDDADTNELGVL
consensus                psediwkkfelvp
```

b. Myc-Box II

```
Human L-      89 IIRRDCMWSGFSARER
Human N-     102 VIIQDCMWSGFSAREK
Human c-     128 IIIQDCMWSGFSAAAK
Ciona        129 KLIKDCMWNGIGHKPH
Asterias     112 ALIQDCMWSSIIAEER
Urchin       136 FLIQDCMWSAIQAEER
Drosophila    68 IRNIDCMWPAMSSCLT
Anopheles    118 QIRHDCMWAGMCADQS
consensus        vii DCMWsgisa er
```

c. Acidic region

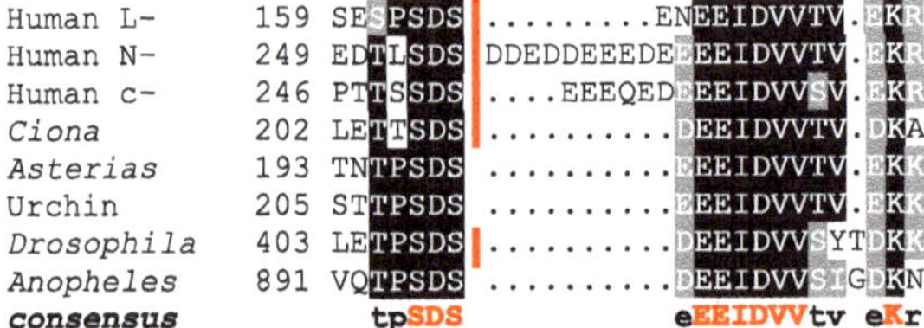

Human L- 159 SESPSDS ....ENEEIDVVTV.EKR
Human N- 249 EDTLSDS DDEDDEEEDEEEEIDVVTV.EKR
Human c- 246 PTTSSDS ....EEEQEDEEEIDVVSV.EKR
Ciona 202 LETTSDS ....DEEIDVVTV.DKA
Asterias 193 TNTPSDS ....EEEIDVVTV.EKK
Urchin 205 STTPSDS ....EEEIDVVTV.EKK
Drosophila 403 LETPSDS ..........DEEIDVVSYTDKK
Anopheles 891 VQTPSDS ..........DEEIDVVSIGDKN
consensus tpSDS eEEIDVVtv eKr

Fig. 2a–c Partial sequence alignments of Myc proteins from different species. Conventions are as for Fig. 1. Species shown are: urchin—*Strongylocentrotus purpuratus* (purple urchin); *Asterias vulgaris* (sea star); others are described in the legend to Fig. 1. Sources of unpublished sequences: *Ciona*—gene name ci0100150934; the BHLHLZ domain of *Anopheles* Myc was identified in a TBLASTN search of the *Anopheles* genome with dMyc as the query; the position of the exon boundaries was predicted based on the position of splice junctions, the amino acid homology at the ends of both exons, and the length of the predicted intron (*Anopheles:* 8,163 bp; *Drosophila:* 8,152 bp for the corresponding intron)

At the amino acid sequence level, Myc proteins are moderately conserved throughout evolution; for example, dMyc and human c-Myc are only 26% identical over the whole sequence (Gallant et al. 1996). However, interspersed in oceans of divergence lie islands of high sequence conservation that correspond to functionally important motifs. Best known are the N-terminally

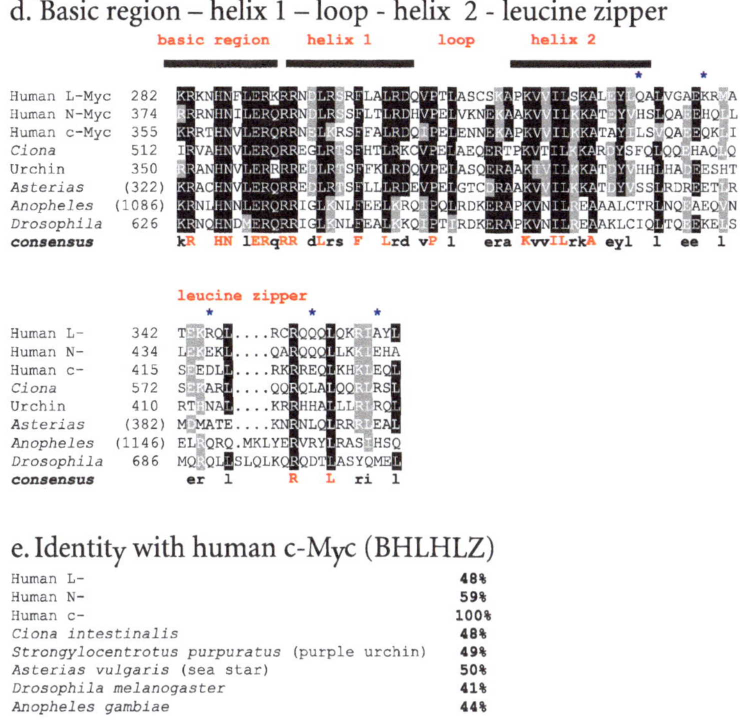

Fig. 2d, e (continued)

located "Myc Box 2", which is part of the transcriptional regulation domain and important for the biological activities of Myc (Fig. 2b; Amati et al. 2001), and the C-terminal BHLHLZ domain, which mediates DNA binding and heterodimerization (Fig. 2d, e; Amati et al. 2001); the presence of these two motifs is a hallmark of all Myc proteins. A second N-terminal motif, known as "Myc Box 1", is also part of the transactivation domain and highly conserved in deuterostome Myc proteins, but much less so in the insect proteins (Fig. 2a). While these motifs have been extensively characterized in vertebrate Myc, considerably less is known about a highly conserved "acidic domain" located in the centre of the protein (Fig. 2c). The corresponding region in the v-Myc oncoprotein is specifically required for the transformation of adult chicken bone marrow cells and peripheral blood macrophages, whereas it is dispensable for the transformation of embryonic chicken cells or quail peripheral blood macrophages (Heaney et al. 1986; Biegalke et al. 1987). The high degree of evolutionary conservation suggests a much broader and more important role for this domain that needs to be defined. Evolutionary constraints on the nucleotide sequence coding for this motif may also explain why the position of the junction between exons 2 and 3 has been conserved in *myc* genes (Fig. 2c).

2.3 Mad/Mnt

In humans and mice, the *mad* family is represented by five genes: *mad1*, *mad3*, *mad4*, *mxi1* and *mnt*. Two family members have been identified in the genome of *C. intestinalis* (P. Gallant, unpublished), whereas *Drosophila* and *Caenorhabditis* only encode one such gene each (*dmnt* and *mdl-1*, respectively; Peyrefitte et al. 2001, Yuan et al. 1998); the same appears to be true for *Anopheles* as well (P. Gallant, unpublished). Thus, early in chordate development a gene duplication involving *mad* seems to have taken place.

Figure 3a shows partial amino acid alignments of the Mad family proteins. The *Ciona* and *Anopheles* proteins are derived from conceptual translations of genomic DNA, and no EST evidence has been published yet; hence only their BHLHLZ region is shown, as the remainder of the protein cannot be predicted with high confidence. Like all members of the Max network, Mad/Mnt proteins are characterized by a BHLHLZ domain. In addition, they contain a region that mediates interaction with the transcriptional corepressor Sin3 known as "Sin3 interaction domain" or SID (Ayer et al. 1995; Eilers et al. 1999; Schreiber-Agus et al. 1995). Based on a comparison of the BHLHLZ regions, the dipteran Mad proteins are most closely related to vertebrate Mnt; the same appears to be true for the worm orthologue (Fig. 3b).

The structure of *mad* genes is less conserved than that of *myc* or *max*. However, in all genes the SID is encoded on a different exon than the BHLHLZ. This opens the possibility for alternative splicing to generate proteins that are able to bind DNA and Max, but lack the interaction with transcriptional corepressors; the resulting proteins could potentially differ radically in their transcriptional properties from SID-containing isoforms. Such alternatively spliced forms have indeed been reported to be produced from the murine *mxi1* locus (Schreiber-Agus et al. 1995) and from the *dmnt* gene (Loo et al. 2005).

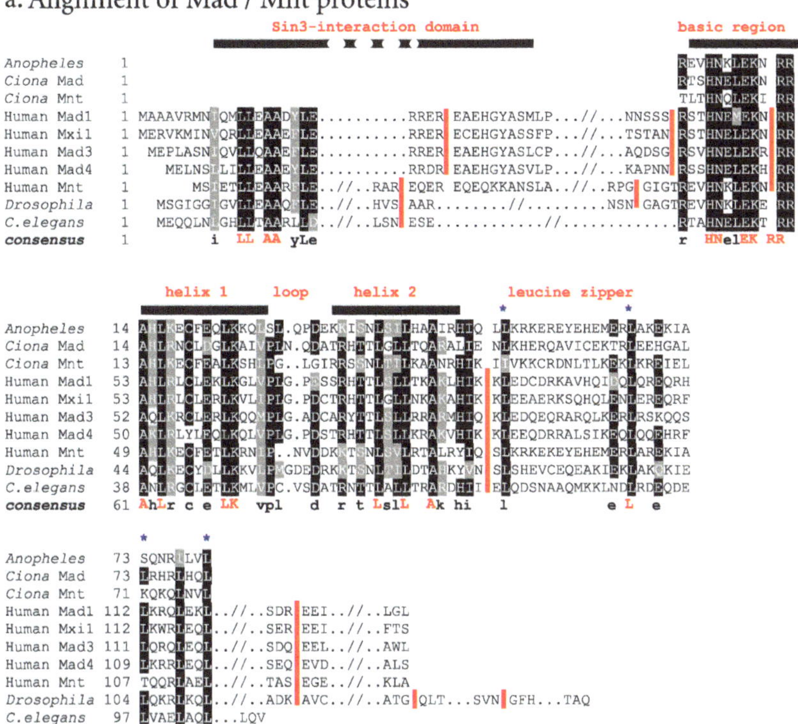

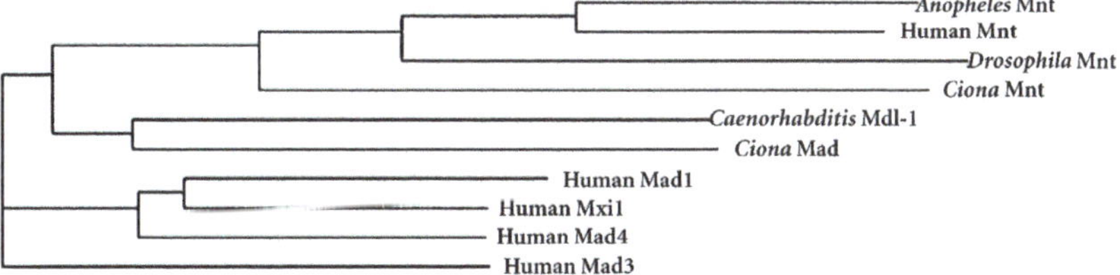

Fig. 3a, b Comparison of Mad/Mnt proteins from different species. **a** Partial amino acid alignment; conventions are as in Fig. 1. **b** Phylogenetic tree of BHLHLZ domains of different Mad/Mnt proteins constructed using CLUSTALW. Species are the same as in Fig. 1. Sources of unpublished sequences: *Anopheles* Mnt—accession number EAA07540 (protein); *Ciona* Mad—gene name ci0100137424; *Ciona* Mnt—gene name ci0100131159

3
Function of the Max Network in Invertebrates

3.1
Drosophila Myc

In invertebrates, the function of Max network components has predominantly been addressed in *Drosophila*. The *dmyc* gene has long been known to the fly-research community under the name of *diminutive* (*dm*), although the identity of *dm* with *dmyc* was only recently recognized (Bridges 1935; Gallant et al. 1996; Schreiber-Agus et al. 1997). While *dmyc* is an essential gene, several hypomorphic viable *dmyc* alleles have been described; flies carrying such mutations are characterized by a number of traits, including reduced body size, slender bristles, a delay in development and female sterility (Bridges 1935; Johnston et al. 1999). The cellular cause for the female sterility is currently unknown, but one of the contributing factors presumably is a defect in the migration and differentiation of somatic follicle cells, in particular of the border cells (J. Maines, personal communication; King 1957; King and Vanoucek 1960). In contrast, the other defects reflect dMyc's role in the control of cellular growth and proliferation: a reduction in *dmyc* activity reduces cellular size and increases the fraction of cells in G1 phase of the cycle (Johnston et al. 1999; T. Hulf et al. 2005), whereas overexpression of dMyc promotes entry into S-phase and increases cellular size and the rate of mass increase (growth) in clones of cells (Johnston et al. 1999). In contrast to vertebrates, the forced expression of dMyc in flies does not accelerate cell division rates, since the G2–M transition is independent of dMyc activity in flies and becomes rate-limiting under conditions of dMyc overexpression where the duration of G1 phase is greatly reduced (Johnston et al. 1999). In endoreplicating cells that lack M-phases, however, forced expression of dMyc induces additional S-phases and results in hyperploidy (Britton et al. 2002; Pierce et al. 2004; Maines et al. 2004). These effects on growth are presumably mediated by the transcriptional regulation of a similar set of target genes as has been proposed for vertebrate Myc, including many genes involved in ribosome function and nucleolar biogenesis (Zaffran et al. 1998; Orian et al. 2003; Hulf et al. 2005). In addition, overexpressed dMyc has been reported to control several cell-cycle regulators at the transcriptional level (Orian et al. 2003), as well as the important regulator of the G1–S transition, cyclin E, at the post-translational level (Prober and Edgar 2000). However, the involvement of these different putative dMyc targets in dMyc-controlled processes has not been addressed genetically.

These initial studies demonstrate a central role for dMyc in the control of growth. What then controls *dmyc* activity itself? So far, three signalling pathways have been implied in this process. The Wnt-family member Wingless was proposed to repress *dmyc* transcription in the presumptive wing margin (Johnston et al. 1999), and Dpp signalling positively regulates dMyc protein levels in the wing imaginal disc (C. Martin-Caballeros, cited in Prober and Edgar 2002). An interesting connection was also made between dMyc and Ras: Activated Ras itself promotes cellular growth, and this effect is mediated in part by an activation of the Raf-MAPK (mitogen-activated protein kinase) module, which results in the accumulation of dMyc protein (Prober and Edgar 2000, 2002). By analogy with the situation in vertebrates, it was speculated that this effect is based on the stabilization of dMyc protein (Sears et al. 1999; Prober and Edgar 2000). A similar process might also occur during normal development, as cells lacking Ras also may have reduced dMyc protein levels (Prober and Edgar 2002). These observations suggest that receptor-tyrosine kinases controlling Ras might also be implied in the regulation of *dmyc*.

Ectopically expressed activated Ras also affects growth by stimulating PI3K activity, but PI3K and dMyc reside on parallel growth-regulatory pathways; forced expression of PI3K does not affect dMyc protein levels, and conversely, forced dMyc expression does not alter the levels of PIP3, the product of PI3K enzymatic activity (Britton et al. 2002; Prober and Edgar 2002). The difference between dMyc and PI3K is illustrated by their different response to environmental conditions. During normal development, PI3K is controlled by nutrient availability, via the activity of the insulin-receptor, and starvation leads to down-regulation of PI3K activity (Britton et al. 2002). If this down-regulation is prevented by constitutive expression of PI3K, larvae become hyper-sensitive to starvation. In contrast, larvae constitutively expressing dMyc survive starvation as well as wild-type larvae, consistent with the idea that nutrient and insulin signalling does not feed into *dmyc* (Britton et al. 2002). The growth-relevant targets downstream of dMyc and PI3K also seem to be different, as co-expressed PI3K and dMyc strongly synergize in the promotion of cellular growth (L. Johnston and P. Gallant, unpublished observations).

While these studies have directly addressed the regulation of *dmyc* protein and mRNA levels, forced dMyc expression has also been shown to overcome proliferation defects caused by genetic lesions in other pathways. Interference with the activity of the Tor kinase (Schmelzle and Hall 2000), either by expression of dominant-negative or wild-type forms of Tor in the wing (both of which function in a dominant-negative fashion), or by overexpression of the tumour suppressors TSC1 and TSC2 in the eye, inhibits growth and reduces organ size; these defects can be reversed by co-expression with dMyc (Tapon et al. 2001; Hennig and Neufeld 2002). Ectopic expression of different transcription factors in the eye primordium interferes with the normal development of the head capsule and results in a striking reduction in head size; this defect can be partially rescued by co-expression with dMyc (Jiao et al. 2001). Finally, certain combinations of mutations in the Pax gene *prd* with partial genomic rescue constructs allow the development of adult male flies that are characterized by small accessory glands; this size defect is rescued by ectopic expression of dMyc (Xue and Noll 2002). These examples further illustrate the ability of dMyc to promote growth and proliferation in different situations. However, additional work is required to determine to what extent and at which level dTOR or Prd, for example, control *dmyc* activity during normal development.

The examples described above indicate that two principal biological activities of Myc proteins have been conserved between flies and vertebrates: the control of growth and proliferation (Elend and Eilers 1999). Indeed, fly and vertebrate Myc proteins are very similar in their molecular function and they can substitute for each other in different assays: When expressed together with dMax in human 293 cells, dMyc activates the expression of a c-Myc responsive reporter construct (Gallant et al. 1996); upon co-expression with human RasV12 dMyc is able to transform rat embryo fibroblasts (Schreiber-Agus et al. 1997); the proliferation defect of mouse embryo fibroblasts that are mutant for *c-myc* is partially rescued by ectopic expression of dMyc (Trumpp et al. 2001). Conversely, different forms of human c-Myc are able to partially rescue the lethality of strong *dmyc* alleles in flies (C. Benassayag et al. 2005). In light of these observations, it is likely that dMyc and human c-Myc fulfill the same molecular tasks, and notably that they control the expression of their target genes in similar ways, by recruiting similar types of transcriptional co-factors as have been described in the vertebrate system, e.g. TRRAP, SNF5, Tip48, Tip49, BAF53, p300/CBP—all of which are also present in the fly genome (McMahon et al. 1998; Cheng et al. 1999; Wood et al. 2000; Park et al. 2002; Vervoorts et al. 2003; Adams et al. 2000; Bellosta et al. 2005).

3.2 Mad and Max

The other components of the Max network have not been extensively studied in flies. No mutations are known for *dmax*, but a null mutation in *dmnt* has recently been identified (Loo et al., manuscript submitted). An initial characterization suggests that overexpression of dMnt inhibits growth and proliferation, and a mutation in *dmnt* has the opposite effect, consistent with the expected properties of an antagonist of dMyc (Loo et al. 2005).

In contrast to flies, *C. elegans* contains two *max* genes (*mxl-1* and *mxl-3*) and one *mad* gene (*mdl-1*), but no *myc* (Yuan et al. 1998). Little is known about the normal function of these genes. Overexpression of dominant-negative forms of Mdl-1 or Mxl-1 (lacking the basic region) or RNA interference with *mxl-1* or *mxl-3* produces no discernible phenotype (Yuan et al. 1998; Maeda et al. 2001; Kamath et al. 2003), whereas RNA interference with *mdl-1* slightly reduces longevity in *daf-2* mutant worms (Murphy et al. 2003). Interestingly, *mdl-1* expression is also negatively regulated by the insulin receptor *daf-2* (Murphy et al. 2003), mutations of which extend lifespan in worms, raising the possibility that Mdl-1 might also contribute to the regulation of lifespan in worms.

Although these experiments do not reveal any involvement in the control of proliferation and growth, Mdl-1 and Mxl-1 do show Mad- and Max-like properties when assayed in a heterologous system. Mdl-1 (and to a lesser extent Mxl-1) is able to interfere with the co-transformation of rat embryo fibroblasts by activated mammalian Ras and c-Myc. The interference by Mdl-1 depends on the SID in Mdl-1, suggesting that Mdl-1 functions like other Mad proteins by recruiting the Sin3-corepressor complex and repressing transcription (Yuan et al. 1998). This result—as well as the sequence similarity—indicates that the (rudimentary) Max network in worms might fulfill similar functions to the vertebrate network. On the other hand, the Max network in worms shows several unique features not found in other metazoans—the absence of a *myc* gene, the existence of two *max*-like genes, the unique genomic structure of the *max* genes and the inability of Mxl-1 to homodimerize (Yuan et al. 1998).

As the phylogenetic relationship between nematodes, arthropods and chordates is still under debate (Hedges 2002), two main hypotheses can be invoked to explain these peculiarities in worms. The first is that worms contain an ancestral form of the Max network; hence, activities executed by Mad:Max complexes are the primary duty of the network, and Myc-like genes have been added later in evolution. The alternative is that *C. elegans* contains a derived Max network that differs in several aspects from an ancestral Max network. As Myc is essential in flies (Bourbon et al. 2002) whereas Mad/Mnt is not (Loo et al. 2005)—suggesting that Myc function is more important for survival—we favour the latter possibility.

4
Speculations and Conclusions

The availability of complete genome sequences enables biologists for the first time to make (reasonably accurate) predictions about the presence *and* absence of certain gene functions in many different species. Based on such information, we can state that components of the Max network exist in all analysed animals, but neither in unicellular organisms nor in plants, suggesting that this network originated early during the evolution of animals. The principal function of the Max network resides in the control of growth and proliferation. These processes are essential for all living cells, and accordingly Myc activity is required for the proliferation of many cells. However, the Max network is not absolutely required in all cell types and it might not be an integral part of the basic cell-cycle machinery or growth apparatus in animals, as indicated by the existence of several vertebrate cell lines that lack core components of the Max network—either Myc (Miyazaki et al. 1995; Mateyak et al. 1997) or Max (Hopewell and Ziff 1995). Rather, it appears that Max network components might relay signals that are typical for multicellular organisms (e.g. patterning signals involved in cell–cell communication) down to the core cell-cycle and growth machinery. The Max network affects the activity of this machinery by modulating, or fine-tuning, the expression of many of its core components (Eisenman 2001). In contrast, Max network components might not be involved in the transmission of nutrient signals (at least in simpler animals), a function that is not specific to metazoans but of equal relevance for unicellular organisms.

The evolutionary conservation and, by inference, the central importance of the Max network is dramatically illustrated by the partial functional interchangeability of Myc proteins from flies and mammals, which further implies that not only core components of the Max networks are conserved (Max, Myc, Mad) but also associated factors that interact with these core components. This high degree of conservation opens new possibilities for the experimental dissection of the Max network, based on one hand on a functional analysis in appropriate model organisms (such as flies) and on the other hand on a bioinformatic analysis of the components making up the Max network. A sequence comparison of components from widely divergent species (in particular flies and mammals) reveals several highly conserved features that did not stand out when only mammals were included in the analysis. Of particular note are the gene structure of Max, which hints at the potential relevance of alternative Max isoforms, and the acidic domain located in the centre of the Myc protein. Clearly, despite intensive research over the last 20 years, the Max network still holds many secrets that will keep biologists busy for some time to come.

Acknowledgements Many thanks to Laura Johnston and Andreas Trumpp for critical reading of the manuscript and for helpful comments. Work in the author's lab is funded by grants from the Swiss National Science Foundation (SNF), the Swiss Cancer League (SKL), and the Zürcher Hochschulverein (FAN).

References

Abe H, Urao T, Ito T, Seki M, Shinozaki K, Yamaguchi-Shinozaki K (2003) Arabidopsis AtMYC2 (bHLH) and AtMYB2 (MYB) function as transcriptional activators in abscisic acid signaling. Plant Cell 15:63–78

Adams MD, Celniker SE, Holt RA, Evans CA, Gocayne JD, Amanatides PG, Scherer SE, Li PW, Hoskins RA, Galle RF, et al (2000) The genome sequence of Drosophila melanogaster. Science 287:2185–2195

Amati B, Frank SR, Donjerkovic D, Taubert S (2001) Function of the c-Myc oncoprotein in chromatin remodeling and transcription. Biochim Biophys Acta 1471:M135–M145

Arsura M, Deshpande A, Hann SR, Sonenshein GE (1995) Variant Max protein, derived by alternative splicing, associates with c-Myc in vivo and inhibits transactivation. Mol Cell Biol 15:6702–6709

Atchley WR, Fitch WM (1995) Myc and Max: molecular evolution of a family of proto-oncogene products and their dimerization partner. Proc Natl Acad Sci U S A 92:10217–10221

Ayer DE, Lawrence QA, Eisenman RN (1995) Mad-Max transcriptional repression is mediated by ternary complex formation with mammalian homologs of yeast repressor Sin3. Cell 80:767–776

Biegalke BJ, Heaney ML, Bouton A, Parsons JT, Linial M (1987) MC29 deletion mutants which fail to transform chicken macrophages are competent for transformation of quail macrophages. J Virol 61:2138–2142

Bellosta P, Hulf T, Diop SB, Usseglio F, Pradel J, Aragnol D, Gallant P (2005) Myc interacts genetically with Tip48/Reptin and Tip49/Pontin to control growth and proliferation during Drosophila development. Proc Nat Acad Sci U S A 102:11799

Benassayag C, Montero L, Colombié N, Gallant P, Cribbs D, Morello D (2005) Human c-Myc isoforms differentialy regulate cell growth and apoptosis in Drosophila. Mol Cell Biol 25:9897

Bishop JM (1983) Cellular oncogenes and retroviruses. Annu Rev Biochem 52:301–354

Blackwood EM, Eisenman RN (1991) Max: a helix-loop-helix zipper protein that forms a sequence-specific DNA-binding complex with Myc. Science 251:1211–1217

Bourbon HM, Gonzy-Treboul G, Peronnet F, Alin MF, Ardourel C, Benassayag C, Cribbs D, Deutsch J, Ferrer P, Haenlin M, et al (2002) A P-insertion screen identifying novel X-linked essential genes in Drosophila. Mech Dev 110:71–83

Bridges CB (1935) Drosophila melanogaster: legend for symbols, mutants, valuations. Drosophila Information Service 3:5–19

Britton JS, Lockwood WK, Li L, Cohen SM, Edgar BA (2002) Drosophila's insulin/PI3-kinase pathway coordinates cellular metabolism with nutritional conditions. Dev Cell 2:239–249

Charron J, Malynn BA, Fisher P, Stewart V, Jeannotte L, Goff SP, Robertson EJ, Alt FW (1992) Embryonic lethality in mice homozygous for a targeted disruption of the N-myc gene. Genes Dev 6:2248–2257

Cheng SW, Davies KP, Yung E, Beltran RJ, Yu J, Kalpana GV (1999) c-MYC interacts with INI1/hSNF5 and requires the SWI/SNF complex for transactivation function. Nat Genet 22:102–105

Davis AC, Wims M, Spotts GD, Hann SR, Bradley A (1993) A null c-myc mutation causes lethality before 10.5 days of gestation in homozygotes and reduced fertility in heterozygous female mice. Genes Dev 7:671–682
Dehal P, Satou Y, Campbell RK, Chapman J, Degnan B, De Tomaso A, Davidson B, Di Gregorio A, Gelpke M, Goodstein DM, et al (2002) The draft genome of Ciona intestinalis: insights into chordate and vertebrate origins. Science 298:2157–2167
Eilers AL, Billin AN, Liu J, Ayer DE (1999) A 13-amino acid amphipathic alpha-helix is required for the functional interaction between the transcriptional repressor Mad1 and mSin3A. J Biol Chem 274:32750–32756
Eisenman RN (2001) Deconstructing myc. Genes Dev 15:2023–2030
Elend M, Eilers M (1999) Cell growth: downstream of Myc—to grow or to cycle? Curr Biol 9:R936–R938
FitzGerald MJ, Arsura M, Bellas RE, Yang W, Wu M, Chin L, Mann KK, DePinho RA, Sonenshein GE (1999) Differential effects of the widely expressed dMax splice variant of Max on E-box vs initiator element-mediated regulation by c-Myc. Oncogene 18:2489–2498
Gallant P, Shiio Y, Cheng PF, Parkhurst SM, Eisenman RN (1996) Myc and Max homologs in Drosophila. Science 274:1523–1527
Gupta MP, Amin CS, Gupta M, Hay N, Zak R (1997) Transcription enhancer factor 1 interacts with a basic helix-loop-helix zipper protein, Max, for positive regulation of cardiac alpha-myosin heavy-chain gene expression. Mol Cell Biol 17:3924–3936
Heaney ML, Pierce J, Parsons JT (1986) Site-directed mutagenesis of the gag-myc gene of avian myelocytomatosis virus 29: biological activity and intracellular localization of structurally altered proteins. J Virol 60:167–176
Hedges SB (2002) The origin and evolution of model organisms. Nat Rev Genet 3:838–849
Hennig KM, Neufeld TP (2002) Inhibition of cellular growth and proliferation by dTOR overexpression in Drosophila. Genesis 34:107–110
Holt RA, Subramanian GM, Halpern A, Sutton GG, Charlab R, Nusskern DR, Wincker P, Clark AG, Ribeiro JM, Wides R, et al (2002) The genome sequence of the malaria mosquito Anopheles gambiae. Science 298:129–149
Hopewell R, Ziff EB (1995) The nerve growth factor-responsive PC12 cell line does not express the Myc dimerization partner Max. Mol Cell Biol 15:3470–3478
Hulf T, Bellosta P, Furrer M, Steiger D, Svensson D, Barbour A, Gallant P (2005) Whole-genome analysis reveals a strong positional bias of conserved dMyc-dependent E-boxes. Mol Cell Biol 25:3401
Hurlin PJ, Steingrimsson E, Copeland NG, Jenkins NA, Eisenman RN (1999) Mga, a dual-specificity transcription factor that interacts with Max and contains a T-domain DNA-binding motif. EMBO J 18:7019–7028
Jiao R, Daube M, Duan H, Zou Y, Frei E, Noll M (2001) Headless flies generated by developmental pathway interference. Development 128:3307–3319
Johansson US, Parsons TJ, Irestedt M, Ericson PGP (2001) Clades within the 'higher land birds', evaluated by nuclear DNA sequences. J Zoolog Syst Evol Res 39:37–51
Johnston LA, Prober DA, Edgar BA, Eisenman RN, Gallant P (1999) Drosophila myc regulates cellular growth during development. Cell 98:779–790

Kamath RS, Fraser AG, Dong Y, Poulin G, Durbin R, Gotta M, Kanapin A, Le Bot N, Moreno S, Sohrmann M, et al (2003) Systematic functional analysis of the Caenorhabditis elegans genome using RNAi. Nature 421:231–237

King MW, Blackwood EM, Eisenman RN (1993) Expression of two distinct homologues of Xenopus Max during early development. Cell Growth Differ 4:85–92

King RC, Burnett RG (1957) Oogenesis in adult Drosophila melanogaster. Growth 21:263–280

King RC, Vanoucek EG (1960) Oogenesis in adult Drosophila melanogaster. X. Studies on the behavior of the follicle cells. Growth 24:333–338

Koskinen PJ, Vastrik I, Makela TP, Eisenman RN, Alitalo K (1994) Max activity is affected by phosphorylation at two NH2-terminal sites. Cell Growth Differ 5:313–320

Lander ES, Linton LM, Birren B, Nusbaum C, Zody MC, Baldwin J, Devon K, Dewar K, Doyle M, FitzHugh W, et al (2001) Initial sequencing and analysis of the human genome. Nature 409:860–921

Loo LW, Secombe J, Little JT, Carlos LS, Yost C, Cheng PF, Flynn EM, Edgar BA, Eisenman RN (2005) The transcriptional repressor dMnt is a regulator of growth in Drosophila melanogaster. Mol Cell Biol 25:7078

Madhavan K, Bilodeau WD, Wadsworth SC (1985) Initial sequencing and analysis of the human genome. Mol Cell Biol 5:7–16

Maeda I, Kohara Y, Yamamoto M, Sugimoto A (2001) Large-scale analysis of gene function in Caenorhabditis elegans by high-throughput RNAi. Curr Biol 11:171–176

Makela TP, Koskinen PJ, Vastrik I, Alitalo K (1992) Alternative forms of Max as enhancers or suppressors of Myc-ras cotransformation. Science 256:373–377

Maines JZ, Stevens LM, Tong X, Stein D (2004) Drosophila dMyc is required for ovary cell growth and endoreplication. Development 131:775

Mateyak MK, Obaya AJ, Adachi S, Sedivy JM (1997) Terminally differentiated skeletal myotubes are not confined to G0 but can enter G1 upon growth factor stimulation. Cell Growth Differ 8:1039–1048

McMahon SB, Van BH, Dugan KA, Copeland TD, Cole MD (1998) The novel ATM-related protein TRRAP is an essential cofactor for the c-Myc and E2F oncoproteins. Cell 94:363–374

Miyamoto MM, Freire NP (2000) Evolution of CpG islands within the myc gene family. Mol Phylogenet Evol 16:475–481

Miyazaki T, Liu ZJ, Kawahara A, Minami Y, Yamada K, Tsujimoto Y, Barsoumian EL, Permutter RM, Taniguchi T (1995) Three distinct IL-2 signaling pathways mediated by bcl-2, c-myc, and lck cooperate in hematopoietic cell proliferation. Cell 81:223–231

Murphy CT, McCarroll SA, Bargmann CI, Fraser A, Kamath RS, Ahringer J, Li H, Kenyon C (2003) Genes that act downstream of DAF-16 to influence the lifespan of Caenorhabditis elegans. Nature 424:277–283

Orian A, Van Steensel B, Delrow J, Bussemaker HJ, Li L, Sawado T, Williams E, Loo LW, Cowley SM, Yost C, et al (2003) Genomic binding by the Drosophila Myc, Max, Mad/Mnt transcription factor network. Genes Dev 17:1101–1114

Oster SK, Ho CS, Soucie EL, Penn LZ (2002) The myc oncogene: MarvelouslY Complex. Adv Cancer Res 84:81–154

Park J, Wood MA, Cole MD (2002) BAF53 forms distinct nuclear complexes and functions as a critical c-Myc-interacting nuclear cofactor for oncogenic transformation. Mol Cell Biol 22:1307–1316

Peyrefitte S, Kahn D, Haenlin M (2001) New members of the Drosophila Myc transcription factor subfamily revealed by a genome-wide examination for basic helix-loop-helix genes. Mech Dev 104:99–104

Pierce SB, Yost C, Britton JS, Loo LW, Flynn EM, Edgar BA, Eisenman RN (2004) dMyc is required for larval growth and endoreplication in Drosophila. Development 131:2317–2327

Prober DA, Edgar BA (2000) Ras1 promotes cellular growth in the Drosophila wing. Cell 100:435–446

Prober DA, Edgar BA (2002) Interactions between Ras1, dMyc, and dPI3K signaling in the developing Drosophila wing. Genes Dev 16:2286–2299

Sarid J, Halazonetis TD, Murphy W, Leder P (1987) Evolutionarily conserved regions of the human c-myc protein can be uncoupled from transforming activity. Proc Natl Acad Sci U S A 84:170–173

Sawai S, Shimono A, Wakamatsu Y, Palmes C, Hanaoka K, Kondoh H (1993) Defects of embryonic organogenesis resulting from targeted disruption of the N-myc gene in the mouse. Development 117:1445–1455

Schmelzle T, Hall MN (2000) TOR, a central controller of cell growth. Cell 103:253–262

Schreiber-Agus N, Chin L, Chen K, Torres R, Rao G, Guida P, Skoultchi AI, DePinho RA (1995) An amino-terminal domain of Mxi1 mediates anti-Myc oncogenic activity and interacts with a homolog of the yeast transcriptional repressor SIN3. Cell 80:777–786

Schreiber-Agus N, Stein D, Chen K, Goltz JS, Stevens L, DePinho RA (1997) Drosophila Myc is oncogenic in mammalian cells and plays a role in the diminutive phenotype. Proc Natl Acad Sci U S A 94:1235–1240

Sears R, Leone G, DeGregori J, Nevins JR (1999) Herpes simplex virus glycoprotein D bound to the human receptor HveA. Mol Cell 3:169–179

Shilo B-Z, Weinberg RA (1981) DNA sequences homologous to vertebrate oncogenes are conserved in Drosophila melanogaster. Proc Natl Acad Sci USA 78:6789–6792

Shors ST, Efiok BJ, Harkin SJ, Safer B (1998) Formation of alpha-Pal/Max heterodimers synergistically activates the eIF2-alpha promoter. J Biol Chem 273:34703–34709

Sollenberger KG, Kao TL, Taparowsky EJ (1994) Structural analysis of the chicken max gene. Oncogene 9:661–664

Spencer CA, Groudine M (1991) Control of c-myc regulation in normal and neoplastic cells. Adv Cancer Res 56:1–48

Tapon N, Ito N, Dickson BJ, Treisman JE, Hariharan IK (2001) The Drosophila tuberous sclerosis complex gene homologs restrict cell growth and cell proliferation. Cell 105:345–355

Tonissen KF, Krieg PA (1994) Analysis of a variant Max sequence expressed in Xenopus laevis. Oncogene 9:33–38

Trumpp A, Refaeli Y, Oskarsson T, Gasser S, Murphy M, Martin GR, Bishop JM (2001) c-Myc regulates mammalian body size by controlling cell number but not cell size. Nature 414:768–773

Urao T, Yamaguchi-Shinozaki K, Mitsukawa N, Shibata D, Shinozaki K (1996) Molecular cloning and characterization of a gene that encodes a MYC-related protein in Arabidopsis. Plant Mol Biol 32:571–576

Vastrik I, Koskinen PJ, Alitalo R, Makela TP (1993) Alternative mRNA forms and open reading frames of the max gene. Oncogene 8:503–507

Venter JC, Adams MD, Myers EW, Li PW, Mural RJ, Sutton GG, Smith HO, Yandell M, Evans CA, Holt RA, et al (2001) The sequence of the human genome. Science 291:1304–1351

Vervoorts J, Luscher-Firzlaff JM, Rottmann S, Lilischkis R, Walsemann G, Dohmann K, Austen M, Luscher B (2003) Stimulation of c-MYC transcriptional activity and acetylation by recruitment of the cofactor CBP. EMBO Rep 4:1–7

Walker CW, Boom JD, Marsh AG (1992) First non-vertebrate member of the myc gene family is seasonally expressed in an invertebrate testis. Oncogene 7:2007–2012

Wanzel M, Herold S, Eilers M (2003) Transcriptional repression by Myc. Trends Cell Biol 13:146–150

Wood MA, McMahon SB, Cole MD (2000) An ATPase/helicase complex is an essential cofactor for oncogenic transformation by c-Myc. Mol Cell 5:321–330

Xue L, Noll M (2002) Dual role of the Pax gene paired in accessory gland development of Drosophila. Development 129:339–346

Yuan J, Tirabassi RS, Bush AB, Cole MD (1998) The C. elegans MDL-1 and MXL-1 proteins can functionally substitute for vertebrate MAD and MAX. Oncogene 17:1109–1118

Zaffran S, Chartier A, Gallant P, Astier M, Arquier N, Doherty D, Gratecos D, Semeriva M (1998) A Drosophila RNA helicase gene, pitchoune, is required for cell growth and proliferation and is a potential target of d-Myc. Development 125:3571–3584

CTMI (2006) 302:255–278

The Mlx Network: Evidence for a Parallel Max-Like Transcriptional Network That Regulates Energy Metabolism

A. N. Billin[1] · D. E. Ayer[2] (✉)

[1]Research and Development, GlaxoSmithKline, 5 Moore Drive, PO Box 13398, Research Triangle Park, NC 27709-3398, USA

[2]Huntsman Cancer Institute, University of Utah, 2000 Circle of Hope, Room 4365, Salt Lake City, Utah 84112-5550, USA
don.ayer@hci.utah.edu

Abstract Recent experiments suggest the existence of a transcriptional network that functions in parallel to the canonical Myc/Max/Mad transcriptional network. Unlike the Myc/Max/Mad network, our understanding of this network is still in its infancy. At the center of this network is a Max-like protein called Mlx; hence we have called this network the Mlx network. Like Max, Mlx interacts with transcriptional repressors and transcriptional activators, namely the Mad family and the Mondo family, respectively. Similar to Max-containing heterodimers, Mlx-containing heterodimers recognize CACGTG E-box elements, suggesting that the transcriptional targets of these two networks may overlap. Supporting this hypothesis, we have observed genetic interactions between the *Drosophila melanogaster* orthologs of Myc and Mondo. In higher eukaryotes, two proteins, MondoA and MondoB/CHREBP/WBSCR14, constitute the Mondo family. At present little is known about the transcriptional targets of MondoA; however, pyruvate kinase is a putative target of MondoB/CHREBP/WBSCR14, suggesting a function for the Mondo family in glucose and/or lipid metabolism. Finally, unlike the predominant nuclear localization of Myc family proteins, both Mondo family members localize to the cytoplasm. Therefore, while the Myc and Mondo families may share some biological functions, it is likely each family is under distinct regulatory control.

1 Introduction

The Myc/Max/Mad transcription factor network controls diverse aspects of cell physiology including cell size, proliferation, differentiation, and death. Extensive efforts have identified important transcriptional target genes controlled by the network as well as the corepressor and coactivator complexes utilized by Myc and Mad to regulate transcription of these targets. Furthermore, an impressively large number of Myc-, Max-, or Mad-interacting proteins have been implicated in modulating the transcriptional activity of the network. Together these data have provided a basic understanding of how the Myc/Max/Mad transcription factor network regulates cell physiology.

Over the last several years, evidence has emerged to support the existence of a novel transcriptional network that in many respects constitutes a mirror image of the Myc/Max/Mad transcriptional network (Fig. 1). Our understanding of this network and its biological function is in its relative infancy and is the subject of this review. The proteins in this novel network share many structural and functional features with the Myc/Max/Mad transcriptional network, including the characteristic basic-region-helix-loop-helix-zipper (bHLHZip) protein and DNA interaction motif. Members of this novel network include the Mondo family of transcriptional activators and the Mad family of transcriptional repressors. Mlx, a Max-like protein, functions as the center of this network and, like Max, appears to act as a partner required by both the Mondo and the Mad families to bind DNA and elicit their respective functions as either transcriptional activators or repressors. For the sake of simplicity, we will refer to this recently identified network as the Mlx network and the canonical network comprising the Myc, Max, and Mad families as the Max network. Given the recent elaboration of the Mlx network, less is known about its biological function. Several lines of evidence suggest that the Mlx network functions analogously to the Max network and, in some cases, the two net-

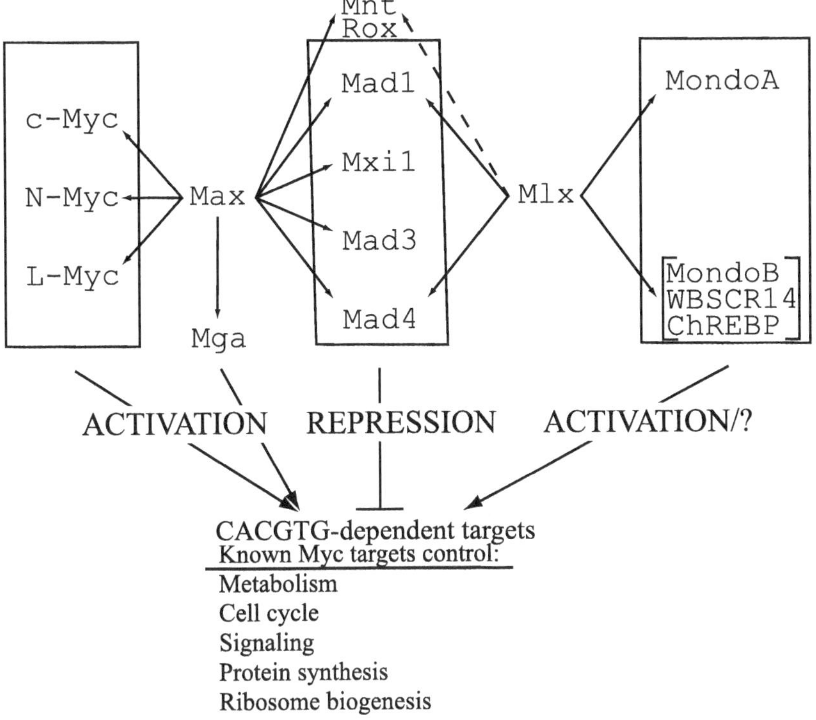

Fig. 1 The Max and Mlx transcriptional networks. A schematic of the two networks is shown. Both Max and Mlx interact with members of the Mad family and Mnt/Rox to repress transcription in a manner that depends on the interaction with the mSin3A/HDAC corepressor complex. The *dashed line* between Mlx and Mnt/Rox indicates that not all investigators have observed this interaction. The Mad family constitutes the repressive or negative arm of both the Max and Mlx networks. By contrast, the Myc family provides the activation or positive arm of the Max network, whereas the Mondo family provides the activation arm of the Mlx network. MondoB and WBSCR14 are identical, whereas ChREBP is the rat ortholog of MondoB/WBSCR14. All Max- and Mlx-containing heterodimers bind the CACGTG class of E-box element, suggesting that the two networks share many transcriptional targets. To date, Myc:Max targets have been extensively characterized. The extent to which these targets with be also regulated by other Max- and Mlx-containing heterodimers remains to be elucidated

works may function in parallel. For example, similar to Myc, one member of the Mondo family appears to regulate glucose metabolism; therefore, the Max and Mlx networks may share overlapping or complementary functions in the regulation of energy homeostasis. However, in contrast to members of the Max network, Mlx and the Mondo family proteins localize to the cytoplasm under most circumstances, suggesting that their transcriptional functions are tightly controlled by subcellular localization. In this chapter we review the discovery of the members of the Mlx network, as well as the regulation of their functions. In addition, we speculate on the potential functional overlap between the Mlx and Max networks.

2 Discovery and Members of the Mlx Transcriptional Network: An Overview

As with many members of the Max transcriptional network, members of the Mlx network were identified by two-hybrid screening. Mlx was identified initially as a binding partner for Mad1 (Billin et al. 1999) and then as a binding partner for Mnt/Rox (Meroni et al. 2000). Mlx is 29% identical and 54% similar to Max, with the similarity restricted to the bHLHZip domain. As such, we proposed the moniker Mlx (for Max-like protein x). Like Max, Mlx itself appears to be transcriptionally inert, but upon heterodimerization with certain Mad family members—and potentially Mnt/Rox—it functions as a transcriptional repressor. The Mlx-containing heterodimeric repressors likely constitute a negative arm of the Mlx network (Fig. 1). The Mondo family comprises the transcriptional activation arm of the Mlx network and was described shortly after the discovery of Mlx (Fig. 1). Two-hybrid screening using Mlx as bait identified MondoA, whereas its paralog MondoB was identified as an expressed sequence tag (Billin et al. 2000). Two-hybrid screening and database mining also identified Williams-Beuren syndrome conserved region 14 (WBSCR14) (de Luis et al. 2000; Cairo et al. 2001). Subsequent analysis has determined that MondoB and WBSCR14 are identical. The rat ortholog of MondoB/WBSCR14 was identified as the carbohydrate response element binding protein or ChREBP (Yamashita et al. 2001). For clarity, we will refer to these proteins collectively as MondoB/WBSCR14/ChREBP; however, when referring to specific experiments we will cite the actual protein used. Like Max-containing heterodimers, each of the Mlx-containing heterodimers can bind the CACGTG E-box element, suggesting that there may be considerable overlap in the transcriptional targets and cellular functions regulated by these two transcription factor networks (Fig. 1).

3 Genes, Proteins, and Expression Patterns

The Mlx gene spans roughly 7 kb with 8 exons. As with *Max* messenger (m)RNA, expression of *Mlx* mRNA is widespread during embryogenesis and in adult tissues (Billin et al. 1999; Meroni et al. 2000). The human and mouse Mlx genes are on chromosomes 17q21.1 and 11, respectively. Human Mlx maps just centromeric to the tumor suppressor BRCA1 in a region that presents a loss of heterogeneity in a number of tumors (Vogelstein and Kinzler 1994), suggesting a tumor suppressor function for Mlx; however, further studies are needed to investigate this intriguing possibility. AceView (www.ncbi.nlm.nih.gov/IEB/Research/Acembly) predicts nine alternatively spliced products encoding seven different protein isoforms. Recent reports have confirmed sequences for three Mlx isoforms differing at the N-terminus of the open reading frame (Billin et al. 1999; Meroni et al. 2000). Mlx-γ has sequences that resemble a bipartite nuclear localization signal, while Mlx-α and -β do not, suggesting that subcellular localization may be one functional difference between the Mlx isoforms. In support of this hypothesis, Mlx-α and -β localized to the cytoplasm, while Mlx-γ localized to the nucleus when overexpressed in HeLa cells (Meroni et al. 2000). However, all three Mlx isoforms localized to the cytoplasm of NIH3T3 cells, suggesting that cell type-specific factors control subcellular localization (our unpublished data).

The genes encoding MondoA and WBSCR14/MondoB are large, spanning approximately 60 and 30 kb, respectively. Like *Mlx*, *MondoA* and *WBSCR14/MondoB/ChREBP* mRNAs are widely expressed during embryogenesis and in adult tissues. Consistent with potential roles in energy metabolism, *MondoA* and *WBSCR14/MondoB/ChREBP* mRNAs are most highly expressed in skeletal muscle and liver, respectively (Billin et al. 2000; Cairo et al. 2001). Each gene comprises 17 exons, providing ample opportunities for alternative splicing. There are currently 5 alternatively spliced complementary (c)DNAs in the public databases for WBSCR14, and AceView predicts 13 different alternatively spliced isoforms for both MondoA and WBSCR14/MondoB. The exact number of spliced isoforms and expressed proteins remains to be determined, but such complexity suggests that alternative splicing will generate considerable functional diversity. Endogenous protein isoforms for WBSCR14/MondoB have not been reported; however, MondoA is expressed primarily as a 135-kDa protein in P19 cells with a minor 110-kDa species also detected (Billin et al. 2000). Thus, the diversity of proteins produced may be significantly less than predicted by the number of spliced mRNAs.

In mouse, both MondoA and WBSCR14/MondoB map to chromosome 5. In human, MondoA maps to 12q24.31 (Billin et al. 2000) and WBSCR14/MondoB maps to 7q11.23 (Billin et al. 2000; de Luis et al. 2000; Cairo et al. 2001). Haploinsufficiency of multiple genes at 7q11.23 is associated with Williams-Beuren syndrome, and hemizygosity has been confirmed for WBSCR14/MondoB in Williams-Beuren patients (de Luis et al. 2000). The clinical features of Williams-Beuren syndrome are complex, including supravalvular aortic stenosis, impaired visual-spatial constructive cognition, mental retardation, and infantile hypercalcemia (Towbin et al. 1999). At present, how or whether WBSCR14/MondoB loss contributes to Williams-Beuren syndrome is not known.

4
Conserved Domains of the Mlx and Mondo Family Proteins

At approximately 900 amino acids in length, members of the Mondo family are among the largest members of the bHLHZip family (Fig. 2a). The Mondo family is conserved across species. The *Drosophila melanogaster* ortholog dmondo is 45% identical and the *Caenorhabditis elegans* ortholog T20B12.6 is 25% identical to MondoA. Similar to the degree of homology between Myc family members, MondoA and WBSCR14/MondoB/ChREBP are 40% identical over their entire open reading frames. The similarity between the Mondo family members is restricted to the N-terminal and C-terminal thirds of the proteins. The middle third of the Mondo proteins is rich in Ser, Thr, and Pro, lacks specific identifiable sequence motifs, and functions as a transcriptional activation domain (Billin et al. 2000; Cairo et al. 2001). The N-terminus contains five blocks of conserved sequence, three of which regulate subcellular localization. The C-terminus contains the bHLHZip domain, followed by a region that is conserved in the Mlx family. The function of each of these conserved domains is discussed in more detail in Sect. 7.

In contrast to the large size disparity between the Mondo and Myc family proteins, Mlx proteins are much more Max-like at approximately 240 amino acids in length (Fig. 2a). Like Max, Mlx is highly conserved among different species with only four amino acid differences between human and mouse Mlx-β. Mlx orthologs are also present in lower eukaryotes with the *D. melanogaster*

Fig. 2a, b Conserved domains of the Mondo and Mlx families. **a** Schematic diagrams of MondoA and Mlx-β. Overall percentage identity between MondoA and the other members of the Mondo family is given (*in parentheses*) directly adjacent to the family member. The percentage identity between different sequence motifs of MondoA to the other members of the Mondo family, Mlx-β and c-Myc are given in the *vertical columns* below each domain. *TAD*, transcriptional activation domain; *bHLHZip*, basic region-helix-loop-helix-zipper; *DCD*, dimerization and cytoplasmic localization domain; the Mondo-specific region contains the five MCR (Mondo conserved region) motifs. **b** Sequence alignment of the basic regions of the Mlx and Mondo families with Mad1, Max, and c-Myc. The conserved residues that target members of these families to the CACGTG subclass of E-box are *highlighted*

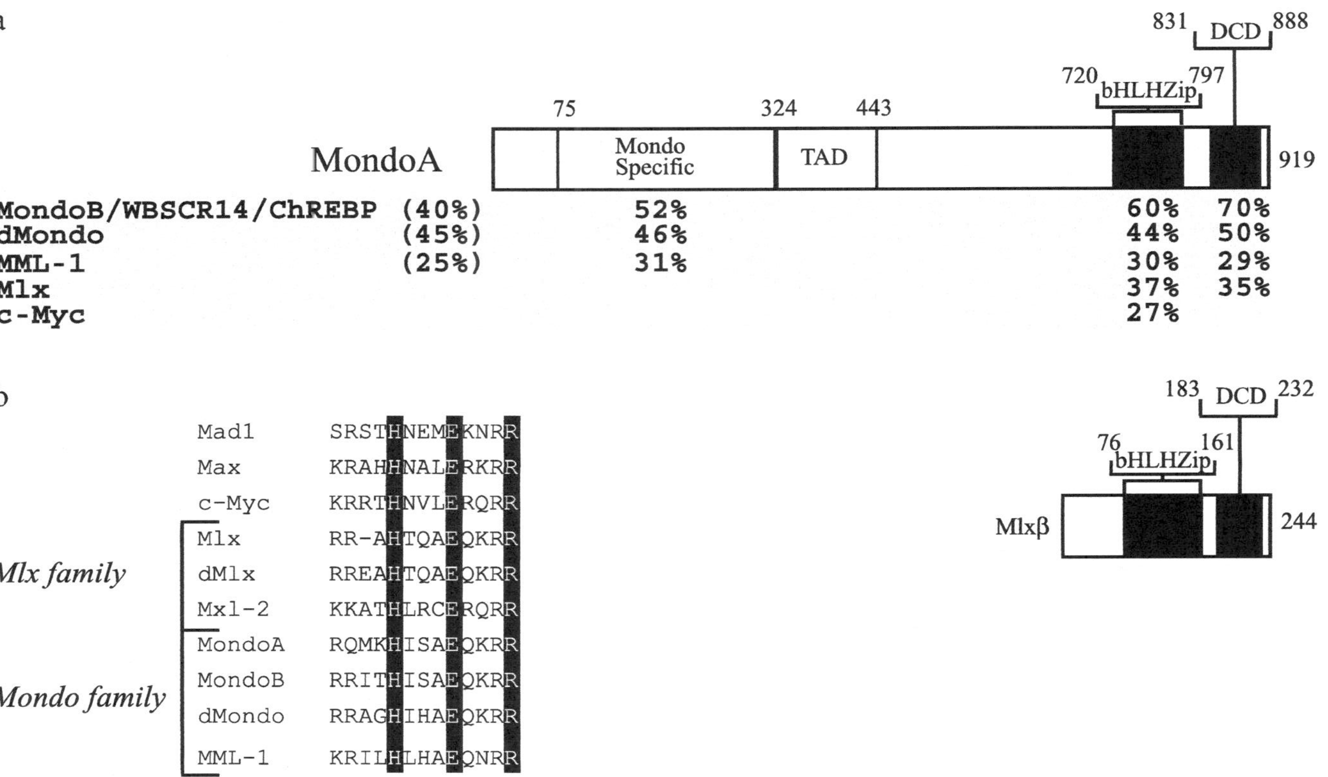
a
831
DCD
888
720
bHLHZip
797
75
324
443
MondoA
Mondo Specific
TAD
919
MondoB/WBSCR14/ChREBP (40%) 52% 60% 70%
dMondo (45%) 46% 44% 50%
MML-1 (25%) 31% 30% 29%
Mlx 37% 35%
c-Myc 27%
b
Mad1 SRSTHNEMEKNRR
Max KRAHHNALERKRR
c-Myc KRRTHNVLERQRR
Mlx RR-AHTQAEQKRR
dMlx RREAHTQAEQKRR
Mxl-2 KKATHLRCERQRR
MondoA RQMKHISAEQKRR
MondoB RRITHISAEQKRR
dMondo RRAGHIHAEQKRR
MML-1 KRILHLHAEQNRR
Mlx family
Mondo family
183
DCD
232
76
bHLHZip
161
Mlxβ
244

ortholog *dmlx* being 45% identical and *C. elegans* ortholog Mxl-2 being 25% identical to Mlx, respectively. Interestingly, the region C-terminal of the bHLHZip domain of the Mlx proteins is similar to the C-terminal region of the Mondo family (Fig. 2a). This domain has been referred to as the WBSCR14-Mlx C-tail (WMC) (Cairo et al. 2001), or a "Zipper-like domain" (Yamashita et al. 2001). Our functional analysis has demonstrated that this domain is required for the cytoplasmic localization of both MondoA and Mlx and also functions as a novel heterodimerization interface. Furthermore, the cytoplasmic localization and dimerization functions are coupled. As such, we have called this domain the dimerization and cytoplasmic localization domain, or DCD, to reflect these two functions (Eilers et al. 2002).

Similar to its location within the Max and Myc proteins, the bHLHZip domains of the Mlx and Mondo proteins are located in the center or close to the C-terminus of the proteins, respectively. Within the larger class of bHLHZip proteins, three residues within the basic region dictate the specificity of E-box DNA binding. For example, the members of the Max network all have His, Glu, and Arg at the respective positions 5, 9, and 13 of their basic regions, which direct binding of the different heterodimers to the CACGTG subclass of E-box (Ferre-D'Amare et al. 1993, 1994; Fig. 2b). Mlx and the Mondo proteins all contain these amino acids similarly positioned within their basic regions, suggesting that Mlx-containing heterodimers also bind CACGTG. In support of this hypothesis, Mad1:Mlx, Mad4:Mlx (Billin et al. 1999), Mnt/Rox:Mlx (Meroni et al. 2000), MondoA:Mlx (Billin et al. 2000), and WBSCR14/MondoB:Mlx heterodimers (Cairo et al. 2001) and ChREBP monomers (Yamashita et al. 2001) all bind CACGTG elements or close relatives. CACGTG binding suggests that members of the Mlx and Max networks may have common transcriptional targets; however, experimental evidence has neither definitively identified nor excluded the possibility of common targets in vivo. Another notable feature of these novel bHLHZip factors is the loop of Mlx that is 6–8 residues longer than those of any other member of the Mlx and Max networks. The loop residues of Max and upstream stimulatory factor (USF) make backbone contacts (Ferre-D'Amare et al. 1993, 1994) outside of the CACGTG core, raising the possibility that Mlx-containing heterodimers may be stabilized on DNA relative to Max-containing heterodimers due to additional contacts between the loop of Mlx and the phosphate backbone. In addition, one cannot rule out the possibility that the longer loop of Mlx might form base-specific contacts and contribute to binding site selection.

Comparison of the available Mondo family sequences using the BLOCKS (http://blocks.fhcrc.org/) or MEME (http://meme.sdsc.edu/) algorithms suggests that the conserved N-terminus of the Mondo family can be divided into five separate sequence motifs. We have termed these sequence motifs MCRI–V for *M*ondo *c*onserved *r*egions (Fig. 3) (Eilers et al. 2002). The functions of MCRI and MCRV are unknown; however, MCRII, III, and IV function in regulating subcellular localization. Furthermore, the spacing between MCRII, III, and IV is also conserved across species, suggesting that they function as a module. By contrast, the locations of MCRI and MCRV are variable, suggesting that they function independently of one another and of the other MCRs. As presented in more detail in Sect. 7, MCRII of MondoA functions as a CRM1-dependent nuclear export signal and MCRIII is a non-consensus binding site for 14-3-3 (Eilers et al. 2002). The high degree of conservation suggests that these domains will function similarly in the other family members. The function of MCRIV is less clear. In ChREBP, MCRIV appears to function as a bipartite nuclear localization signal (Kawaguchi et al. 2001), but several residues that this signal comprises are not conserved in MondoA, suggesting that MCRIV has different functions in the two proteins.

Structural features of the Mondo family suggest that it may function analogously to the Myc family. In the N-terminus of the Myc proteins, Myc Box II is highly conserved, required for its function as a transforming oncogene and binds a number of proteins involved in modifying chromatin structure (McMahon et al. 1998; Wood et al. 2000). By including Myc family members with the Mondo family in the MEME analysis, a Myc Box II-like sequence is identified in MondoA and WBSCR14/MondoB/ChREBP that lies between MCRIV and MCRV (Fig. 3). Interestingly, this putative homology is not found in *dmondo* or T20B12.6. The motif identified by the MEME algorithm is compared against each sequence in the query and each alignment is assigned a P-value. For MondoA and WBSCR14/MondoB the resulting P-values are very low, suggesting that the identification of the Myc Box II-like motif in the Mondo family is likely to be significant. It will be of interest to determine the contribution of this domain to the function of the Mondo family members.

5 Functions of the Mlx Transcriptional Network

Max can interact with all of the members of the Myc and Mad families as well as Mnt/Rox and Mga. By contrast, Mlx appears to only interact with Mad1, Mad4, and Mnt/Rox, but not with members of the Myc family or Max (Fig. 1) (Billin et al. 1999; Meroni et al. 2000). There are conflicting reports as to whether Mlx can homodimerize and bind DNA, which likely result from differences in the experimental conditions used to measure interaction, including variations in the two-hybrid approach, varying levels of Mlx-interacting proteins present

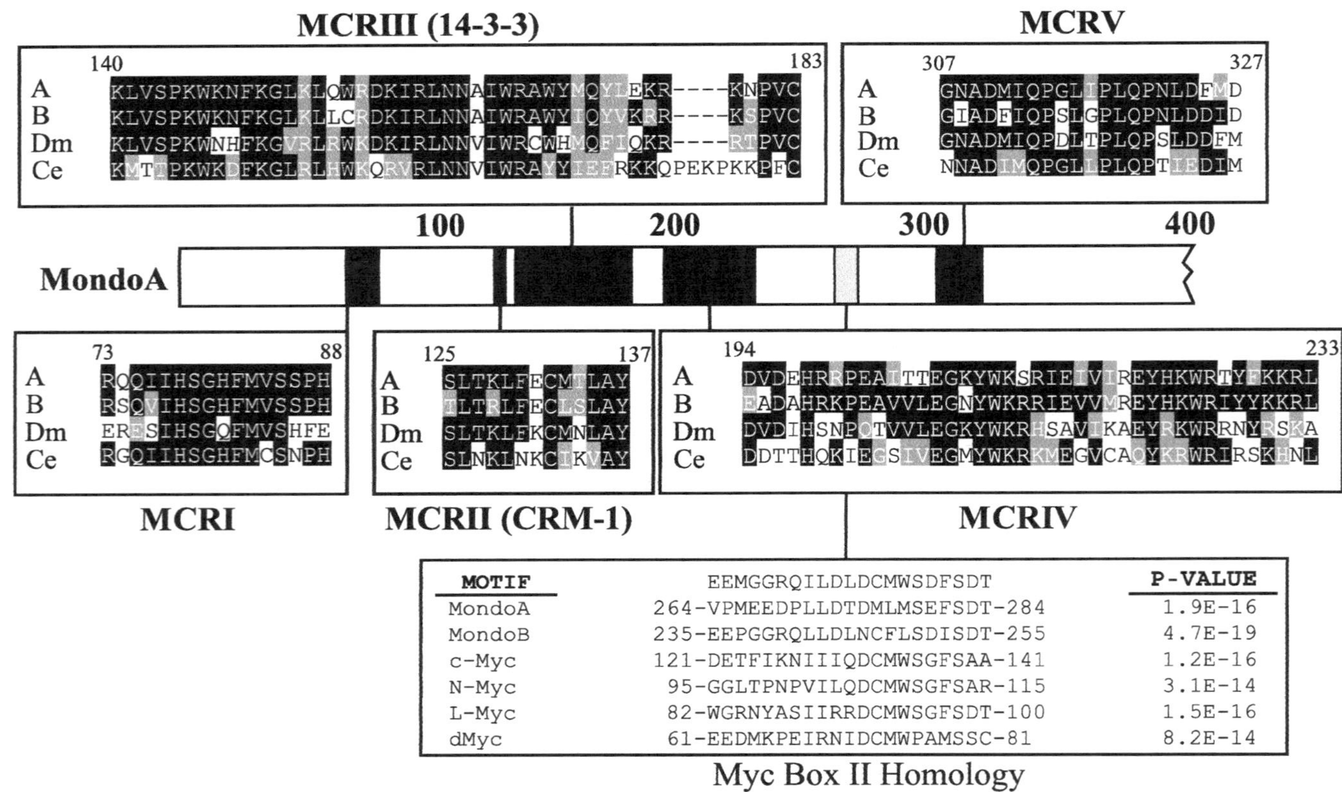
MCRIII (14-3-3)
MCRV
MCRI
MCRII (CRM-1)
MCRIV
MondoA
100
200
300
400
A
B
Dm
Ce
140
183
307
327
73
88
125
137
194
233
MOTIF
P-VALUE
EEMGGRQILDLDCMWSDFSDT
MondoA 264-VPMEEDPLLDTDMLMSEFSDT-284 1.9E-16
MondoB 235-EEPGGRQLLDLNCFLSDISDT-255 4.7E-19
c-Myc 121-DETFIKNIIIQDCMWSGFSAA-141 1.2E-16
N-Myc 95-GGLTPNPVILQDCMWSGFSAR-115 3.1E-14
L-Myc 82-WGRNYASIIRRDCMWSGFSDT-100 1.5E-16
dMyc 61-EEDMKPEIRNIDCMWPAMSSC-81 8.2E-14
Myc Box II Homology

◄

Fig. 3 Conserved sequence motifs within the N-terminal Mondo specific region. Sequence alignment of the Mondo family using the BLOCKS or MEME algorithms identifies 5 blocks of sequence. We have termed these domains the Mondo Conserved Regions or MCRs I-V. Their relative position within the N-terminus of MondoA is shown. Alignments of MondoA (A), WBSCR14/MondoB/ChREBP (B), dmondo (Dm) and MML-1/T20B12.6 (Ce) are given with identical residues shaded black and similar residues shaded grey. The numbering given corresponds to positions in MondoA. Current data suggests that MCRII and MCRIII of MondoA function as a CRM1 dependent nuclear export signal and a 14-3-3 binding site, respectively. In ChREBP, a portion of MCRIV functions as a nuclear localization signal. The MEME algorithm also identifies a sequence in MondoA and WBSCR14/MondoB/ChREBP that resembles Myc Box II. The relative position of this domain is shown aligned with Myc family members from human and *D. melanogaster*

in reticulocyte lysate and pulldown assays, and in the amounts of protein used in the DNA binding assays. Regardless, given the sequence similarity between Mlx and Max, the more restricted dimerization specificity of Mlx is surprising. A recent structural comparison of Myc:Max and Mad:Max heterodimers with Max homodimers revealed that the observed propensity of Max to form heterodimers over homodimers can be attributed to structural differences in the coiled-coil leucine zipper regions (Nair and Burley 2003). It will be of interest to determine (1) whether partner selection by Mlx is dictated by similar mechanisms and (2) the contribution of the DCD to dimerization selectivity and strength. Interestingly, as determined by both two-hybrid and gel shift assays using recombinant proteins, it appears as if Max and Mlx have similar intrinsic affinities for Mad1. However, when interaction is measured with proteins synthesized in vitro or by immunoprecipitation from cell lysates, a clear preference for Mad1:Max over Mad1:Mlx heterodimers is observed, suggesting that the interaction between Mad1 and Mlx may be controlled by posttranscriptional modification (our unpublished data).

The function of the Mad family proteins when bound to Max or Mlx is similar if not identical. For example, Mad1:Mlx heterodimers bind CACGTG E-box elements and repress transcription from CACGTG-dependent reporters. As expected, transcriptional repression by Mad1:Mlx is dependent upon dimerization, DNA binding, and interaction with the mSin3A/HDAC corepressor complex (Billin et al. 1999). Transfection of Mnt/Rox also results in repression from CACGTG-dependent reporter genes, but the level of repression is not significantly altered in the presence of added Mlx (Meroni et al. 2000), suggesting that Mnt/Rox binding and repression may be mediated by endogenous Max or Mlx. Therefore, in addition to being a sequence analog of Max on many levels, Mlx appears to be a functional analog of Max.

Given the number of similarities to Max, it was proposed that Mlx might also function as the center of a transcriptional network, mediating both transcriptional activation and repression activities of its binding partners (Billin et al. 1999; Meroni et al. 2000). Similar to their role in the Max network, Mad proteins seem to constitute the transcriptional repression arm of a putative Mlx transcriptional network. The finding that Mlx activated transcription when transfected alone and that this activation required an intact leucine zipper and basic region (Billin et al. 1999) further supported the hypothesis that Mlx could also interact with transcriptional activators. As detailed in the following sections, MondoA and ChREBP localize to the cytoplasm, but can activate transcription from CACGTG-dependent reporters when targeted to the nucleus. Transcriptional activation by MondoA depends on heterodimerization with Mlx (Billin et al. 2000), supporting the hypothesis that the Mondo family comprises the transcriptional activation arm of the Mlx transcriptional network. Transcriptional activation by ChREBP did not require cotransfection of Mlx, suggesting that this activity was mediated by interactions with endogenous Mlx (Yamashita et al. 2001) or, alternatively, is truly independent of Mlx. In contrast to MondoA and ChREBP, WBSCR14/MondoB repressed transcription weakly from E-box containing promoters; however, the subcellular localization of WBSCR14/MondoB was not determined in these experiments (Cairo et al. 2001).

6
Transcriptional Targets of the Mlx Network

The Max network plays an important role in cell growth and division. Many of the direct and indirect transcriptional targets of Myc:Max have been identified, and to a large extent their activities are consistent with effector functions in both of these processes (reviewed in Grandori and Eisenman 1997; Dang 1999; most recent primary reports, Fernandez et al. 2003; O'Connell et al. 2003; Orian et al. 2003). For example, Myc:Max heterodimers regulate metabolism, cell-cycle signaling, protein synthesis, and ribosome biogenesis (Fig. 1). At present the targets of the other Max-containing heterocomplexes are less well understood. Two recent papers in which the basic region of Mad1 replaced that of Myc suggest a large degree of overlap between Myc:Max and Mad:Max transcriptional targets (James and Eisenman 2002; Nikiforov et al. 2003).

By contrast, only a few of the transcriptional targets of the Mondo family are known. For example, MondoA can activate transcription from synthetic promoters dependent on CACGTG binding sites (Billin et al. 2000). Furthermore, MondoA can activate the lactate dehydrogenase A promoter (our unpublished data) and ChREBP can activate the L-type pyruvate kinase promoter (Yamashita et al. 2001), both in a CACGTG-dependent manner. Interestingly, these promoters are also activated by Myc (Shim et al. 1997; Collier et al. 2003), demonstrating that the Mondo family can indeed regulate bona fide Myc transcriptional targets. While the full spectrum of Mondo targets remains to be illuminated, these initial findings support the general hypothesis that the Mondo and Myc families share at least some overlapping functions.

7 Regulation of Subcellular Localization

With a few exceptions (for example, Craig et al. 1993; Wakamatsu et al. 1993; Okano et al. 1999), the majority of current evidence suggests that members of the Max network localize primarily to the nucleus. As such, it seems unlikely that regulated nuclear entry or exit contributes significantly to the function of the canonical Max network. By contrast, MondoA and ChREBP appear to localize to the cytoplasm of most cell types examined (Billin et al. 2000; Yamashita et al. 2001; Eilers et al. 2002), suggesting that nuclear entry and accumulation, presumably in response to extracellular signaling cues, controls their function as transcriptional activators. ChREBP accumulates in the nucleus in response to carbohydrate (Yamashita et al. 2001); however, MondoA does not (our unpublished data), suggesting that MondoA requires either different or additional signals for nuclear accumulation.

While the events that trigger the nuclear accumulation of MondoA:Mlx heterodimers remain elusive, the domains responsible for the cytoplasmic localization of both MondoA and Mlx have been identified and their functions characterized. The C-terminal DCD functions as a nonautonomous cytoplasmic localization domain in both MondoA and Mlx that is required to keep each protein monomer in the cytoplasm. In addition, MCRII and MCRIII of MondoA constitute a potent cytoplasmic localization domain that functions in conjunction with the DCD to sequester MondoA in the cytoplasm. Unlike the DCD domain, the MCRII and MCRIII module can override the nuclear localization activity of a variety of strong nuclear localization signals, suggesting that it functions autonomously. Finally, Mlx-β contains a nuclear retention signal within its N-terminus and MondoA contains a nuclear localization signal that maps to its central region (Eilers et al. 2002).

While multiple domains control the subcellular localization of MondoA:Mlx, the concerted activity of these domains in controlling nuclear accumulation can be explained by a relatively conceptually simple two-step model (Fig. 4). First, in addition to functioning as a cytoplasmic localization

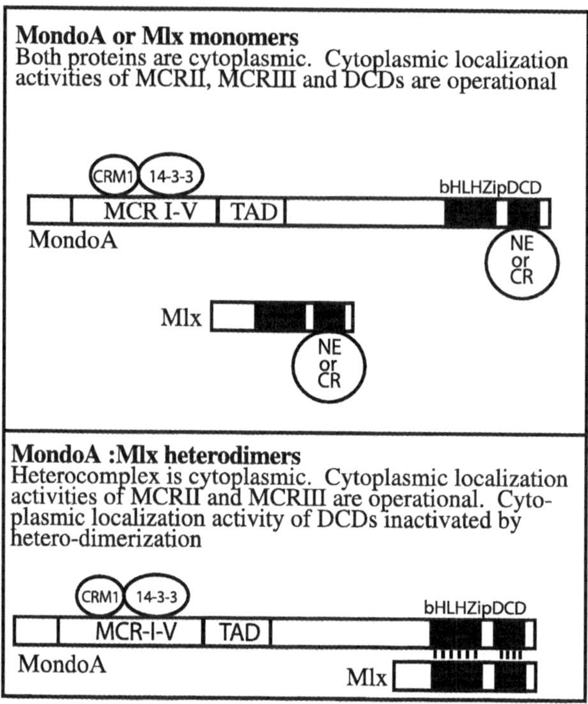

Fig. 4 Complex mechanisms control the subcellular localization of MondoA and Mlx. MondoA and Mlx monomers are retained in the cytoplasm by the DCD at the C-terminus of both proteins and by MCRII and MCRIII of MondoA. The DCD may contribute to cytoplasmic localization by interacting with nuclear export signals (*NE*) or with cytoplasmic retention signals (*CR*). It is also possible that the DCD domain functions by masking nuclear export signals (not shown). When MondoA and Mlx form heterodimers, the cytoplasmic localization activity of the DCD is canceled by dimerization mediated by both the leucine zipper and the DCD. Dimerization forms a transcription factor that is capable of nuclear entry, DNA binding, and presumably transcription activation; however, the heterodimer is retained in the cytoplasm by MCRII and MCRIII. We propose that the heterodimer accumulates in the nucleus in response to an extracellular cue that must impinge on MCRII and III and abrogate both CRM1 and 14-3-3 binding. The identity of this putative signal is currently unknown

domain—perhaps via interactions with cytoplasmic anchors or nuclear export machinery—the DCDs of MondoA and Mlx constitute a novel heterodimerization interface. Heterodimerization between MondoA and Mlx, mediated by both the leucine zipper and the DCD, inactivates the cytoplasmic localization function of the DCD. As such, heterodimerization between MondoA and Mlx in the cytoplasm renders the heterocomplex capable of DNA binding and functioning as a transcriptional activator; however, due to the activity of MCRII and MCRIII the complex remains in the cytoplasm (Fig. 4). Therefore, we have proposed that the second regulatory step is a signal-dependent inactivation of the cytoplasmic localization function of MCRII and MCRIII that allows nuclear accumulation of the heterocomplex. It seems most likely that dimerization between MondoA and Mlx is constitutive, suggesting that the proposed signaling event must be rate limiting for nuclear entry.

MCRII and MCRIII are highly conserved across species, suggesting conserved functions in regulating subcellular localization. MCRII functions as a CRM1-dependent nuclear export signal, whereas MCRIII is, at least in part, a non-canonical binding site for 14-3-3 proteins (Eilers et al. 2002). Elimination of either one of these functions by point mutagenesis alone has little effect on the cytoplasmic localization of a nuclear localization signal (NLS)-tagged fusion protein that encodes both MCRII and MCRIII, but mutation of both domains simultaneously results in the dramatic localization of the fusion protein to the nucleus. Therefore, CRM1 and 14-3-3 appear to function together to regulate the cytoplasmic localization of MondoA. Current evidence suggests that 14-3-3 proteins control the subcellular localization of a number of substrates by indirect mechanisms (for example, Brunet et al. 2002). As compared to wildtype, mutants of MondoA that cannot bind 14-3-3, but retain interactions with CRM1, accumulate in the nucleus to a greater degree in response to the nuclear export inhibitor leptomycin B (Eilers et al. 2002). This suggests that 14-3-3 may contribute to the subcellular localization of MondoA by either stabilizing CRM1 binding or by masking a yet-to-be-identified nuclear localization signal.

ChREBP was initially identified by its CACGTG DNA binding activity detected in nuclear extracts prepared from the livers of starved rats that had been refed a high carbohydrate diet (Yamashita et al. 2001). This activity was also detected when the rats were refed a high fat diet, but to a lesser extent. This CACGTG binding activity was purified and the protein responsible identified as the rat homolog of WBSCR14/MondoB. While little information is currently available on the regulation of endogenous ChREBP, a green fluorescent protein (GFP)-ChREBP fusion protein accumulates in the cytoplasm following transfection but translocates to the nucleus in response to increased glucose levels (Kawaguchi et al. 2001). Once in the nucleus, ChREBP can activate transcription from the glycolytic enzyme L-type pyruvate kinase promoter in a CACGTG- and glucose-dependent manner, supporting a role in regulating the transcriptional response to increased carbohydrate. Recent experiments demonstrate that two protein kinase A (PKA) phosphorylation sites, one adjacent to MCRIV (Ser196) and one within the basic region of

ChREBP (Thr666), regulate ChREBP nuclear accumulation and DNA binding activity, respectively (Kawaguchi et al. 2001). In addition, an AMP-activated protein kinase (AMPK) site located approximately 100 amino acids upstream of the basic region (Ser568) also appears to regulate the DNA binding activity of ChREBP (Kawaguchi et al. 2002). All of these sites were identified using database searching. Furthermore, recombinant ChREBP or ChREBP peptides were phosphorylated by recombinant PKA or AMPK in vitro. The relevance of each of these sites to ChREBP function was established using a variety of assays and a collection of ChREBP mutants containing substitutions that either mimic or abrogate phosphorylation. It has yet to be determined whether these sites are phosphorylated in vivo by the appropriate kinase, and therefore these results must be interpreted somewhat cautiously.

How does each of the proposed phosphorylation events regulate the activity of ChREBP? In low glucose, when ChREBP localizes to the cytoplasm, it is proposed that Ser196 and Thr666 are phosphorylated by PKA. Ser196 is adjacent to MCRIV, which contains a bipartite NLS, suggesting that its phosphorylation blocks the function of the NLS. Thr666 is located within the basic region, and substitution of this residue with Asp blocks DNA binding, suggesting that phosphorylation of this residue inhibits DNA binding (Kawaguchi et al. 2001). There is a fair amount of variation at this position within the other members of the bHLHZip family (Fig. 3), suggesting that it does not form base-specific hydrogen bonds with the E-box element; this contention is supported by the high resolution crystal structures of Max homodimers and Myc:Max heterodimers (Ferre-D'Amare et al. 1993; Nair and Burley 2003). As such, phosphorylation of Thr666 may inhibit DNA binding by charge repulsion with the DNA phosphate backbone. High glucose is proposed to activate the phosphatase PP2A, via an indirect mechanism, thereby leading to the dephosphorylation of Ser196 and Thr666 (Kawaguchi et al. 2001). Therefore, elevation in glucose levels unmasks the activity of the NLS resulting in nuclear accumulation, DNA binding, and the subsequent activation of target genes.

In contrast to Ser196 and Thr666, Ser568 is thought to be unphosphorylated under both low and high glucose conditions. However, in response to fatty acid treatment, Ser568 appears to be phosphorylated by AMPK, which results in reduced DNA binding (Kawaguchi et al. 2002). Such a downregulation of ChREBP DNA binding and transcriptional activity may account for the so-called "glucose sparing" effect where key mediators of glycolysis and lipogenesis are downregulated by high fatty acid concentrations. It is currently unknown how phosphorylation of Ser568 inhibits DNA binding from its location about 100 amino acids distant from the ChREBP-DNA interface.

At present, the information concerning the regulation of the subcellular localization of MondoA and ChREBP is complementary. For MondoA, many of the specific mechanisms that control subcellular localization have been elucidated, but the signal that triggers nuclear accumulation is unknown. By contrast, the signal that triggers nuclear accumulation of ChREBP is known, as are many of the regulatory phosphorylation sites, but the mechanistic details are missing. Given the high overall sequence similarity between ChREBP and MondoA, one expects that their activity will be controlled by some of the same mechanisms.

In spite of the high overall similarity between ChREBP and MondoA and the likelihood they are regulated by many of the same mechanisms, there appear to be key differences. For example, the NLS within MCRIV of ChREBP is only partially conserved in MCRIV of MondoA. In support of the idea that this domain will have alternate functions in the different proteins, a fusion between MCRIV of MondoA and a heterologous NLS localized completely to the cytoplasm (our unpublished data), suggesting that this domain in MondoA functions as a cytoplasmic localization domain as opposed to an NLS. Finally, despite the high overall sequence similarity between MondoA and ChREBP, the phosphorylation sites that appear to regulate ChREBP activity are not conserved in MondoA; perhaps explaining why the subcellular localization of MondoA is unaffected by glucose levels. In fact, the residues that correspond to Ser196 and Thr666 in ChREBP are both positively charged lysines in MondoA, suggesting completely different modes of regulation.

8 Function of the Max and Mlx Transcriptional Networks in Lower Eukaryotes

In higher eukaryotes, Max can interact with at least nine members of the bHLHZip family, while Mlx has at least four bHLHZip binding partners. Furthermore, each Max- or Mlx-containing heterocomplex can mediate transcriptional effects through the same CACGTG E-box element, potentially regulating many of the same target genes (Fig. 1). Together, these findings suggest that the ultimate biological outputs generated by these transcriptional networks are a result of both functional redundancy and transcriptional crosstalk between the different possible heterocomplexes. Both of these issues complicate the interpretation of the phenotypes of mice bearing targeted deletions of different family members or expressing exogenous transgenes. As such, a number of investigators have begun to study the Myc/Max/Mad/Mondo/Mlx families in *D. melanogaster* and *C. elegans*. In both organisms, the transcriptional network is much simpler, providing an excellent opportunity to study its function without the aforementioned complications.

A single representative of each of the mammalian transcription factor families is present in *D. melanogaster*, i.e., there are single Myc, Mondo, Mad/Mnt, Max, and Mlx orthologs. To date, dMyc has been most extensively studied and, as in mammalian cells, plays a role in growth and proliferation (Johnston et al. 1999). Data concerning the function of dMondo dMad, dMax, and dMlx has not yet been published; however, we present data below demonstrating that dMyc and dMondo interact genetically. As expected, dMyc associates with dMax (Gallant et al. 1996); however, it will be necessary to determine the spectrum of interactions allowed among the other sequence orthologs of the Max and Mlx networks and the transcriptional activity of the different heterocomplexes to obtain a clear picture of the extent of functional conservation. Based on the available evidence, *D. melanogaster* appears to have canonical Max and Mlx networks, providing a simplified and genetically tractable model system to study their functions.

The *C. elegans* genome also expresses clear orthologs of Max, Mad, Mlx, and Mondo. Published data suggest that the transcriptional repression arm of the Max network is largely intact in worm. For example, the Max and Mad orthologs MXL-1 and MDL-1 preferentially form heterodimers that bind CACGTG sites. Amazingly, overexpression of MDL-1 in mammalian cells blocks Myc+Ras cotransformation in a manner similar to Mad1, demonstrating functional conservation (Yuan et al. 1998). In contrast to *D. melanogaster* and other higher organisms, there is no clear Myc ortholog in the *C. elegans* genome, suggesting that the transcriptional activation arm of the Max network is missing in this organism. We have identified a putative Mondo ortholog, T20B12.6, which contains each of the MCRs in its N-terminus. Interestingly, T20B12.6 is also 26% identical and 42% similar to the C-terminal half of human c-Myc. Therefore, T20B12.6 has sequence characteristics of both Myc and Mondo family proteins, and we have elected to call this protein, MML-1 for Myc and Mondo-like 1. Given the possibility that the Myc and Mondo families have some functions in common, this finding raises the intriguing possibility that MML-1 carries out all of the Mondo and Myc functions in *C. elegans*. The current data regarding the function of Max and Mlx networks in *C. elegans* are limited, but one model is that MXL-1:MDL-1 heterodimers constitute the transcriptional repression arm of the network, while the Mlx ortholog, MXL-2, interacts solely with MML-1 to constitute the transcriptional activation arm of the network. Biochemical methods as well as the powerful forward and reverse genetic tools available in *C. elegans* should shed light on the functions of each heterocomplex.

9 Does the Mondo Family Function in Parallel with the Myc Family?

Both Myc:Max and MondoA:Mlx heterodimers bind to CACGTG E-box elements and activate transcription, suggesting that they may have overlapping or complementary biological functions. This finding also raises the exciting possibility that the Mlx network may function in parallel, or in conjunction, with the Max network. To date, the information about the biological role(s) of the Mondo family is limited; however, that which is available is consistent with a role for the Mondo family in controlling growth and energy metabolism. For example, overexpression of WBSCR14 leads to a reduction in cell colony number; however, it was not tested whether cell growth or apoptosis was affected in this experiment (Cairo et al. 2001). In addition, both Myc and ChREBP can regulate a number of genes involved in cellular energy homeostasis (examples include, Shim et al. 1997; Osthus et al. 2000; Yamashita et al. 2001; Riu et al. 2002). Thus, the case for functional similarities between the Myc family and the Mondo family has some experimental support; however, the full extent of functional overlap between the two families remains to be determined. We have recently gained some insight into this question using *D. melanogaster* as a model system.

During embryogenesis *dmlx* is ubiquitously expressed, whereas *dmondo* expression is restricted to the amnioserosa and dorsal pharyngeal musculature. Both genes are expressed during larval, pupal, and adult stages (Peyrefitte et al. 2001). To investigate the biological role of the *mondo* genes and determine whether the *mondo* and *myc* families function in related biological pathways, we first obtained a P-element insertion in *dmondo* from the Berkeley *Drosophila* Genome Project and analyzed it molecularly. The P-element, k05106, is located in the sixth intron of the *dmondo* gene, about 10 kb from the transcription initiation site at cytological band 39C1-D1. Precise excision of the P-element insertion resulted in flies that appeared wildtype with no apparent associated lethality. Thus, the P-element insertion, and not other genetic alterations, is responsible for the observed phenotypes. This P-element insertion drastically reduces the level of *dmondo* expression and is therefore likely to be a hypomorphic allele of *dmondo* (our unpublished data). We have called this allele of *dmondo*, *dmon1*.

Homozygotes of *dmon1* complete embryogenesis; however, homozygous *dmon1* adult female flies hatched at approximately one-half the expected frequency and 1–2 days later than heterozygotes. The remaining females die during pupation, as indicated by the presence of many pupal cases containing melanized dead female pupae. In contrast, very few male homozygotes survived. About half failed to remove themselves from the pupal case and those that did hatch were weak and usually became mired in the food (our unpublished data). Thus, *dmon1* homozygotes are sub-viable, suggesting a role for *dmon1* in normal fly development and/or physiology. These data are also consistent with published complementation mapping studies performed in the 39C1-D1 region. Hewes et al. noted reduced viability of the k05106 homozygotes. Further, they noted that k05106 in *trans* to a deletion for the 39C1-D1 region, Df(2L)TW1, do not survive to adulthood (Hewes et al. 2000). This further argues for the k05106 insertion acting as a strong hypomorph. Taken together, these data suggest that *dmondo* is critical for full viability of *D. melanogaster*.

To determine whether the *mondo* and *myc* families have overlapping or complementary biological functions, we tested for genetic interaction between the *diminutive* (*dm*) allele of *dmyc* (Gallant et al. 1996; Schreiber-Agus et al. 1997) and the *dmon1* allele. Animals homozygous for *dm* are fully viable but smaller than normal, and females are sterile due to defective oogenesis (Lindsley and Zimm 1992; Gallant et al. 1996). We tested for genetic interaction between *dm* and *dmon1* by determining whether the double mutant combination *dm/dm*; *dmon1/dmon1* was viable. Only females were examined because of the small number of surviving homozygous *dmon1* males. Of 619 females scored, no double homozygotes were recovered. However, animals in the other progeny classes were recovered, including animals heterozygous for *dm* and homozygous for *dmon1* (Fig. 5). Thus, double homozygosity for *dm*

dMyc *(dm)* and dMondo *(dmon1)* Interact Genetically

Cross	Genotypes of Progeny	Expected	Observed
dm/FM7 ; dmon1/CyO	dm/dm ; dmon1/CyO	207	258
X	dm/FM7 ; dmon1/CyO	207	294
dm/Y ; dMondo/CyO	dm/FM7 ; dmon1/dmon1	103	67
	dm/dm ; dmon1/dmon1	103	0

Fig. 5 dm and dmon1 interact genetically. The indicated cross was performed and female progeny were scored for the appropriate markers to determine genotype. If no lethal interaction occurs, then animals homozygous for *dm* and *dmon1* should be present at the same frequency as animals heterozygous for *dm* and *dmon1*. No flies homozygous for both *dm* and *dmon1* were recovered, revealing a synthetic lethal genetic interaction between *dm* and *dmon1* and suggesting that dMyc and dMondo function is parallel or redundant pathways

and *dmon1* results in synthetic lethality. Recently, genomic binding sites for dMax, dMnt, and dMyc have been discovered using genome-wide isolation and sequencing of bound chromatin (Orian et al. 2003). Sites for dMnt have been identified in the *dmlx* promoter, and dMyc binding sites have been identified in the *dmondo* promoter. This biochemical data, along with the lethal genetic interaction between *dmondo* and *dmyc*, supports the hypothesis that these two transcription factor networks act together to regulate the expression of at least one essential gene.

10 Future Directions

Much of the current evidence suggests that the Mondo proteins function as Myc analogs, but with their subcellular localization highly regulated. Furthermore, the interaction of Mlx with members of the Mad and Mondo families suggests that, like Max, Mlx functions as the center of a transcription factor network. Given the many similarities between the Max and Mlx networks, it is likely that they share at least some functions in parallel. A great deal has been learned over the last several years about the functions of this extension of the Mlx network; however, our understanding of the role of the Mlx network in cellular physiology remains meager relative to volumes of information currently available about the Max network. As such, several key questions concerning the function of the Mlx network remain to be answered. Most importantly, which functions of the Mondo proteins overlap with functions of Myc, and which are novel? For example, do Mondo family members regulate all Myc targets or only a subset? Given current evidence, it is possible that the greatest functional overlap is at promoters of glycolytic and perhaps other metabolic enzymes. In a similar vein, it will be also important to (1) examine the role of the Mondo family using models of cellular transformation and (2) determine the contribution of the Myc Box II homology to Mondo function. Similarly, it will be important to determine whether the functions of MondoA and/or WBSCR14/MondoB are altered in human cancers. The contribution, if any, of WBSCR14/MondoB loss to Williams-Beuren syndrome must also be investigated. The restricted interaction of Mlx with Mad1 and Mad4 suggests that they may have functions different from those of Mxi1 and Mad3 and may function differently as a heterodimer with Mlx relative to their function as a heterodimer with Max.

While we have presented a case for the Mondo family functioning similarly to the Myc family, the Mondo family also has many unique features that are worthy of further study. Given that ChREBP accumulates in the nucleus in response to glucose while MondoA does not, discovering the extracellular cue that drives MondoA nuclear accumulation is a priority. Furthermore, while some aspects of MondoA and WBSCR14/MondoB/ChREBP regulation are clearly different, the overall sequence similarity between these proteins suggests that certain aspects of their function and regulation are likely to be shared. For example, MCRI and MCRV and the DCD domains are highly conserved within the family, but how these two MRCs contribute to function and how the DCD functions as a cytoplasmic localization domain are currently unknown. How the DCD contributes to dimerization affinity, specificity, and potentially DNA binding site selection also provides an interesting experimental direction. Similarly, the restricted dimerization specificity observed for Mlx also provides an interesting structural problem. Finally, multiple lines of evidence suggest nuclear functions for both MondoA and WBSCR14/MondoB/ChREBP, but given that under most circumstances they localize to the cytoplasm, one cannot rule out additional functions in this subcellular compartment.

Acknowledgements We apologize to our many colleagues whose work could not be cited directly. We thank the members of the Ayer lab for critical reading of this manuscript. The work in the Ayer lab is funded by grants from the National Institutes of Health, the American Cancer Society, and the Huntsman Cancer Foundation. D.E.A. is a Scholar of the Leukemia and Lymphoma Society.

References

Billin AN, Eilers AL, Queva C, Ayer DE (1999) Mlx, a novel max-like BHLHZip protein that interacts with the max network of transcription factors. J Biol Chem 274:36344–36350

Billin AN, Eilers AL, Coulter KL, Logan JS, Ayer DE (2000) MondoA, a novel basic helix-loop-helix-leucine zipper transcriptional activator that constitutes a positive branch of a max-like network. Mol Cell Biol 20:8845–8854

Brunet A, Kanai F, Stehn J, Xu J, Sarbassova D, Frangioni JV, Dalal SN, DeCaprio JA, Greenberg ME, Yaffe MB (2002) 14-3-3 transits to the nucleus and participates in dynamic nucleocytoplasmic transport. J Cell Biol 156:817–828

Cairo S, Merla G, Urbinati F, Ballabio A, Reymond A (2001) WBSCR14, a gene mapping to the Williams-Beuren syndrome deleted region, is a new member of the Mlx transcription factor network. Hum Mol Genet 10:617–627

Collier JJ, Doan TT, Daniels MC, Schurr JR, Kolls JK, Scott DK (2003) c-Myc is required for the glucose-mediated induction of metabolic enzyme genes. J Biol Chem 278:6588–6595

Craig RW, Buchan HL, Civin CI, Kastan MB (1993) Altered cytoplasmic/nuclear distribution of the c-myc protein in differentiating ML-1 human myeloid leukemia cells. Cell Growth Differ 4:349–357

Dang CV (1999) c-Myc target genes involved in cell growth, apoptosis, and metabolism. Mol Cell Biol 19:1–11

de Luis O, Valero MC, Jurado LA (2000) WBSCR14, a putative transcription factor gene deleted in Williams-Beuren syndrome: complete characterisation of the human gene and the mouse ortholog. Eur J Hum Genet 8:215–222

Eilers AL, Sundwall E, Lin M, Sullivan AA, Ayer DE (2002) A novel heterodimerization domain, CRM1, and 14-3-3 control subcellular localization of the MondoA-Mlx heterocomplex. Mol Cell Biol 22:8514–8526

Fernandez PC, Frank SR, Wang L, Schroeder M, Liu S, Greene J, Cocito A, Amati B (2003) Genomic targets of the human c-Myc protein. Genes Dev 17:1115–1129

Ferre-D'Amare AR, Prendergast GC, Ziff EB, Burley SK (1993) Recognition by Max of its cognate DNA through a dimeric b/HLH/Z domain. Nature 363:38–45

Ferre-D'Amare AR, Pognonec P, Roeder RG, Burley SK (1994) Structure and function of the b/HLH/Z domain of USF. EMBO J 13:180–189

Gallant P, Shiio Y, Cheng PF, Parkhurst SM, Eisenman RN (1996) Myc and Max homologs in Drosophila. Science 274:1523–1527

Grandori C, Eisenman RN (1997) Myc target genes. Trends Biochem Sci 22:177–181

Hewes RS, Schaefer AM, Taghert PH (2000) The cryptocephal gene (ATF4) encodes multiple basic-leucine zipper proteins controlling molting and metamorphosis in Drosophila. Genetics 155:1711–1723

James L, Eisenman RN (2002) Myc and Mad bHLHZip domains possess identical DNA-binding specificities but only partially overlapping functions in vivo. Proc Natl Acad Sci U S A 99:10429–10434

Johnston LA, Prober DA, Edgar BA, Eisenman RN, Gallant P (1999) Drosophila myc regulates cellular growth during development. Cell 98:779–790

Kawaguchi T, Takenoshita M, Kabashima T, Uyeda K (2001) Glucose and cAMP regulate the L-type pyruvate kinase gene by phosphorylation/dephosphorylation of the carbohydrate response element binding protein. Proc Natl Acad Sci U S A 98:13710–13715

Kawaguchi T, Osatomi K, Yamashita H, Kabashima T, Uyeda K (2002) Mechanism for fatty acid "sparing" effect on glucose-induced transcription: regulation of carbohydrate-responsive element-binding protein by AMP-activated protein kinase. J Biol Chem 277:3829–3835

Lindsley DL, Zimm GG (1992) The genome of Drosophila melanogaster. Academic Press, San Diego, pp 1–804

McMahon SB, Van Buskirk HA, Dugan KA, Copeland TD, Cole MD (1998) The novel ATM-related protein TRRAP is an essential cofactor for the c-Myc and E2F oncoproteins. Cell 94:363–374

Meroni G, Cairo S, Merla G, Messali S, Brent R, Ballabio A, Reymond A (2000) Mlx, a new Max-like bHLHZip family member: the center stage of a novel transcription factors regulatory pathway? Oncogene 19:3266–3277

Nair SK, Burley SK (2003) X-ray structures of Myc-Max and Mad-Max recognizing DNA. Molecular bases of regulation by proto-oncogenic transcription factors. Cell 112:193–205

Nikiforov MA, Popov N, Kotenko I, Henriksson M, Cole MD (2003) The mad and myc basic domains are functionally equivalent. J Biol Chem 278:11094–11099

O'Connell BC, Cheung AF, Simkevich CP, Tam W, Ren X, Mateyak MK, Sedivy JM (2003) A large scale genetic analysis of c-Myc-regulated gene expression patterns. J Biol Chem 278:12563–12573

Okano HJ, Park WY, Corradi JP, Darnell RB (1999) The cytoplasmic Purkinje onconeural antigen cdr2 down-regulates c-Myc function: implications for neuronal and tumor cell survival. Genes Dev 13:2087–2097

Orian A, van Steensel B, Delrow J, Bussemaker HJ, Li L, Sawado T, Williams E, Loo LWM, Cowley SM, Yost C, Pierce S, Edgar BA, Parkhurst SM, Eisenman RN (2003) Genomic binding by the Drosophila Myc, Max, Mad/Mnt transcription factor network. Genes Dev17:1101–1114

Osthus RC, Shim H, Kim S, Li Q, Reddy R, Mukherjee M, Xu Y, Wonsey D, Lee LA, Dang CV (2000) Deregulation of glucose transporter 1 and glycolytic gene expression by c-Myc. J Biol Chem 275:21797–21800

Peyrefitte S, Kahn D, Haenlin M (2001) New members of the Drosophila Myc transcription factor subfamily revealed by a genome-wide examination for basic helix-loop-helix genes. Mech Dev 104:99–104

Riu E, Ferre T, Mas A Hidalgo A, Franckhauser S, Bosch F (2002) Overexpression of c-myc in diabetic mice restores altered expression of the transcription factor genes that regulate liver metabolism. Biochem J 368:931–937

Schreiber-Agus N, Stein D, Chen K, Goltz JS, Stevens L, DePinho RA (1997) Drosophila Myc is oncogenic in mammalian cells and plays a role in the diminutive phenotype. Proc Natl Acad Sci U S A 94:1235–1240

Shim H, Dolde C, Lewis BC, Wu CS, Dang G, Jungmann RA, Dalla-Favera R, Dang CV (1997) c-Myc transactivation of LDH-A: implications for tumor metabolism and growth. Proc Natl Acad Sci U S A 94:6658–6663

Towbin JA, Casey B, Belmont J (1999) The molecular basis of vascular disorders. Am J Hum Genet 64:678–684

Vogelstein B, Kinzler KW (1994) Has the breast cancer gene been found? Cell 79:1–3

Wakamatsu Y, Watanabe Y, Shimono A, Kondoh H (1993) Transition of localization of the N-Myc protein from nucleus to cytoplasm in differentiating neurons. Neuron 10:1–9

Wood MA, McMahon SB, Cole MD (2000) An ATPase/helicase complex is an essential cofactor for oncogenic transformation by c-Myc. Mol Cell 5:321–330

Yamashita H, Takenoshita M, Sakurai M, Bruick RK, Henzel WJ, Shillinglaw W, Arnot D, Uyeda K (2001) A glucose-responsive transcription factor that regulates carbohydrate metabolism in the liver. Proc Natl Acad Sci U S A 98:9116–9121

Yuan J, Tirabassi RS, Bush AB, Cole MD (1998) The C. elegans MDL-1 and MXL-1 proteins can functionally substitute for vertebrate MAD and MAX. Oncogene 17:1109–1118

Subject Index

Current Topics in Microbiology and Immunology

Volumes published since 1989 (and still available)

Vol. 258: **Takada, Kenzo (Ed.):** Epstein-Barr Virus and Human Cancer. 2001. 38 figs. IX, 233 pp. ISBN 3-540-41506-8

Vol. 259: **Hauber, Joachim, Vogt, Peter K. (Eds.):** Nuclear Export of Viral RNAs. 2001. 19 figs. IX, 131 pp. ISBN 3-540-41278-6

Vol. 260: **Burton, Didier R. (Ed.):** Antibodies in Viral Infection. 2001. 51 figs. IX, 309 pp. ISBN 3-540-41611-0

Vol. 261: **Trono, Didier (Ed.):** Lentiviral Vectors. 2002. 32 figs. X, 258 pp. ISBN 3-540-42190-4

Vol. 262: **Oldstone, Michael B.A. (Ed.):** Arenaviruses I. 2002. 30 figs. XVIII, 197 pp. ISBN 3-540-42244-7

Vol. 263: **Oldstone, Michael B. A. (Ed.):** Arenaviruses II. 2002. 49 figs. XVIII, 268 pp. ISBN 3-540-42705-8

Vol. 264/I: **Hacker, Jörg; Kaper, James B. (Eds.):** Pathogenicity Islands and the Evolution of Microbes. 2002. 34 figs. XVIII, 232 pp. ISBN 3-540-42681-7

Vol. 264/II: **Hacker, Jörg; Kaper, James B. (Eds.):** Pathogenicity Islands and the Evolution of Microbes. 2002. 24 figs. XVIII, 228 pp. ISBN 3-540-42682-5

Vol. 265: **Dietzschold, Bernhard; Richt, Jürgen A. (Eds.):** Protective and Pathological Immune Responses in the CNS. 2002. 21 figs. X, 278 pp. ISBN 3-540-42668X

Vol. 266: **Cooper, Koproski (Eds.):** The Interface Between Innate and Acquired Immunity, 2002. 15 figs. XIV, 116 pp. ISBN 3-540-42894-X

Vol. 267: **Mackenzie, John S.; Barrett, Alan D. T.; Deubel, Vincent (Eds.):** Japanese Encephalitis and West Nile Viruses. 2002. 66 figs. X, 418 pp. ISBN 3-540-42783X

Vol. 268: **Zwickl, Peter; Baumeister, Wolfgang (Eds.):** The Proteasome-Ubiquitin Protein Degradation Pathway. 2002. 17 figs. X, 213 pp. ISBN 3-540-43096-2

Vol. 269: **Koszinowski, Ulrich H.; Hengel, Hartmut (Eds.):** Viral Proteins Counteracting Host Defenses. 2002. 47 figs. XII, 325 pp. ISBN 3-540-43261-2

Vol. 270: **Beutler, Bruce; Wagner, Hermann (Eds.):** Toll-Like Receptor Family Members and Their Ligands. 2002. 31 figs. X, 192 pp. ISBN 3-540-43560-3

Vol. 271: **Koehler, Theresa M. (Ed.):** Anthrax. 2002. 14 figs. X, 169 pp. ISBN 3-540-43497-6

Vol. 272: **Doerfler, Walter; Böhm, Petra (Eds.):** Adenoviruses: Model and Vectors in Virus-Host Interactions. Virion and Structure, Viral Replication, Host Cell Interactions. 2003. 63 figs., approx. 280 pp. ISBN 3-540-00154-9

Vol. 273: **Doerfler, Walter; Böhm, Petra (Eds.):** Adenoviruses: Model and Vectors in VirusHost Interactions. Immune System, Oncogenesis, Gene Therapy. 2004. 35 figs., approx. 280 pp. ISBN 3-540-06851-1

Vol. 274: **Workman, Jerry L. (Ed.):** Protein Complexes that Modify Chromatin. 2003. 38 figs., XII, 296 pp. ISBN 3-540-44208-1

Vol. 275: **Fan, Hung (Ed.):** Jaagsiekte Sheep Retrovirus and Lung Cancer. 2003. 63 figs., XII, 252 pp. ISBN 3-540-44096-3

Vol. 276: **Steinkasserer, Alexander (Ed.):** Dendritic Cells and Virus Infection. 2003. 24 figs., X, 296 pp. ISBN 3-540-44290-1

Vol. 277: **Rethwilm, Axel (Ed.):** Foamy Viruses. 2003. 40 figs., X, 214 pp. ISBN 3-540-44388-6

Vol. 278: **Salomon, Daniel R.; Wilson, Carolyn (Eds.):** Xenotransplantation. 2003. 22 figs., IX, 254 pp. ISBN 3-540-00210-3

Vol. 279: **Thomas, George; Sabatini, David; Hall, Michael N. (Eds.):** TOR. 2004. 49 figs., X, 364 pp. ISBN 3-540-00534X

Vol. 280: **Heber-Katz, Ellen (Ed.):** Regeneration: Stem Cells and Beyond. 2004. 42 figs., XII, 194 pp. ISBN 3-540-02238-4

Vol. 281: **Young, John A. T. (Ed.):** Cellular Factors Involved in Early Steps of Retroviral Replication. 2003. 21 figs., IX, 240 pp. ISBN 3-540-00844-6

Vol. 282: **Stenmark, Harald (Ed.):** Phosphoinositides in Subcellular Targeting and Enzyme Activation. 2003. 20 figs., X, 210 pp. ISBN 3-540-00950-7

Vol. 283: **Kawaoka, Yoshihiro (Ed.):** Biology of Negative Strand RNA Viruses: The Power of Reverse Genetics. 2004. 24 figs., IX, 350 pp. ISBN 3-540-40661-1

Vol. 284: **Harris, David (Ed.):** Mad Cow Disease and Related Spongiform Encephalopathies. 2004. 34 figs., IX, 219 pp. ISBN 3-540-20107-6

Vol. 285: **Marsh, Mark (Ed.):** Membrane Trafficking in Viral Replication. 2004. 19 figs., IX, 259 pp. ISBN 3-540-21430-5

Vol. 286: **Madshus, Inger H. (Ed.):** Signalling from Internalized Growth Factor Receptors. 2004. 19 figs., IX, 187 pp. ISBN 3-540-21038-5

Vol. 287: **Enjuanes, Luis (Ed.):** Coronavirus Replication and Reverse Genetics. 2005. 49 figs., XI, 257 pp. ISBN 3-540-21494-1

Vol. 288: **Mahy, Brain W. J. (Ed.):** Foot-and-Mouth-Disease Virus. 2005. 16 figs., IX, 178 pp. ISBN 3-540-22419X

Vol. 289: **Griffin, Diane E. (Ed.):** Role of Apoptosis in Infection. 2005. 40 figs., IX, 294 pp. ISBN 3-540-23006-8

Vol. 290: **Singh, Harinder; Grosschedl, Rudolf (Eds.):** Molecular Analysis of B Lymphocyte Development and Activation. 2005. 28 figs., XI, 255 pp. ISBN 3-540-23090-4

Vol. 291: **Boquet, Patrice; Lemichez Emmanuel (Eds.)** Bacterial Virulence Factors and Rho GTPases. 2005. 28 figs., IX, 196 pp. ISBN 3-540-23865-4

Vol. 292: **Fu, Zhen F (Ed.):** The World of Rhabdoviruses. 2005. 27 figs., X, 210 pp. ISBN 3-540-24011-X

Vol. 293: **Kyewski, Bruno; Suri-Payer, Elisabeth (Eds.):** CD4+CD25+ Regulatory T Cells: Origin, Function and Therapeutic Potential. 2005. 22 figs., XII, 332 pp. ISBN 3-540-24444-1

Vol. 294: **Caligaris-Cappio, Federico, Dalla Favera, Ricardo (Eds.):** Chronic Lymphocytic Leukemia. 2005. 25 figs., VIII, 187 pp. ISBN 3-540-25279-7

Vol. 295: **Sullivan, David J.; Krishna Sanjeew (Eds.):** Malaria: Drugs, Disease and Post-genomic Biology. 2005. 40 figs., XI, 446 pp. ISBN 3-540-25363-7

Vol. 296: **Oldstone, Michael B. A. (Ed.):** Molecular Mimicry: Infection Induced Autoimmune Disease. 2005. 28 figs., VIII, 167 pp. ISBN 3-540-25597-4

Vol. 297: **Langhorne, Jean (Ed.):** Immunology and Immunopathogenesis of Malaria. 2005. 8 figs., XII, 236 pp. ISBN 3-540-25718-7

Vol. 298: **Vivier, Eric; Colonna, Marco (Eds.):** Immunobiology of Natural Killer Cell Receptors. 2005. 27 figs., VIII, 286 pp. ISBN 3-540-26083-8

Vol. 299: **Domingo, Esteban (Ed.):** Quasispecies: Concept and Implications. 2006. 44 figs., XII, 401 pp. ISBN 3-540-26395-0

Vol. 300: **Wiertz, Emmanuel J.H.J.; Kikkert, Marjolein (Eds.):** Dislocation and Degradation of Proteins from the Endoplasmic Reticulum. 2006. 19 figs., VIII, 168 pp. ISBN 3-540-28006-5

Vol. 301: **Doerfler, Walter; Böhm, Petra (Eds.):** DNA Methylation: Basic Mechanisms. 2006. 24 figs., VIII, 324 pp. ISBN 3-540-29114-8